Lecture Notes in Mathematics 1513

L. S. Block W. A. Coppel

Dynamics in One Dimension

Springer-Verlag
Berlin Heidelberg New York
London Paris Tokyo
Hong Kong Barcelona
Budapest

Authors

Louis Stuart Block
Department of Mathematics
University of Florida
Gainesville, Florida 32611, USA

William Andrew Coppel
Department of Theoretical Physics
Institute of Advanced Studies
Australian National University
GPO Box 4
Canberra 2601, Australia

Mathematics Subject Classification (1991): 26A18, 54H20, 58F08

ISBN 3-540-55309-6 Springer-Verlag Berlin Heidelberg New York
ISBN 0-387-55309-6 Springer-Verlag New York Berlin Heidelberg

Printed in Germany

Typesetting: Camera ready by author
Printing and binding: Druckhaus Beltz, Hemsbach/Bergstr.
46/3140-543210 - Printed on acid-free paper

Preface

There has recently been an explosion of interest in one-dimensional dynamics. The extremely complicated — and yet orderly — behaviour exhibited by the logistic map, and by unimodal maps in general, has attracted particular attention. The ease with which such maps can be explored with a personal computer, or even with a pocket calculator, has certainly been a contributing factor. The unimodal case is extensively studied in the book of Collet and Eckmann [49], for example.

It is not so widely known that a substantial theory has by now been built up for *arbitrary* continuous maps of an interval. It is quite remarkable how many strong, general properties can be established, considering that such maps may be either real-analytic or nowhere differentiable. The purpose of the present book is to give a clear, connected account of this subject. Thus it updates and extends the survey article of Nitecki [96]. The two books [112], [113] by Šarkovskii and his collaborators contain material on the same subject. However, they are at present available only in Russian and in general omit proofs. Here complete proofs are given. In many cases these have previously been difficult of access, and in some cases no complete proof has hitherto appeared in print.

Our standpoint is topological. We do not discuss questions of a measure-theoretical nature or connections with ergodic theory. This is not to imply that such matters are without interest, merely that they are outside our scope. [A forthcoming book by de Melo and van Strien discusses these matters, and also the theory of smooth maps.] The material here could indeed form the basis for a course in topological dynamics, with many of the general concepts of that subject appearing in a concrete situation and with much greater effect.

Several of the results included here were first established for piecewise monotone maps. There exist also other results which are valid for piecewise monotone maps, but which do not hold for arbitrary continuous maps. Although we include some results of this nature, we do not attempt to give a full account of the theory of piecewise monotone maps.

The final chapter of the book deals with extensions to maps of a circle of the preceding results for maps of an interval. In contrast to the earlier chapters, the results here are merely stated, with references to the literature for the proofs. [Complete proofs are given in a forthcoming book by Alsedà, Llibre and Misiurewicz, which also discusses the material in our

Chapters 1,7 and 8.] We do not discuss at all some results which have been established for other one-dimensional structures. The pre-eminent importance of the interval and the circle appears to us adequate justification for our title. The list of references at the end of the book, although extensive, has no pretence to completeness.

This book has its origin in a course of lectures which the older author gave at the Australian National University in 1984. The first four chapters are based on the xeroxed notes for that course. However, the older author acknowledges that without the assistance of the younger author the book could never have reached its present greatly expanded form. We accept responsibility equally for the final product.

Our manuscript was originally submitted as a whole volume for the series *Dynamics Reported.* After its submission responsibility for publication of this series passed from Wiley and Teubner to Springer-Verlag. The resulting changes in format would not have presented insurmountable difficulties if the authors had been experts with TEX or LATEX. Since we were not, we decided instead to produce a good camera-ready manuscript, following the instructions to authors provided by Springer-Verlag for its *Lecture Notes in Mathematics* series. We are extremely grateful to the Managing Editors of *Dynamics Reported*, Professors U. Kirchgraber and H.O. Walther, for the time and care they devoted to our manuscript, for obtaining valuable referees' reports, and finally for generously agreeing to its appearance in the *Lecture Notes in Mathematics* rather than in *Dynamics Reported.*

We thank Professor Xiong Jincheng for contributing some unpublished results (Propositions VI.53 and VI.54), and the referees for several useful suggestions. We take this opportunity to thank also the numerous typists who have assisted us over a period of eight years. W.A.C. is grateful to the University of Florida for support during a visit to Gainesville in 1987. L.S.B. would like to thank the Australian National University for its hospitality during visits in 1988 and 1990. These visits considerably accelerated progress on the book. L.S.B. also thanks the University of Göttingen for its hospitality during a visit in 1988, and Zbigniew Nitecki for helpful conversations during that visit. Finally he thanks Ethan Coven for many helpful conversations over the past few years.

We dedicate this book to our families, in gratitude for their support.

Louis Block
Andrew Coppel

Contents

Introduction

This book is primarily concerned with the asymptotic behaviour of sequences (x_n) defined iteratively by $x_{n+1} = f(x_n)$, where f is an arbitrary continuous map of an *interval* into itself. The sequence (x_n) is the *trajectory* of the initial point x_0 under the map f.

An important reason for studying this problem, in addition to its intrinsic interest, comes from higher-dimensional dynamics. The extremely complicated behaviour of some 3-dimensional flows, or 2-dimensional diffeomorphisms, is also observed in non-invertible 1-dimensional maps. We hope to gain a better understanding of this behaviour by studying 1-dimensional maps, since they are much more amenable to mathematical analysis. Many remarkable properties of such maps have been established in recent years.

If in the trajectory (x_n) we have $x_p = x_0$ for some $p > 0$, then $x_{n+p} = x_n$ for every $n > 0$. Thus the trajectory (x_n) is *periodic*. It is said to have *period p* if p is the least positive integer such that $x_p = x_0$. It turns out that if a continuous map f has a periodic trajectory with a given period p, then it necessarily has periodic trajectories with certain other periods. A complete description of all possible sets of periods, for the periodic trajectories of a continuous map of an interval, is given by a theorem of Šarkovskii, which is stated and proved in Chapter I. An interesting feature of this proof is the use of directed graphs.

In Chapter II we begin the study of nonperiodic trajectories. A simple example is a trajectory (x_n) with $x_n = x_2$ for all $n > 2$ and either $x_2 < x_0 < x_1$ or $x_1 < x_0 < x_2$. A map possessing such a trajectory is said to be *turbulent*. It turns out that all trajectories of non-turbulent maps are subject to rather stringent restrictions, whereas turbulent maps possess some trajectories which behave wildly. This wild behaviour is established by using the *shift map* of a symbol space. Thus, even though we are primarily interested in maps of an interval, we are naturally led to consider maps of other spaces. In Chapter II we also study the effects of slightly perturbing the given map f, and we give some results which hold for continuously differentiable or piecewise monotone maps, but not for all continuous maps.

The notions of stable and unstable manifold, of a periodic point, are important in the theory of smooth diffeomorphisms. For continuous maps of an interval, the stable set – or basin of attraction – of a periodic point may not be a manifold or have nice properties. However, as we show in Chapter III, the *unstable manifold* exists and is a well-behaved object. Moreover, one also has *left* and *right* unstable manifolds.

In Chapter III we also study *homoclinic points*. This term was first used by Poincaré, for diffeomorphisms, to describe a point belonging to both the stable and unstable manifolds of a periodic point. In our situation we demand instead that the point hit the periodic point after finitely many iterations, in addition to belonging to its unstable manifold. We show that there is a close relationship between turbulence and the existence of homoclinic points.

Periodicity represents the most precise type of repetitive behaviour. Several other types are studied in *topological dynamics*. In Chapter IV we discuss ordinary *recurrence* (or 'Poisson stability') and *nonwandering*ness. In Chapter V we consider *strong recurrence* (often also called 'recurrence'), *regular recurrence*, and *chain recurrence*. Chain recurrence is the weakest, and the most recently introduced, of these types of repetitive behaviour. Our treatment of it has some novelty, since we adopt a purely topological definition instead of the usual metric one.

Many of the results of Chapters IV and V are valid for continuous maps of any compact metric space. However, there are also results which are specific to maps of an interval. We mention, in particular, a remarkable characterization of ω-*limit points* due to Šarkovskii. These results give a strength to the theory for an interval which is lacking in the general case.

We define a map of an interval into itself to be *chaotic* if some iterate of the map is turbulent or, equivalently, if there exists a periodic point whose period is not a power of 2. It is shown in Chapter VI that there is a marked distinction between the behaviour of chaotic and non-chaotic maps. Exaggerated claims about a new theory of chaos have been appearing in the popular press. In fact there is no generally accepted definition of chaos. It is our view that any definition for more general spaces should agree with ours in the case of an interval. This requirement is not satisfied by some of the definitions used in the literature. The definition given above is strictly 1-dimensional. However, we show that a map is chaotic if and only if some iterate has the shift map as a *factor*, and we propose this as a general definition. Other definitions which meet our requirement are certainly possible, notably that some iterate is *topologically mixing* (as shown in Chapter VI) or that the map has positive *topological entropy* (as shown later in Chapter VIII), but they do not really call for the use of a new word. Ultimately it will probably be necessary to distinguish between different types of chaotic behaviour, in the same way as for recurrence.

To characterize a periodic trajectory we need to know not only its period but also its *type*, i.e. the way in which its points are ordered on the real line. It may be asked if Šarkovskii's theorem on periods can be strengthened to take account of types. That is, if a map has a periodic trajectory of a given type, does it necessarily have periodic trajectories of certain other types? In a sense this question is completely answered by a theorem of Baldwin, which says that a periodic trajectory of type P *forces* a periodic trajectory of type Q if and only if the *linearization* of P has a trajectory of type Q. We prove Baldwin's theorem in Chapter VII, but we do not investigate in detail the rather complicated partial ordering of types which forcing induces.

A periodic trajectory is said to be *primary* if it forces no periodic trajectory with the same period. In Chapter VII we also characterize completely the primary trajectories, and we prove that a map is chaotic if and only if it has a periodic trajectory which is not primary.

Chapter VIII is devoted to the important concept of topological entropy. After establishing the main results which hold for any compact topological space, we devote our attention to results which hold for a compact interval. The most profound of these is a theorem of Misiurewicz, one of whose consequences is the result, already implied, that a map is chaotic if and only if it has positive topological entropy.

Finally, in Chapter IX we summarize, with references only for the proofs, extensions to maps of a *circle* of the foregoing results for maps of an interval. In the literature some results have also been given for 1-dimensional branched manifolds, and in particular for '*Y*', but these lie outside our scope. [See, for example, L. Alsedà, J. Llibre and M. Misiurewicz, *Trans. Amer. Math. Soc.*, **313** (1989), 475-538, and a series of papers by A.M. Blokh in *Teor. Funktsii Funktsional. Anal. i Prilozhen.*]

An introduction, such as this, frequently concludes with some remarks on prerequisites. A most attractive feature of our subject is that the only knowledge demanded of the reader would be contained in a first course on real analysis. For the reader possessing this knowledge we present a variety of interesting and nontrivial results which were unknown thirty years ago! We hope that some readers may be stimulated to make additional contributions of their own, even if it means that our book will become outdated.

I
Periodic Orbits

1 ŠARKOVSKII'S THEOREM

By an *interval* we will always mean, except in Chapter VIII, a connected subset of the real line which contains more than one point. Thus an interval may be open, half-open or closed, but not degenerate, and an endpoint of an interval need not belong to the interval. However, the phrase 'nondegenerate interval' will sometimes be used for emphasis. We will denote by $\langle a,b \rangle$ the closed interval with endpoints a and b, when we do not know (or care) whether $a < b$ or $a > b$.

Let $f: I \to I$ be a continuous map of the interval I into itself. Having performed the map f once we can perform it again, and again, and again. That is, we consider the *iterates* f^n defined inductively by

$$f^1 = f, \quad f^{n+1} = f \circ f^n \qquad (n \geq 1).$$

We also take f^0 to be the identity map, defined by $f^0(x) = x$ for every $x \in I$. Evidently f^n is also a continuous map of I into itself. We are interested in the behaviour of the *trajectory* of x, i.e. the sequence $f^n(x)$ $(n \geq 0)$, for arbitrary $x \in I$. It is convenient to make a distinction between the trajectory of x and the *orbit* of x, which is the set of points $\{f^n(x) : n \geq 0\}$.

There is a very simple graphical procedure for following trajectories. In the (x,y) – plane draw the curve $y = f(x)$ and the straight line $y = x$. To obtain the trajectory with initial point x_0 we go vertically to $y = f(x)$, then horizontally to $y = x$. This gives $x_1 = f(x_0)$, and the process is repeated *ad infinitum* (see Figure 1).

A point $c \in I$ is said to be a *fixed point* of f if $f(c) = c$. Thus the fixed points are given by the intersections of the curve $y = f(x)$ and the straight line $y = x$. If the interval I is compact it necessarily contains at least one fixed point. For if $I = [a,b]$ we have

$$f(a) - a \geq 0 \geq f(b) - b,$$

and so the assertion follows from the intermediate value theorem for continuous functions.

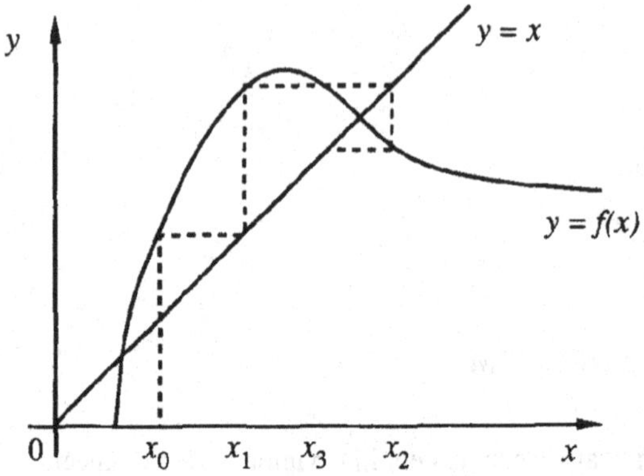

Fig. 1

A point $c \in I$ is said to be a *periodic point* of f with *period m* if $f^m(c) = c, f^k(c) \neq c$ for $1 \leq k < m$. The orbit of c then consists of the m distinct points $c, f(c), \ldots, f^{m-1}(c)$ and the trajectory of c consists of the same points repeated periodically. By abuse of language the orbit of c will also be said to be periodic. A fixed point is a periodic point of period 1.

If $f: I \rightarrow \mathbb{R}$ is a continuous map of an interval into the real line, then all or some of the iterates may be defined on a subinterval of I and we may still talk about periodic points. Throughout this chapter, unless otherwise stated, f will denote an arbitrary continuous map of an arbitrary interval I into the real line.

We are going to study first the periodic orbits of f. Our main objective will be the proof of the following striking theorem, due to Šarkovskii [102].

THEOREM 1 *Let the positive integers be totally ordered in the following way:*

$$3 \prec 5 \prec 7 \prec 9 \prec \ldots \prec 2.3 \prec 2.5 \prec \ldots \prec 2^2.3 \prec 2^2.5 \prec \ldots\ldots \prec 2^3 \prec 2^2 \prec 2 \prec 1.$$

If f has a periodic orbit of period n and if $n \prec m$, then f also has a periodic orbit of period m.

However, on the way we will derive a number of results of independent interest.

LEMMA 2 *If J is a compact subinterval such that $J \subseteq f(J)$, then f has a fixed point in J.*

Proof If $J = [a,b]$ then for some $c,d \in J$ we have $f(c) = a, f(d) = b$. Thus $f(c) \leq c, f(d) \geq d$, and the result follows again from the intermediate value theorem. □

LEMMA 3 *If J, K are compact subintervals such that $K \subseteq f(J)$, then there is a compact subinterval $L \subseteq J$ such that $f(L) = K$.*

Proof Let $K = [a,b]$ and let c be the greatest point in J for which $f(c) = a$. If $f(x) = b$ for some $x \in J$ with $x > c$, let d be the least. Then we can take $L = [c,d]$. Otherwise $f(x) = b$ for some $x \in J$ with $x < c$. Let c' be the greatest and let $d' \leq c$ be the least $x \in J$ with $x > c'$ for which $f(x) = a$. Then we can take $L = [c',d']$. □

LEMMA 4 *If $J_0, J_1, \ldots, J_m$ are compact subintervals such that $J_k \subseteq f(J_{k-1})$ $(1 \leq k \leq m)$, then there is a compact subinterval $L \subseteq J_0$ such that $f^m(L) = J_m$ and $f^k(L) \subseteq J_k$ $(1 \leq k < m)$.*

If also $J_0 \subseteq J_m$, then there exists a point y such that $f^m(y) = y$ and $f^k(y) \in J_k$ $(0 \leq k < m)$.

Proof The first assertion holds for $m = 1$, by Lemma 3. We assume that $m > 1$ and that it holds for all smaller values of m. Then we can choose $L' \subseteq J_1$ so that $f^{m-1}(L') = J_m$ and $f^k(L') \subseteq J_{k+1}$ $(1 \leq k < m-1)$. We now choose $L \subseteq J_0$ so that $f(L) = L'$.

The second assertion follows from the first, by Lemma 2. □

As a first application of these ideas we prove

PROPOSITION 5 *Between any two points of a periodic orbit of period $n > 1$ there is a point of a periodic orbit of period less than n.*

Proof Let $a < b$ be two adjacent points of the orbit of period n. Since there is one more point of the orbit to the left of b than to the left of a we must have $f^m(a) > a, f^m(b) < b$ for some m such that $1 \leq m < n$. It follows at once that $f^m(c) = c$ for some c such that $a < c < b$, assuming that f^m is defined throughout $[a,b]$. However, the same conclusion can be reached without this assumption. For if $J_k = \langle f^k(a), f^k(b) \rangle$ is the closed interval with endpoints $f^k(a)$ and $f^k(b)$ then $J_k \subseteq f(J_{k-1})$ $(1 \leq k \leq m)$. But $J_0 \subseteq J_m$, since $f^m(a) \geq b, f^m(b) \leq a$. The result now follows from Lemma 4. □

The method of argument used here can be refined. Suppose again that f has a periodic orbit of period $n > 1$. Let $x_1 < \ldots < x_n$ be the distinct points of this orbit and set $I_j = [x_j, x_{j+1}]$ $(1 \leq j < n)$. With the periodic orbit we associate a *directed graph*, or *digraph*, in the following way. The vertices of the directed graph are the subintervals $I_1, \ldots, I_{n-1}$ and there is an arc $I_j \to I_k$ if I_k is contained in the closed interval $\langle f(x_j), f(x_{j+1}) \rangle$ with endpoints $f(x_j)$ and $f(x_{j+1})$.

For example, suppose c is a periodic point of period 3 with $f(c) < c < f^2(c)$. The corresponding directed graph has two vertices, namely the intervals $I_1 = [f(c), c]$ and $I_2 = [c, f^2(c)]$, connected in the following way:

$$\circlearrowleft I_1 \rightleftarrows I_2$$

A number of properties of our digraphs follow from the definition. First, for any vertex I_j there is always at least one vertex I_k for which $I_j \to I_k$. Moreover, it is always possible to choose $k \neq j$, unless $n = 2$. The proof is trivial.

Secondly, for any vertex I_k there is at least one vertex I_j for which $I_j \to I_k$. Moreover, it is always possible to choose $j \neq k$, unless n is even and $k = n/2$. We will prove this by contradiction. Suppose there is no $j \neq k$ for which $I_j \to I_k$. Then if $i \neq k, f(x_i) \leq x_k$ implies $f(x_{i+1}) \leq x_k$ and $f(x_i) \geq x_{k+1}$ implies $f(x_{i+1}) \geq x_{k+1}$. If $f(x_{k+1}) \geq x_{k+1}$ it follows that $f(x_i) \geq x_{k+1}$ for $k < i \leq n$, which is impossible because no proper subset of the orbit can be mapped into itself by f. Hence $f(x_{k+1}) \leq x_k$, and similarly $f(x_k) \geq x_{k+1}$. Thus $I_k \to I_k$. Moreover $f(x_i) \leq x_k$ for $k < i \leq n$ and $f(x_i) \geq x_{k+1}$ for $1 \leq i \leq k$, which implies $n = 2k$.

Thirdly, the digraph always contains a loop. For, since $f(x_1) > x_1$ and $f(x_n) < x_n$ we have $f(x_j) > x_j$ and $f(x_{j+1}) < x_{j+1}$ for some j $(1 \leq j < n)$. Then $f(x_j) \geq x_{j+1}$ and $f(x_{j+1}) \leq x_j$, and hence $I_j \to I_j$.

A cycle $J_0 \to J_1 \to \ldots \to J_{n-1} \to J_0$ of length n in the digraph will be said to be a *fundamental cycle* if J_0 contains an endpoint c such that $f^k(c)$ is an endpoint of J_k for $1 \leq k < n$. A fundamental cycle always exists and is unique. For, without loss of generality take $c = x_1$, so that $J_0 = I_1$. Suppose $J_0, \ldots, J_{i-1}$ have been defined. If $J_{i-1} = [a,b]$, so that $f^{i-1}(c)$ is either a or b, we must take J_i to be the uniquely determined interval $I_k \subseteq \langle f(a), f(b) \rangle$ which has $f^i(c)$ as one endpoint. Then $J_n = J_0$ and we obtain a cycle of length n.

In the fundamental cycle some vertex must occur at least twice among $J_0, \ldots, J_{n-1}$, since the digraph has only $n-1$ vertices. On the other hand, any vertex occurs at most twice, since an interval I_k has only two endpoints. If the fundamental cycle contains the vertex I_k twice then it can be decomposed into two cycles of smaller length, each of which contains I_k only once and consequently is primitive.

Here a cycle in a digraph is said to be *primitive* if it does not consist entirely of a cycle of smaller length described several times. Straffin [118], who first showed the relevance of directed graphs in this connection, observed that the existence of a primitive cycle of length m enables one to deduce the existence of a periodic orbit of period m.

LEMMA 6 *Suppose f has a periodic point of period $n > 1$. If the associated digraph contains a primitive cycle $J_0 \to J_1 \to \dots \to J_{m-1} \to J_0$ of length m, then f has a periodic point y of period m such that $f^k(y) \in J_k$ $(0 \le k < m)$.*

Proof By Lemma 4 there exists a point y such that $f^m(y) = y$ and $f^k(y) \in J_k$ $(0 \le k < m)$. Since the cycle is primitive and distinct intervals J_k have at most one endpoint in common it follows that y has period m, unless possibly $y = x_i$ for some i and n is a divisor of m. However, this is possible only if the cycle is a multiple of the fundamental cycle since, given J_{k-1}, the requirements $f^k(y) \in J_k$ and $J_{k-1} \to J_k$ uniquely determine J_k. □

The principle embodied in Lemma 6 is remarkably powerful. Consider the previous example of the digraph associated to a periodic point of period 3. Corresponding to the loop $I_1 \to I_1$ there is a fixed point of f and corresponding to the primitive cycle $I_1 \to I_2 \to I_1$ there is a point of period 2. Moreover, for any positive integer $m > 2$ there is a point of period m, corresponding to the primitive cycle $I_1 \to I_2 \to I_1 \to I_1 \to \dots \to I_1$ of length m. Thus there are orbits of period n for every $n \ge 1$, in agreement with Theorem 1. In fact our proof of Theorem 1 will make essential use of this principle.

PROPOSITION 7 *If f has a periodic point of period > 1, then it has a fixed point and a periodic point of period 2.*

Proof The first assertion follows at once from the fact that the digraph of a periodic orbit always contains a loop. [More simply, if f has no fixed point then either $f(x) > x$ for all x or $f(x) < x$ for all x, and hence f has no periodic point.]

To prove the second assertion, let n be the least positive integer greater than 1 such that f has a periodic point of period n. We will assume $n > 2$ and deduce a contradiction. In fact the fundamental cycle decomposes into two cycles of smaller length, each of which is primitive. Since at least one of these has length greater than 1, it follows from Lemma 6 that there is a periodic point with period strictly between 1 and n. □

Proposition 7 was first proved in Coppel [52]. We next use Lemma 6 to obtain a result due to Štefan [117].

PROPOSITION 8 *Suppose f has a periodic orbit of odd period $n > 1$, but no periodic orbit of odd period strictly between 1 and n. If c is the midpoint of the orbit of odd period n, then the points of this orbit have the order*

$$f^{n-1}(c) < f^{n-3}(c) < \ldots < f^2(c) < c < f(c) < \ldots < f^{n-2}(c)$$

or the reverse order

$$f^{n-2}(c) < \ldots < f(c) < c < f^2(c) < \ldots < f^{n-3}(c) < f^{n-1}(c).$$

In either case the associated digraph is given by Figure 2, where $J_1 = \langle c, f(c)\rangle$ and $J_k = \langle f^{k-2}(c), f^k(c)\rangle$ for $1 < k < n$.

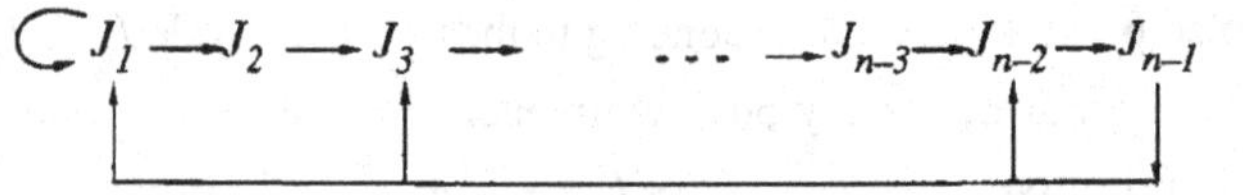

Fig. 2

Proof The fundamental cycle decomposes into two smaller primitive cycles, one of which has odd length. This length must be 1, since f has no orbit of odd period strictly between 1 and n. Thus the fundamental cycle has the form

$$J_1 \to J_1 \to J_2 \to \ldots \to J_{n-1} \to J_1 ,$$

where $J_i \neq J_1$ for $1 < i < n$. If we had $J_i = J_k$, where $1 < i < k < n$, then by omitting the intermediate vertices we would obtain a smaller primitive cycle. Moreover, by excluding the loop at J_1 if necessary, we can arrange that its length is odd. Since this is contrary to hypothesis we conclude that $J_1 ,\ldots, J_{n-1}$ are all distinct and thus a permutation of $I_1 ,\ldots, I_{n-1}$. Similarly we cannot have $J_i \to J_k$ if $k > i + 1$ or if $k = 1$ and $i \neq 1, n-1$.

Suppose $J_1 = I_h = [a,b]$. Since J_1 is directed only to J_1 and J_2, the interval J_2 is adjacent to J_1 on the real line and f maps one endpoint of J_1 into an endpoint of J_1 and the other endpoint of J_1 into an endpoint of J_2. Since the endpoints are not fixed points, there are just two possibilities: either

$$x_h = a,\ x_{h+1} = f(a),\ x_{h-1} = f^2(a) ,$$

or

$$x_{h+1} = b,\ x_h = f(b),\ x_{h+2} = f^2(b) .$$

We consider only the first case, the argument in the second case being similar.

For $n = 3$ the result now follows immediately. Suppose $n > 3$. If $f^3(a) < f^2(a)$ then $J_2 \to J_1$, which is forbidden. Hence $f^3(a) > f^2(a)$. Since J_2 is not directed to J_k for $k > 3$ it follows that $J_3 = [f(a), f^3(a)]$ is adjacent to J_1 on the right. If $f^4(a) > f^3(a)$ then $J_3 \to J_1$, which is forbidden. Hence $f^4(a) < f^2(a)$ and, since J_3 is not directed to J_k for $k > 4$, $J_4 = [f^4(a), f^2(a)]$ is adjacent to J_2 on the left. Proceeding in this way we see that the order of the intervals J_i on the real line is given by

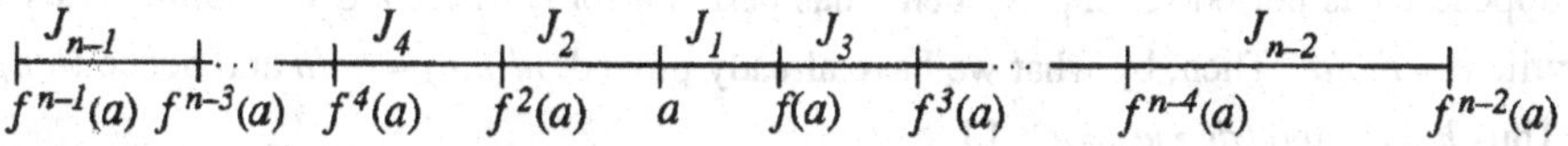

Since the endpoints of J_{n-1} are mapped into a and $f^{n-2}(a)$ we have $J_{n-1} \to J_k$ if and only if k is odd. It is readily verified now that there are no other arcs in the digraph besides those already found. □

An orbit of odd period $n > 1$ with either of the two configurations described in Proposition 8 will be called a *Stefan orbit.*

The next result follows immediately, using Lemma 6.

PROPOSITION 9 *If f has a periodic point of odd period $n > 1$, then it has periodic points of arbitrary even order and periodic points of arbitrary odd order $> n$.*

Proof We may suppose that n is minimal, so that the associated digraph is given by Proposition 8. If $m < n$ is even then $J_{n-1} \to J_{n-m} \to J_{n-m+1} \to \dots \to J_{n-1}$ is a primitive cycle of length m. If $m > n$ is even or odd then

$$J_1 \to J_2 \to \dots \to J_{n-1} \to J_1 \to J_1 \to \dots \to J_1$$

is a primitive cycle of length m. □

The following lemma relates the period of a periodic point of f to its period as a periodic point of some iterate. The result will later be used repeatedly, sometimes without explicit reference.

LEMMA 10 *If c is a periodic point of f with period n then, for any positive integer h, c is a periodic point of f^h with period n/(h,n), where (h,n) denotes the greatest common divisor of h and n.*

Conversely, if c is a periodic point of f^h with period m then c is a periodic point of f with period mh/d, where d divides h and is relatively prime to m.

Proof Suppose c has period n for f and let $m = n/(h,n)$. Then $f^{mh}(c) = c$. On the other hand, if $f^{kh}(c) = c$ then n divides kh and hence m divides k.

Suppose c has period m for f^h. Then c has period n for f, where n divides mh. Thus we can write $n = mh/d$. Then, by what we have already proved, $n/(h,n) = nd/h$ and hence $(h,n) = h/d$. Thus $h = de$ and $(de,me) = e$. □

We are now in a position to prove Šarkovskii's theorem.

Proof of Theorem 1 We give the proof initially for $f{:}I \to I$. Write $n = 2^d q$, where q is odd. Suppose first that $q = 1$ and $m = 2^e$, where $0 \le e < d$. By Proposition 7 we may assume $e > 0$. The map $g = f^{m/2}$ has a periodic point of period 2^{d-e+1}, by Lemma 10, and hence also a periodic point of period 2, by Proposition 7. This point has period m for f, by Lemma 10 again.

Suppose next that $q > 1$. The remaining cases to be considered are $m = 2^d r$, where either (i) r is even, or (ii) r is odd and $r > q$. The map $g = f^{2^d}$ has a periodic point of period q and hence also has a periodic point of period r, by Proposition 9. In case (i) this point has period $m = 2^d r$ for f. In case (ii) its f-period is $2^e r$ for some $e \le d$. If $e = d$ we are finished. If $e < d$ we can replace n by $2^e r$. Since $m = 2^e(2^{d-e} r)$ it then follows from case (i) that f also has a periodic point of period m.

We now give the proof for $f{:}I \to \mathbb{R}$. Let x_1 and x_n denote respectively the least and greatest points of a periodic orbit of f of period n. Then $K{:}= [x_1, x_n] \cup f[x_1, x_n]$ is a compact interval. Define a continuous map $g{:}K \to K$ by setting $g(x) = f(x_1)$ if $x \le x_1$, $g(x) = f(x)$ if $x \in [x_1, x_n]$, and $g(x) = f(x_n)$ if $x \ge x_n$. Since g has a periodic orbit of period n, g also has a periodic orbit of period m, by what we have already proved. Since this orbit of period m is contained in the interval $[x_1, x_n]$, it is also a periodic orbit of f. □

We draw attention to the fact that if, in the statement of Theorem 1, the orbit of period n is contained in a subinterval J, then J also contains an orbit of period m.

2 SUPPLEMENTARY REMARKS

Šarkovskii's theorem is best possible in the sense that, for each positive integer n, there is a continuous map f of a compact interval into itself which has a periodic point of period n but no periodic point of period m for any $m \prec n$. Also, there exists a continuous map f with periodic points of period 2^d for every $d \geq 0$, but of no other periods. These claims will now be established by means of examples.

We prove first a converse of Straffin's primitive cycle principle.

LEMMA 11 *Suppose f has a periodic orbit $x_1 < \ldots < x_n$ of period $n > 1$ and let $\mathbb{G}$ be the associated digraph. Suppose also that f is strictly monotonic on each subinterval $I_j = [x_j, x_{j+1}]$ $(1 \leq j < n)$. If f has an orbit of period m in the open interval (x_1, x_n), then either $\mathbb{G}$ contains a primitive cycle of length m, or m is even and $\mathbb{G}$ contains a primitive cycle of length $m/2$.*

Proof If the point $c \in (x_1, x_n)$ has period m for f then, for each k, there is a unique vertex J_k of $\mathbb{G}$ such that $f^{k-1}(c) \in J_k$. Moreover $J_1 \to \ldots \to J_m \to J_1$ is a cycle of length m in $\mathbb{G}$, since f is monotonic on each interval J_k. This cycle is a multiple of a primitive cycle of length ℓ, where ℓ divides m and $1 \leq \ell \leq m$. The map f^ℓ of a subinterval of J_1 onto J_1, determined by this primitive cycle, is strictly monotonic. If it is increasing the orbit of period m can 'close' only if $m = \ell$. If it is decreasing the orbit can close only if $m = \ell$ or $m = 2\ell$. □

EXAMPLE 12 Let $n = 2h + 1$ be an odd integer >1. Let $I = [0,2h]$ and let $f: I \to I$ be the piecewise linear map defined by

$$f(0) = 2h,\ f(h-1) = h + 1,\ f(h) = h - 1\ ,\ f(2h-1) = 0,\ f(2h) = h,$$

(see Figure 3). Then h is a periodic point of period n, since

$$f^{2k-1}(h) = h - k, f^{2k}(h) = h + k \quad (1 \leq k \leq h),$$

and $f^{2h+1}(h) = h$. If we put

$$J_{2k-1} = [h-k, h-k+1], J_{2k} = [h+k-1, h+k] \quad (1 \leq k \leq h)\ ,$$

the associated digraph is given by Figure 2. Since the digraph does not contain any primitive cycle of length m, where m is odd and $1 < m < n$, it follows from Lemma 11 that f does not have any periodic orbit with such a period m.

In Example 12 we constructed a continuous map f of the interval $I = [0,2h]$ to itself with a periodic point of odd period $n > 1$, but no periodic points of odd period strictly between 1 and n. By mapping an arbitrary compact interval J linearly onto the interval I, we can construct a continuous map of J to itself with the same properties. A similar remark applies to the examples which follow. This method of constructing new examples will be put on a more general basis in the discussion of topological conjugacy later in the chapter.

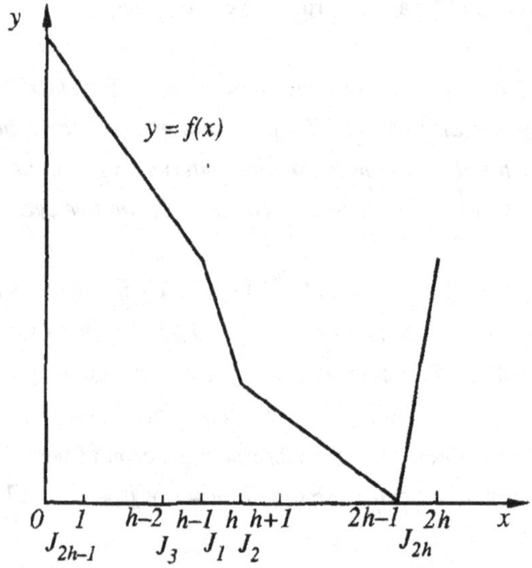

Fig. 3

EXAMPLE 13 Let f be a continuous map of the compact interval $[a,a+h]$ into itself. Let g be the continuous map of the compact interval $[a,a+3h]$ into itself defined by

$$g(x) = f(x) + 2h \quad \text{for } a \le x \le a + h,$$

$$g(x) = x - 2h \quad \text{for } a + 2h \le x \le a + 3h,$$

and by linearity for $a + h < x < a + 2h$ (see Figure 4).

Then g maps the intervals $[a,a+h]$ and $[a+2h,a+3h]$ into one another. Moreover $g^2(x) = f(x)$ for $a \le x \le a + h$ and $g^2(x) = f(x-2h) + 2h$ for $a + 2h \le x \le a + 3h$. The map g has a unique fixed point c in the interval $(a+h,a+2h)$, and a trajectory which starts from any other point of this interval ultimately leaves it. It follows that if x is a periodic point of f with period n, then x and $x + 2h$ are periodic points of g with period $2n$. Moreover, except for the fixed point c, all periodic points of g are accounted for in this way.

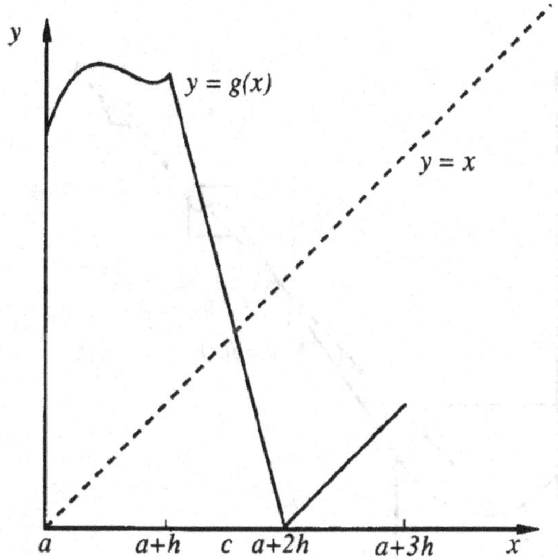

Fig. 4

Consequently, if f has (unique) periodic orbits of periods $1, 2, ..., 2^d$ and no other periodic orbits, then g has (unique) periodic orbits of periods $1, 2, ..., 2^{d+1}$ and no other periodic orbits.

Similarly, if f has a periodic orbit of period $n = 2^d q$, where $q > 1$ is odd and $d \geq 0$, but no periodic orbit of period $m \prec n$, then g has a periodic orbit of period $2n$ but no periodic orbit of period $m \prec 2n$.

Starting, for $d = 0$, with Example 12 if $q > 1$ and a constant map if $q = 1$, it follows by induction on d that for each $n \geq 1$ there is a continuous map f with a periodic orbit of period n, but no periodic orbit of period m for any $m \prec n$.

EXAMPLE 14 Let $I = [0,1]$ and, for each $k \geq 0$, let $J_k = [1 - 1/3^k, 1 - 2/3^{k+1}]$. Let f_k be a continuous map of J_k to itself. We define a continuous map $f: I \to I$ by setting $f(1) = 1$, $f(x) = f_k(x)$ if $x \in J_k$ and by linearity elsewhere (see Figure 5).

Evidently the map f has a periodic orbit of period n if and only if some f_k has a periodic orbit of period n. In particular, if we choose f_k to have an orbit of period 2^k but no orbit of period 2^{k+1}, then f has orbits of period 2^d for every $d \geq 0$, and of no other periods. In this example the set of all periodic points of f is closed. (Another such map f, for which the set of all periodic points is not closed, is given in Example VI. 30.)

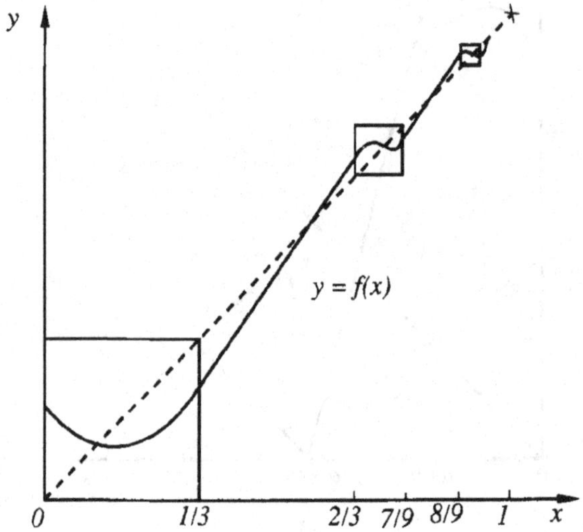

Fig. 5

This completes the proof that Šarkovskii's theorem is best possible. Examples 12 and 13 are given in Štefan [117] and Example 14 in Delahaye [60].

The original proof by Šarkovskii [102] of Theorem 1 differs from that given here and uses ideas which will be discussed in Chapter II. Our proof follows those of Ho and Morris [68] and Block *et al.* [37].

The next result, due to Cosnard [54], shows that if a map has a periodic orbit of odd period $n > 1$, then it has a Stefan orbit of period n. It is not claimed, and in general is not true, that every orbit of period n is a Stefan orbit.

PROPOSITION 15 *If f has a periodic orbit of odd period $n > 1$, then f has a periodic point c of period n such that either*

$$f^{n-1}(c) < f^{n-3}(c) < ... < f^2(c) < c < f(c) < ... < f^{n-2}(c)$$

or

$$f^{n-2}(c) < ... < f(c) < c < f^2(c) < ... < f^{n-3}(c) < f^{n-1}(c).$$

Proof By Proposition 8, the result holds if f does not have an orbit of odd period strictly between 1 and n. Hence it is sufficient to show that if there is a point c of period n satisfying, say, the first set of inequalities above then there is a point x of period $n+2$ such that

$$f^{n+1}(x) < f^{n-1}(x) < ... < f^2(x) < x < f(x) < ... < f^n(x).$$

Evidently there is a fixed point $a \in (c, f(c))$, a point $b \in (a, f(c))$ such that $f(b) = c$, and a point $d \in (c,a)$ such that $f(d) = b$. Put

$$J_0 = [d,a],\ J_1 = [a,b],\ J_2 = [c,d],\ J_3 = [b,f(c)],$$
$$J_4 = [f^2(c), c],\ J_5 = [f(c), f^3(c)],\ ...,\ J_{n+1} = [f^{n-1}(c), f^{n-3}(c)].$$

Then $J_{k+1} \subseteq f(J_k)$ for $0 \le k \le n$ and $J_0 \subseteq f(J_{n+1})$. Hence, by Lemma 4, there exists a point x such that $f^{n+2}(x) = x$ and $f^k(x) \in J_k$ $(0 \le k \le n+1)$. Since we must actually have $f^k(x) \in \operatorname{int} J_k$, the result follows. □

With any directed graph there is associated an *adjacency matrix*. If the digraph has $n-1$ vertices $J_1, \ldots, J_{n-1}$ the $(n-1) \times (n-1)$ adjacency matrix $A = (a_{ik})$ is defined by $a_{ik} = 1$ if there is an arc $J_i \to J_k$ and $= 0$ otherwise. [Conversely, any matrix of zeros and ones determines in this way a directed graph.] By relabelling the vertices, the rows and columns of the adjacency matrix are subjected to the same permutation. In spite of this lack of uniqueness we will still refer to *the* adjacency matrix of a digraph. The *characteristic polynomial* of a digraph is the uniquely determined polynomial det $(\lambda I - A)$. The relationship between the cycle structure of a digraph and its characteristic polynomial is considered in Cvetković *et al.* [59], especially Theorems 1.2 and 3.1.

For example, the digraph of a Stefan orbit of odd period $n > 1$ (Figure 2) has the adjacency matrix

$$\mathrm{St}_n = \begin{bmatrix} 1 & 1 & 0 \ldots & 0 & 0 \\ 0 & 0 & 1 \ldots & 0 & 0 \\ & & \ldots & \ldots & \\ 0 & 0 & 0 \ldots & 0 & 1 \\ 1 & 0 & 1 \ldots & 1 & 0 \end{bmatrix}.$$

LEMMA 16 *The characteristic polynomial* $\Psi_n(\lambda)$ *of* St_n *is given by*

$$\Psi_n(\lambda) = (\lambda^n - 2\lambda^{n-2} - 1)/(\lambda + 1)\,. \tag{1}$$

Proof Let $\omega_n(\lambda)$ denote the characteristic polynomial of the matrix obtained from St_n by replacing the element 1 in the top left hand corner by 0. Then it is easily verified that

$$\Psi_n(\lambda) = \omega_n(\lambda) - \lambda\omega_{n-2}(\lambda)\,.$$

On the other hand, $\omega_n(\lambda)$ satisfies the recurrence relation $\omega_n(\lambda) = \lambda^2\omega_{n-2}(\lambda) - 1$ and hence

$$\begin{aligned}\omega_n(\lambda) &= \lambda^{n-1} - (1 + \lambda^2 + \ldots + \lambda^{n-3})\\ &= (\lambda^{n+1} - 2\lambda^{n-1} + 1)/(\lambda^2 - 1)\,.\end{aligned}$$

The claimed formula for $\Psi_n(\lambda)$ now follows immediately. □

The general concept of topological conjugacy, referred to in the discussion following Example 12, will now be introduced. Let X and Y be metric spaces, and let $f{:}X \to X$ and $g{:}Y \to Y$ be continuous maps. A *homeomorphism* $h{:}X \to Y$ is a continuous, one-to-one map of X onto Y with a continuous inverse. (Continuity of the inverse is automatically satisfied if X is compact.) The maps f and g are said to be *topologically conjugate* if there exists a homeomorphism $h{:}X \to Y$ such that

$$h \circ f(x) = g \circ h(x) \quad \text{for every } x \in X,$$

i.e. the diagram

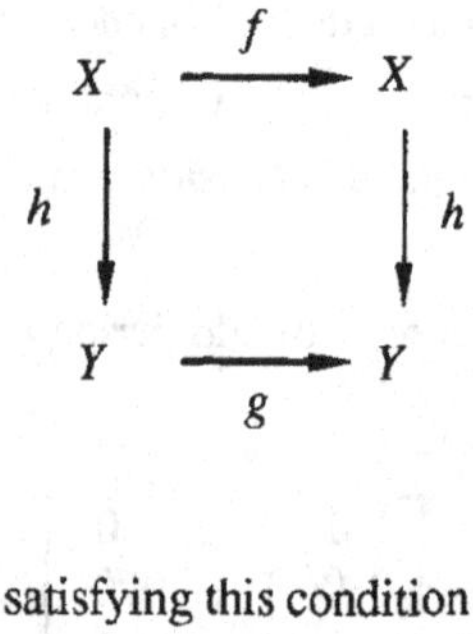

commutes. A homeomorphism h satisfying this condition is called a (*topological*) *conjugacy*.

For later use we introduce also some related notions. An *endomorphism* $h{:}X \to Y$ is a continuous map of X into Y. The maps f and g are said to be *topologically semi-conjugate* if there exists an endomorphism h of X *onto* Y such that $h \circ f = g \circ h$. The endomorphism h is then said to be a *semi-conjugacy* and g is said to be a *factor* of f.

The significance of these concepts is that topologically conjugate maps possess essentially the same properties, and that many properties are also inherited by factors. It follows easily from the definition that topological conjugacy is an equivalence relation, i.e. the following statements hold:

(i) f is topologically conjugate to f,
(ii) if f is topologically conjugate to g, then g is topologically conjugate to f,
(iii) if f_1 is topologically conjugate to f_2 and f_2 is topologically conjugate to f_3, then f_1 is topologically conjugate to f_3.

LEMMA 17 *Suppose $f{:}X \to X$ and $g{:}Y \to Y$ are topologically conjugate maps, and let $h{:}X \to Y$ be the corresponding conjugacy. Then a point $x \in X$ is a periodic point of f of period n if and only if the point $h(x) \in Y$ is a periodic point of g of period n.*

Proof From $h \circ f = g \circ h$ we obtain by induction $h \circ f^k = g^k \circ h$ for every positive integer k. Since h is one-to-one, it follows that $f^k(x) = x$ if and only if $g^k[h(x)] = h(x)$. Hence x has period n for f if and only if $h(x)$ has period n for g. □

We now return to maps of an interval. If $X = I$ and $Y = J$ are intervals on the real line, then a function $h{:}I \to J$ is a homeomorphism if and only if h is continuous, onto, and either strictly increasing or strictly decreasing. It follows from Lemma 17 that any such function h may be used to replace the map $f{:}I \to I$, where I is the interval considered in Example 12, by a map $g{:}J \to J$ of an arbitrary compact interval J by taking $g = h \circ f \circ h^{-1}$.

Let $f{:}I \to \mathbb{R}$ and $g{:}J \to \mathbb{R}$ be continuous maps of intervals into the real line. We say that a periodic orbit P of f and a periodic orbit Q of g have the same *type* if there exists a one-to-one and order-preserving map h of the points of P onto the points of Q such that $h \circ f(x) = g \circ h(x)$ for every $x \in$ P.

As we have defined it, the directed graph associated with a periodic orbit, and hence also the corresponding adjacency matrix, depends only on the way in which f permutes the ordered points of the periodic orbit, i.e. only on the type of the periodic orbit, and not on its behaviour elsewhere. Hence we will sometimes denote the adjacency matrix corresponding to a periodic orbit P by A_P and its characteristic polynomial by $\chi_P(\lambda)$. This notation will receive more formal justification in Chapter VII, when the type of a periodic orbit is defined without reference to any continuous map.

In the same chapter we will also define the notions of strongly simple orbit and double cover. The following remarks will enable us to determine the characteristic polynomial of the digraphs corresponding to such periodic orbits. This will be used in Chapter VIII to obtain a lower bound for the topological entropy of an arbitrary periodic orbit.

Let P = $\{x_1 < \ldots < x_n\}$ be a periodic orbit of period $n = qm$, where $q > 1$ and $m > 1$, such that each of the m blocks $\{x_{q(k-1)+1}, \ldots, x_{qk}\}$ $(k=1, \ldots, m)$ of q consecutive points is a periodic orbit of f^m (with period q) of the same type as R. Suppose, moreover, that the blocks are permuted by f like a periodic orbit (with period m) of the same type as Q, and that with at most one exception f maps each block monotonically onto its image. By choosing as vertices of the digraph first the intervals whose endpoints belong to the same block, the blocks being ordered according to the way in which they are permuted by f, and finally the intervals whose endpoints lie in different blocks, the adjacency matrix A_P of P can be given the form

$$A_P = \begin{bmatrix} 0 & I & 0 & \ldots & 0 & 0 \\ 0 & 0 & I & \ldots & 0 & 0 \\ & & \cdots & & \cdots & \\ 0 & 0 & 0 & \ldots & I & 0 \\ A_R & 0 & 0 & \ldots & 0 & 0 \\ * & * & * & & * & A_Q \end{bmatrix},$$

where I here denotes not an interval but the $(q-1)\times(q-1)$ unit matrix. We will show that the corresponding characteristic polynomial is

$$\chi_P(\lambda) = \chi_R(\lambda^m)\,\chi_Q(\lambda)\,. \tag{2}$$

In fact this is an immediate consequence of the following purely algebraic result.

LEMMA 18 *The characteristic polynomial of the $m\times m$ block matrix*

$$B = \begin{bmatrix} 0 & I & 0 & \ldots & 0 \\ 0 & 0 & I & \ldots & 0 \\ & & \cdots & \cdots & \\ 0 & 0 & 0 & \ldots & I \\ A & 0 & 0 & \ldots & 0 \end{bmatrix},$$

where A is an $h\times h$ matrix and I is the $h\times h$ unit matrix, is $\chi(\lambda^m)$, where $\chi(\lambda)$ is the characteristic polynomial of A.

Proof If λ_j is an eigenvalue of A with corresponding eigenvector x_j, and if $\mu_j^m = \lambda_j$, then μ_j is an eigenvalue of B with corresponding eigenvector

$$\begin{bmatrix} x_j \\ \mu_j x_j \\ \cdots \\ \mu_j^{m-1} x_j \end{bmatrix}.$$

Hence if A has distinct eigenvalues λ_j ($j = 1, \ldots, h$), then B has distinct eigenvalues $\mu_j \omega^k$ ($j = 1, \ldots, h$; $k = 0, 1, \ldots, m-1$), where $\mu_j^m = \lambda_j$ and $\omega = \exp(2\pi i/m)$. Since

$$(\lambda - \mu_j)(\lambda - \mu_j\omega) \ldots (\lambda - \mu_j\omega^{m-1}) = \lambda^m - \lambda_j,$$

it follows that at least in this case the characteristic polynomial of B is $\chi(\lambda^m)$. But any matrix A is the limit of a sequence of matrices with distinct eigenvalues and so, by continuity, the same holds also in the general case. □

In particular, let $P = \{x_1 < \ldots < x_n\}$ be a periodic orbit of period $n = 2m$ such that the m pairs $\{x_{2k-1}, x_{2k}\}$ ($k = 1, \ldots, m$) of adjacent points are orbits of period 2 of f^m, and f permutes these pairs like an orbit Q of period m. Then the adjacency matrix A_P of P can be given the form

$$A_P = \begin{bmatrix} 0\ 1\ 0 \ldots 0\ 0 & \\ 0\ 0\ 1 \ldots 0\ 0 & \\ \ldots \quad \ldots & \\ 0\ 0\ 0 \ldots 1\ 0 & \\ 1\ 0\ 0 \ldots 0\ 0 & \\ *\ *\ * \ldots * & A_Q \end{bmatrix},$$

and hence the characteristic polynomial of P is

$$\chi_P(\lambda) = (\lambda^m - 1)\,\chi_Q(\lambda). \tag{3}$$

A directed graph can be associated not only with a periodic orbit, but also with any invariant finite set of points. Let $P = \{x_1, \ldots, x_n\}$, where $x_1 < \ldots < x_n$, be a finite set of points such that $f(P) \subseteq P$. The vertices of the directed graph are again the intervals $I_j = [x_j, x_{j+1}]$ ($j = 1, \ldots, n-1$) and there is an arc $I_j \to I_k$ if and only if $I_k \subseteq \langle f(x_j), f(x_{j+1})\rangle$. Corresponding to the directed graph there is an adjacency matrix, as already observed. The directed graph associated with an invariant finite set of points (in particular, with a periodic orbit) is often called its *Markov graph*, because of the relation between its adjacency matrix and topological Markov chains.

The existence of periodic points can be deduced also from the Markov graph of an invariant finite set. The following lemma is an immediate consequence of Lemma 4.

LEMMA 19 *Let P be a finite set of points such that f(P) ⊆ P. If the corresponding directed graph contains a cycle* $J_0 \to J_1 \to \ldots \to J_{m-1} \to J_0$ *of length m, then there is a fixed point y of* f^m *such that* $f^k(y) \in J_k$ *for* $0 \le k < m$. □

The significance of this extended use of directed graphs is suggested by the following striking algebraic property, discovered by I.M. Isaacs (see Coppel [53]).

PROPOSITION 20 *Let* $P = \{x_1, \ldots, x_n\}$, *where* $x_1 < \ldots < x_n$, *be a finite subset of the real line, and suppose both f and g map P into itself. Put* $I_j = [x_j, x_{j+1}]$ *and let A(f), A(g) be the adjacency matrices of the associated Markov graphs with the vertices* $I_1, \ldots, I_{n-1}$ *in their natural order.*

Then the composite map g ∘ f also maps P into itself, and the corresponding adjacency matrix is determined by

$$A(g \circ f) \equiv A(f)\,A(g) \pmod 2.$$

Proof Let $e_1, \ldots, e_n$ be the standard basis for the n-dimensional vector space V over the field F_2 with two elements. Then $e_1+e_2, e_2+e_3, \ldots, e_{n-1}+e_n$ form a basis for the subspace V_0 of all vectors with coordinate sum zero. Any map π of the set $E = \{e_1, \ldots, e_n\}$ into itself admits a unique extension to a linear transformation T of V. Moreover T maps V_0 into itself. The matrix A^* of this linear transformation relative to the basis $e_1+e_2, \ldots, e_{n-1}+e_n$ is determined in the following way. If π maps e_i into e_j and e_{i+1} into e_k, then T maps $e_i + e_{i+1}$ into

$$e_j + e_k = \sum_{s=l}^{m-1} (e_s + e_{s+1}),$$

where $l = \min(j,k)$ and $m = \max(j,k)$. Hence the i-th column of A^* has ones from the l-th row to the $(m-1)$-th row, and zeros elsewhere.

Under the bijection $x_i \to e_i$ the map f of P determines a unique map $\pi = \pi(f)$ of E. From what has been said the adjacency matrix A(f) is just the transpose of the matrix A^*. Since $h = g \circ f$ implies $\pi(h) = \pi(g) \circ \pi(f)$ and $A_h^* = A_g^* A_f^*$, the result follows. □

It will be noted that the inversion of order in the statement of Proposition 20 results from retaining both the analyst's definition of functional composition and the graph-theorist's definition of adjacency matrix.

COROLLARY 21 *Let* $P = \{x_1, \ldots, x_n\}$, *where* $x_1 < \ldots < x_n$, *be a finite subset of the real line, and suppose f permutes the elements of* P. *If* $A(f)$ *is the adjacency matrix of the associated Markov graph, then*

$$A(f)^h \equiv I \pmod 2,$$

where h is the least common multiple of the periods of all cycles in the cycle decomposition of the permutation f. Thus $A(f)$ *is a non-singular matrix.* □

II
Turbulence

1 PRELIMINARY RESULTS

We now study in more detail both periodic and non-periodic trajectories. The notion of turbulence, which is introduced here, sharpens and unifies a number of results concerning continuous maps of an interval.

Throughout f will denote an arbitrary continuous map of an interval I into itself. The map f will be said to be *turbulent* if there exist compact subintervals J,K with at most one common point such that

$$J \cup K \subseteq f(J) \cap f(K) . \tag{1}$$

It will be said to be *strictly turbulent* if the subintervals J,K can be chosen disjoint.

The terminology here was suggested by Theorem 1 of Lasota and Yorke [75]. It follows at once from the definition that if f is (strictly) turbulent then f^n is (strictly) turbulent for every $n > 1$. The following example shows that a map may be turbulent, but not strictly turbulent.

EXAMPLE 1 Let f: $[0,1] \to [0,1]$ be the piecewise linear map defined by

$$f(0) = 0,\ f(1/2) = 1,\ f(1) = 0.$$

Then (1) holds with $J = [0,1/2]$ and $K = [1/2,1]$, but it is readily seen that (1) does not hold for any disjoint compact subintervals J,K.

It is evident that the map f: $I \to I$ is turbulent if there exist points $a, b, c \in I$ such that

$$f(b) = f(a) = a,\ f(c) = b \tag{2}$$

and either $a < c < b$ or $b < c < a$. The following lemma shows that this sufficient condition for turbulence is also necessary.

LEMMA 2 *If f is turbulent, then there exist points $a, b, c \in I$ such that* (2) *holds and either*

$$a < c < b,$$
$$f(x) > a \text{ for } a < x < b,$$
$$x < f(x) < b \text{ for } a < x < c,$$

or

$$b < c < a,$$
$$f(x) < a \text{ for } b < x < a,$$
$$b < f(x) < x \text{ for } c < x < a.$$

Proof Let $J = [\alpha,\beta]$ and $K = [\gamma,\delta]$, where $\beta \leq \gamma$, be compact subintervals such that (1) holds. If $\beta = \gamma$ we may assume that $f(\beta) \neq \beta$, since otherwise we can choose $J_0 \subset J$ so that $f(J_0) = f(J)$ and $J_0 \cap K = \emptyset$.

Let a' be the least fixed point of f in J and let b' be the greatest point of K for which $f(b') = a'$. Suppose first that $f(c') = b'$ for some $c' \in (a',b')$. Then we can take a to be the greatest fixed point of f in $[a',c')$, b to be the least point of $(c',b']$ for which $f(b) = a$, and c to be the least point of $(a,c']$ for which $f(c) = b$.

Suppose next that $f(x) < b'$ for $a' \leq x \leq b'$. Then f takes the value δ in the intervals $[\alpha,a')$ and $(b',\delta]$, and $f(x) > x \geq \alpha$ for $\alpha \leq x < a'$. Thus $f(a'') = a''$ for some $a'' \in (b',\delta]$, $f(b'') = a''$ for some $b'' \in [\alpha,a')$, and $f(c'') = b''$ for some $c'' \in (a',\beta]$. Then, as before, we can take a to be the least fixed point of f in $(c'',a'']$, b to be the greatest point of $[b'',c'')$ for which $f(b) = a$, and c to be the greatest point of $[c'',a)$ for which $f(c) = b$. □

We study next the relation between turbulence and the existence of periodic orbits.

LEMMA 3 *If f is turbulent, then f has periodic points of all periods.*

Proof Let J,K be compact subintervals with at most one common point such that (1) holds. By Lemma 2 we may assume that if J and K are not disjoint then their common point is not periodic.

Since $J \subseteq f(J)$, f has a fixed point. By Lemma I.4, for any given $n > 1$ there exists a point $x \in J$ such that $f^n(x) = x$ and $f^k(x) \in K$ for $0 < k < n$. Evidently x has period n. □

LEMMA 4 *If f has a periodic point of odd period $n > 1$, then f^2 is strictly turbulent and f^m is also strictly turbulent for every $m \geq n$.*

Proof We may suppose n chosen so that f does not have an orbit of odd period strictly between 1 and n. Then, by Proposition I.8, the orbit of period n has the form

$$f^{n-1}(x) < f^{n-3}(x) < \ldots < f^2(x) < x < f(x) < \ldots < f^{n-2}(x),$$

or its mirror image. Without loss of generality, assume it has the form displayed. Choose d between x and $f(x)$, so that $f(d) = x$ and hence $d < f^2(d)$. Choose a between $f^{n-1}(x)$ and $f^{n-3}(x)$, so close to $f^{n-1}(x)$ that $f^2(a) > d$. Choose b between a and $f^{n-3}(x)$, so close to $f^{n-3}(x)$ that $f^2(b) < a$, and choose c between $f^{n-3}(x)$ and d, so close to $f^{n-3}(x)$ that $f^2(c) < a$. Then $J = [a,b]$ and $K = [c,d]$ are disjoint and

$$f^2(J) \supseteq [f^2(b), f^2(a)] \supseteq [a,d],$$
$$f^2(K) \supseteq [f^2(c), f^2(d)] \supseteq [a,d].$$

Thus f^2 is strictly turbulent.

Now choose $e \in (f^{n-1}(x), f^{n-3}(x))$ so that $f(e) = d$. If we set

$$\tilde{J} = [f^{n-1}(x), e], \ \tilde{K} = [x,d],$$

then $f^n(\tilde{J}) \supseteq [f^{n-1}(x), f^{n-2}(x)]$ and

$$f^n(\tilde{K}) \supseteq f^{n-1}[x,d] \supseteq [f^{n-1}(x), f^{n-2}(x)].$$

Thus f^n is strictly turbulent. Moreover since

$$f\,[f^{n-1}(x), f^{n-2}(x)] \supseteq [f^{n-1}(x), f^{n-2}(x)],$$

f^m is strictly turbulent for every $m \geq n$. □

COROLLARY 5 *If f has a periodic point of period $2^k q$, where $q > 1$ is odd and $k \geq 0$, then $f^{2^{k+1}}$ is strictly turbulent.* □

The next example shows that f^2 may be strictly turbulent, although f has no orbit of odd period > 1.

EXAMPLE 6 Let f: $[-1,1] \to [-1,1]$ be the piecewise linear map defined by

$$f(-1) = f(-2/3) = 1, \ f(0) = 0, \ f(1/2) = -1, \ f(1) = 0.$$

If $J = [0,1/3]$ and $K = [1/2,1]$, then $f^2(J) = f^2(K) = [0,1]$. But f has no periodic point of odd period, except the fixed point at the origin.

It will now be shown that the trajectories of non-turbulent maps are subject to considerable restrictions.

LEMMA 7 *Suppose f is not turbulent and let $c \in I$, $n > 1$.*
If $f^n(c) \le c < f(c)$, then $f(x) > x$ for every $x \in [f^n(c), c]$.
If $f(c) < c \le f^n(c)$, then $f(x) < x$ for every $x \in [c, f^n(c)]$.

Proof It will be sufficient to establish the first assertion. Assume, on the contrary, that the interval $[f^n(c), c]$ contains a fixed point of f, and let z be the greatest. Then $f(x) > x$ for $z < x \le c$. For a unique integer m $(2 \le m \le n)$ we have $z < f^j(c)$ $(0 \le j < m)$, $f^m(c) \le z$. Then $c < f^{m-1}(c)$, since $f^m(c) \le z < f^{m-1}(c)$. If we had $f^j(c) < f^{m-1}(c) \le f^{j+1}(c)$ for some j $(0 \le j \le m-2)$, then $J = [z, f^j(c)]$ and $K = [f^j(c), f^{m-1}(c)]$ would satisfy (1), which is contrary to hypothesis. Consequently $f^j(c) < f^{m-1}(c)$ implies $f^{j+1}(c) < f^{m-1}(c)$. Since $c < f^{m-1}(c)$, this yields the contradiction $f^{m-1}(c) < f^{m-1}(c)$. □

We will say that a point $x \in I$ is a *U-point* if $f(x) > x$ and a *D-point* if $f(x) < x$. The letters U and D are, of course, abbreviations for 'up' and 'down'.

LEMMA 8 *Suppose f is not turbulent and let $c \in I$. Then all U-points in the orbit of c lie to the left of all D-points in its orbit. If the orbit contains a fixed point, then it lies to the left of all D-points and to the right of all U-points.*

Proof If $f^j(c) \le f^k(c) < f^{k+1}(c)$ for some j,k with $j > k$ then $f^j(c) < f^{j+1}(c)$, by Lemma 7. Similarly if $f^{k+1}(c) < f^k(c) \le f^j(c)$ for some j,k with $j > k$ then $f^{j+1}(c) < f^j(c)$. It follows that any U-point $f^j(c)$ lies to the left of any D-point $f^k(c)$, regardless of whether $k > j$ or $k < j$. If $f^h(c)$ is a fixed point then $f^i(c) = f^h(c)$ for all $i > h$ and the same argument applies. □

2 MAIN THEOREMS

The preceding results have some rather substantial consequences. The following theorem is a composite of Propositions 2.2 and 3.4 in Li *et al.* [76].

THEOREM 9 *Suppose that, for some $c \in I$ and some $n > 1$,*

$$f^n(c) \le c < f(c).$$

If n is odd, then f has a periodic point of period q, for some odd q satisfying $1 < q \le n$.

If n is even, then at least one of the following alternatives holds:

(i) *f has a periodic point of period q, for some odd q satisfying* $1 < q \le n/2+1$,

(ii) $f^k(c) < f^j(c)$ *for all even k and all odd j with* $0 \le j, k \le n$.

Proof If $n = 2$ then (ii) holds. We assume that $n > 2$ and the theorem holds for all smaller values of n (also when the inequalities for the points in the trajectory of c are all reversed). We may assume also that f is not turbulent, since otherwise f has periodic points of arbitrary period.

Let $x_k = f^k(c)$ $(k \ge 0)$. Then x_n is not a fixed point, by Lemma 8. Hence each point x_k $(0 \le k \le n)$ is either a U-point or a D-point, and both types occur. Moreover, if x_i is the greatest U-point and x_j the least D-point then $x_i < x_j$, again by Lemma 8. Since $x_{i+1} \ge x_j$ and $x_{j+1} \le x_i$, the interval (x_i, x_j) contains a fixed point of f. Let z be the least fixed point of f in this interval. Then $f(x) > x$ for $x_n \le x < z$, by Lemma 7. If n is odd, or if n is even and (ii) does not hold, there is an $h < n$ such that x_h and x_{h+1} are both $\le x_i$ or both $\ge x_j$. Evidently $x_h \ne x_i, x_j$. Let J_h be the interval $[x_h, x_i]$ if $x_h < x_i$ and the interval $[x_j, x_h]$ if $x_h > x_j$. Also for $k = 0,1,\ldots,n$ $(k \ne h)$ let J_k be the interval $[x_k, z]$ if $x_k \le x_i$ and the interval $[z, x_k]$ if $x_k \ge x_j$. Then it is easily verified that $J_{k+1} \subseteq f(J_k)$ $(0 \le k < n)$. Since $J_0 \subseteq J_n$, there exists a point $y \in J_0$ such that $f^n(y) = y, f^k(y) \in J_k$ $(1 \le k < n)$, by Lemma I.4. Then y has period m, where m divides n. Assume $m = 1$. Then y is a fixed point and it belongs to the intersection of all J_k. But J_h and J_i have at most the endpoint x_i in common, and J_h and J_j have at most the endpoint x_j in common. Hence we must have $1 < m \le n$. If n is odd then m is also odd, and thus the theorem is proved for this case.

Suppose n is even. The periodic point y is a U-point, since $y \in [x_0, z)$. If y_s is the greatest U-point and y_t the least D-point in the orbit of y then $y_s < y_t$, by Lemma 8. Moreover $f(x) > x$ for $y \le x \le y_s$, by Lemma 7, and hence $z > y_s$. On the other hand $z < y_t$, since y_t is a D-point. Since $f^h(y)$ and $f^{h+1}(y)$ lie on the same side of z, by construction, the alternative (ii) does not hold for the orbit of y.

Thus we are reduced to showing that if c itself is periodic with even period $n > 2$ then either (i) or (ii) holds.

Suppose first that $n = 4$. If $x_3 < x_0 < x_1$, or $x_0 < x_1 < x_2$, or $x_0 < x_3 < x_2$, then f has a periodic point of period 3 by what we have already proved (with the inequalities reversed in the last case). If none of these possibilities occurs, then (ii) holds.

Suppose next that $n/2$ is odd. For any k let $k' = k + n/2$. Assume that, for some k, x_k and $x_{k'}$ lie on the same side of z. If x_k is closer than $x_{k'}$ to z, then either $x_{k'} < x_k < x_{k+1}$ or $x_{k'} > x_k > x_{k+1}$. Hence f has a periodic point of odd period q $(1 < q \leq n/2)$, by what we have already proved. If $x_{k'}$ is closer than x_k to z the same conclusion holds, since $x_k = x_{k+n}$. Thus we may assume that x_k and $x_{k'}$ are on opposite sides of z for every k. Without loss of generality we may also assume that c is the greatest U-point in its orbit. Then $x_{1+n/2} < x_0 < x_1$. Hence, by the induction hypothesis, f has a periodic point of odd period q $(1 < q \leq n/2)$ unless $x_0, x_2, ..., x_{1+n/2}$ are all less than z and $x_1, x_3, ..., x_{n/2}$ all greater than z. But then $x_{n/2}, x_{2+n/2}, ..., x_{n-1}$ are all greater than z and $x_{1+n/2}, x_{3+n/2}, ..., x_{n-2}$ are all less than z. Thus (ii) holds.

Suppose finally that $n/2$ is even and $n > 4$. As in the previous case we may assume that x_k and $x_{k'+1}$ are on opposite sides of z for every k. But then x_k and x_{k+2} are on the same side of z for every k, and hence (ii) holds. □

A trajectory $\{f^n(c)\}$ will be said to be *alternating* if either $f^k(c) < f^j(c)$ for all even k and all odd j, or $f^k(c) > f^j(c)$ for all even k and all odd j.

By Theorem 9, if the map f has a periodic trajectory, of period $n > 1$, which is not alternating then it has a periodic orbit of odd period $q > 1$. Moreover the upper bounds for q given in Theorem 9 are sharp. For odd n this is already shown by Example I.12. For even n it is shown by the next examples, due to Li *et al.* [76].

Let $c = x_0, x_1, ..., x_{n-1}$ be n distinct points on the real line and let I be the smallest interval containing these points. We define a continuous map $f: I \to I$ by setting $f(x_k) = x_{k+1}$ $(0 \leq k < n-1)$, $f(x_{n-1}) = x_0$, and by taking f to be strictly monotonic on the open subintervals between adjacent points x_k. Then c is periodic with period n. The non-existence of periodic points with the periods mentioned in the examples follows from Lemma I.11.

EXAMPLE 10 $n \geq 10$ even and $m = n/2$ odd. If

$$x_{n-1} < x_{m-1} < x_{n-3} < x_{m-3} < ... < x_2 < x_m < x_0$$

$$< x_1 < x_{m+1} < x_3 < x_{m+3} < ... < x_{m-2} < x_{n-2}$$

then the trajectory of c is not alternating, but f has no periodic point of period $m-2$.
[It may be verified that the adjacency matrix of the corresponding digraph has characteristic polynomial

$$(\lambda^m - 1)(\lambda^m - 2\lambda^{m-2} - 1)/(\lambda + 1).]$$

EXAMPLE 11 $n \geq 8$ even and $m = n/2$ even. If

$$x_m < x_{n-1} < x_{m-2} < x_{n-3} < \ldots < x_2 < x_{m+1} < x_0$$
$$< x_1 < x_{m+2} < x_3 < x_{m+4} < \ldots < x_{n-2} < x_{m-1}$$

then the trajectory of c is not alternating, but f has no periodic point of period $m-1$.
[It may be verified that the adjacency matrix of the corresponding digraph has characteristic polynomial

$$(\lambda^n - 2\lambda^{n-2} - 2\lambda^{m-1} - 1)/(\lambda + 1).]$$

The next result sharpens Proposition 3.6 of Li *et al.* [76].

THEOREM 12 *Suppose that f^2 is not turbulent. If, for some $c \in I$ and some $n > 1$, we have either $f^n(c) \leq c < f(c)$ or $f(c) < c \leq f^n(c)$, then the trajectory of c is alternating.*

Proof Again set $x_k = f^k(c)$ $(k \geq 0)$. By Lemma 8, since f is not turbulent, the map f has a fixed point z such that all U-points of the trajectory of c lie to the left of z and all D-points to the right of z. If the trajectory contains a fixed point then z may be taken to be this fixed point.

Suppose first that $n = 2$ and, without loss of generality, $x_2 \leq x_0 < x_1$. If the trajectory of x_0 is not alternating, then $x_2 < x_0$ and there is a least integer h such that either $x_h \leq x_{h+1} \leq z$ or $x_h \geq x_{h+1} \geq z$. Evidently $h > 1$ and the first or second alternative holds according as h is even or odd. Assume first that $h = 2$. Then

$$[z, x_1] \subseteq f[x_2, x_0] \cap f[x_0, z]$$

and hence

$$[x_2, z] \subseteq f^2[x_2, x_0] \cap f^2[x_0, z],$$

contrary to the hypothesis that f^2 is not turbulent. Hence we can assume that $h > 2$ and that there is no counterexample to the theorem for any smaller value of h.

If for some j with $2 < j \leq h$ we have $x_j \leq x_{j-2} < x_{j-1}$ or $x_j \geq x_{j-2} > x_{j-1}$, then for the trajectory of $x'_0 = x_{j-2}$ we have $h' = h - j + 2 < h$. Hence, by the induction hypothesis, the trajectory of x'_0 is alternating, which gives a contradiction. Therefore x_{j-2} lies outside the interval $<x_{j-1}, x_j>$ for $2 < j \leq h$. Taking $j = 3, \ldots, h$ in succession we obtain

$$x_2 < x_4 < \ldots < x_h < x_{h-1} < \ldots < x_3 < x_1 \quad \text{if } h \text{ is even,}$$
$$x_2 < x_4 < \ldots < x_{h-1} < x_h < \ldots < x_3 < x_1 \quad \text{if } h \text{ is odd.}$$

We will show that $x_h < x_0$ if h is even and $x_{h-1} < x_0$ if h is odd.

Assume, on the contrary, that $x_{k-2} < x_0 \leq x_k$ for some even k $(2 < k \leq h)$. Then $x_0 \neq x_k$, since $x_1 \neq x_{k+1}$ (even if $k = h$, since $x_{h+1} \leq z$). Hence

$$[x_2, x_k] \subseteq f^2[x_{k-2}, x_0] \cap f^2[x_0, x_k],$$

if $k = h$ as well as if $k < h$, which is a contradiction.

If h is even then $x_2 < x_h < x_0 < z$ and $x_{h+1} \leq z$, hence

$$[x_2, z] \subseteq f^2[x_h, x_0] \cap f^2[x_0, z],$$

which is a contradiction. If h is odd we obtain a contradiction similarly. This completes the proof for $n = 2$.

We now suppose that $n > 2$ and that the theorem holds for all smaller values of n. Moreover we restrict attention to the case $x_n \leq x_0 < x_1$. By what we have already proved we may suppose $x_0 < x_2$. By Lemma 4, the map f does not have a periodic point of odd period > 1. Hence, by Theorem 9, n is even and $x_k < x_j$ for all even k and odd j with $0 \leq j, k \leq n$. Since $x_n \leq x_2 < x_3$, it follows from the induction hypothesis that the trajectory of x_2 is alternating. Since x_0 is a U-point and x_1 a D-point, this implies that the trajectory of x_0 is also alternating, as we wished to prove. □

The next example shows that in Theorem 12 we cannot replace "If f^2 is not turbulent" by "If f has no orbit of odd period > 1".

EXAMPLE 13 Let f: [0,4] → [0,4] be the piecewise linear map defined by

$$f(0) = 2, f(1) = 4, f(2) = 3, f(3) = 2, f(4) = 0.$$

For $x_0 = 1$ we have $x_2 < x_0 < x_1$, but the trajectory of x_0 is not alternating, since $x_3 < x_4$. On the other hand, since $f^2(x) = x$ for $2 \leq x \leq 3$, an orbit of odd period > 1 cannot have a point in [2,3] and therefore also not in [0,1/2]. Since f[1/2,2] = [3,4] and f[3,4] = [0,2], it follows that f has no orbit of odd period > 1.

We are now in a position to strengthen Lemma 4.

THEOREM 14 *If n is not a power of 2, the following statements are equivalent:*

(i) *f has a periodic point of period n,*
(ii) *f^n is strictly turbulent,*
(iii) *f^n is turbulent.*

Proof Suppose first that n is odd. This implies $n > 1$, since n is not a power of 2. Then (i) $\Rightarrow$ (ii), by Lemma 4, and (ii) $\Rightarrow$ (iii) is trivial. It remains to show that (iii) $\Rightarrow$ (i). By Lemma 2 we may suppose that there exist points $a < c < b$ such that $f^n(c) = b, f^n(b) = f^n(a) = a$. If $f(a) \neq a$ then f has a periodic point of period n, by Šarkovskii's theorem. Thus we may suppose that $f(a) = a$. If $f(c) < c$ then f has a periodic point of period n, by Theorem 9 (n odd) and Šarkovskii's theorem. Thus we may suppose that $f(c) > c$. Since $a = f^{2n}(c) < c$ and $f(a) = a$, it follows from Lemma 7 that f is turbulent. Hence f has periodic points of every period, by Lemma 3.

Suppose next that $n = 2^d q$, where $q > 1$ is odd and $d \geq 1$. Then the equivalence of (i) - (iii) for f follows from their equivalence for f^{2^d}, by Lemma I.10 and Šarkovskii's theorem. □

Theorem 14 is due to Blokh [41] with a different, but equivalent, definition of turbulence. It is an immediate consequence of Theorem 14 and Lemma 3 that the following conditions are equivalent:

(i) f has a periodic point whose period is not a power of 2,
(ii) f^m is strictly turbulent for some positive integer m,
(iii) f^n is turbulent for some positive integer n.

The map f will be said to be *chaotic* if one, and hence all three, of these conditions is satisfied. This term was first used by Li and Yorke [77] in a different sense, although without giving any formal definition. It follows at once from our definition that, for any $k > 0, f$ is chaotic if and only if f^k is chaotic.

Let $\mathbb{K}$ denote the set of all continuous maps $f: I \to I$ which are chaotic. Also let $\mathbb{P}_n$, $\mathbb{T}_n$, $\mathbb{S}_n$ denote respectively the set of all continuous maps $f: I \to I$ such that f has a periodic point of period n, f^n is turbulent, f^n is strictly turbulent.

By definition, $\mathbb{K} = \bigcup_{n>0} \mathbb{T}_n$ and, by Theorem 14,

$$\mathbb{P}_n = \mathbb{T}_n = \mathbb{S}_n \quad \text{if } n \text{ is not a power of 2.}$$

Moreover, by Lemmas 3 and 4, the *Šarkovskii stratification*

$$\mathbb{P}_3 \subset \mathbb{P}_5 \subset \mathbb{P}_7 \subset \ldots \subset \mathbb{P}_6 \subset \mathbb{P}_{10} \subset \ldots \subset \mathbb{P}_{12} \subset \mathbb{P}_{20} \subset \ldots$$
$$\subset \mathbb{P}_8 \subset \mathbb{P}_4 \subset \mathbb{P}_2 \subset \mathbb{P}_1$$

admits the following refinement:

$$\mathbb{S}_1 \subset \mathbb{T}_1 \subset \mathbb{P}_3 \subset \mathbb{P}_5 \subset \mathbb{P}_7 \subset \ldots \subset \mathbb{S}_2 \subset \mathbb{T}_2 \subset \mathbb{P}_6 \subset \mathbb{P}_{10} \subset \ldots$$
$$\subset \mathbb{S}_4 \subset \mathbb{T}_4 \subset \mathbb{P}_{12} \subset \mathbb{P}_{20} \subset \ldots \subset \mathbb{K} \subset \ldots \subset \mathbb{P}_8 \subset \mathbb{P}_4 \subset \mathbb{P}_2 \subset \mathbb{P}_1 .$$

This *turbulence stratification* provides an intuitive picture of many of our results.

3 SYMBOLIC DYNAMICS

To justify the terms 'turbulent' and 'chaotic' it should be shown that such maps behave more wildly than non-chaotic maps. The benign nature of non-chaotic maps will be brought out in Chapter VI. Here we devote our attention to establishing the vicious nature of chaotic maps. This will be done using *symbolic dynamics*.

Let Σ denote the set of all infinite sequences $\alpha = (a_1, a_2, \ldots)$, where $a_k = 0$ or 1. We can define a metric on Σ by setting $d(\alpha, \alpha) = 0$ and $d(\alpha, \beta) = 2^{-m}$ if $\alpha \neq \beta$ and m is the least positive integer such that $a_m \neq b_m$. In fact the triangle inequality holds in the stronger form

$$d(\alpha, \gamma) \leq \max\, [d(\alpha, \beta), d(\beta, \gamma)].$$

It follows that a sequence $\{\alpha_n\}$ converges if and only if $d(\alpha_n, \alpha_{n+1}) \to 0$ as $n \to \infty$. Moreover, any sequence $\{\alpha_n\}$ contains a convergent subsequence. For there exist infinitely many n for which α_n has the same first term a_1. Let n_1 be the least such n and set $\beta_1 = \alpha_{n_1}$. Among these n there exist infinitely many for which α_n also has the same second term a_2. Let n_2 be the least such $n > n_1$ and set $\beta_2 = \alpha_{n_2}$. Continuing in this way, we obtain a convergent subsequence $\{\beta_j\}$. Thus Σ is a compact metric space.

A compact set X in a metric space is said to be a *Cantor set* if it has no isolated point and is totally disconnected, *i.e.* for any two distinct points α,β there exist disjoint closed subsets A,B of X which contain α,β respectively. It is easily seen that a compact subset of the real line is a

Cantor set if and only if it has no isolated point and contains no interval. It will now be shown that Σ is a Cantor set.

It is evident that no point of Σ is isolated. If α,β are distinct points of Σ, define A to be the set of all $\gamma \in \Sigma$ such that $d(\alpha, \gamma) < d(\alpha, \beta)$ and B to be the set of all $\gamma \in \Sigma$ such that $d(\beta, \gamma) < d(\alpha, \beta)$. Then A and B are closed sets containing α and β respectively. Moreover, by virtue of the strong triangle inequality, they are disjoint. Thus Σ is totally disconnected.

The *shift* operator σ is defined by $\sigma((a_1, a_2, ...)) = (a_2, a_3, ...)$. It is a 2-1 map of Σ onto itself. Moreover σ is continuous, since $d[\sigma(\alpha), \sigma(\beta)] \le 2d(\alpha, \beta)$. The sequence $\alpha = (a_1, a_2, ...)$, where $a_k = 1$ if k is divisible by n and $= 0$ otherwise, is periodic with period n. Thus σ has periodic points with period n, for every positive integer n.

These interpretations of Σ and σ are understood in the statement of the following proposition.

PROPOSITION 15 *If f is strictly turbulent, then there exists a compact set $X \subseteq I$ such that $f(X) = X$ and a continuous map h of X onto Σ such that each point of Σ is the image of at most two points of X and*

$$h \circ f(x) = \sigma \circ h(x) \quad \textit{for every } x \in X,$$

that is, the accompanying diagram commutes:

$$\begin{array}{ccc} X & \xrightarrow{f} & X \\ {\scriptstyle h}\downarrow & & \downarrow{\scriptstyle h} \\ \Sigma & \xrightarrow[\sigma]{} & \Sigma \end{array}$$

Conversely, if there exist such a set X and map h, then f^4 is strictly turbulent.

Proof Let I_0, I_1 be disjoint compact subintervals such that

$$I_0 \cup I_1 \subseteq f(I_0) \cap f(I_1).$$

Let $I_{a_1a_2}$ be a subinterval of I_{a_1} of minimal length such that $f(I_{a_1a_2}) = I_{a_2}$, where $a_1, a_2 = 0$ or 1. Proceeding inductively, let $I_{a_1...a_k}$ be a subinterval of $I_{a_1...a_{k-1}}$ of minimal length such that $f(I_{a_1...a_k}) = I_{a_2...a_k}$, where $a_1, ..., a_k = 0$ or 1. It is readily seen that

$$I_{a_1 \dots a_k} \cap I_{b_1 \dots b_k} = \varnothing \quad \text{if } (a_1, \dots, a_k) \neq (b_1, \dots, b_k).$$

For any infinite sequence $\alpha = (a_1, a_2, \dots)$ of 0's and 1's let

$$I_\alpha = \bigcap_{k=1}^{\infty} I_{a_1 \dots a_k}.$$

Then I_α is either a compact interval or a single point. Moreover $I_\alpha \cap I_\beta = \varnothing$ if $\alpha \neq \beta$, and hence I_α is an interval for at most countably many values of α. Since f is continuous,

$$f(I_\alpha) = \bigcap_{k=2}^{\infty} f(I_{a_1 \dots a_k}) = \bigcap_{k=2}^{\infty} I_{a_2 \dots a_k} = I_{\sigma(\alpha)},$$

where $\sigma(\alpha) = (a_2, a_3, \dots)$ as before. Thus if I_α is a point, so also is $I_{\sigma(\alpha)}$.

Let

$$\hat{X} = \bigcup_{\alpha \in \Sigma} I_\alpha.$$

We will show that $\hat{X}$ is a closed, and hence compact, subset of $I_0 \cup I_1$. Suppose $x_n \to x$, where $x_n \in I_{\alpha_n}$. By restricting attention to a subsequence we may assume that $\alpha_n \to \alpha = (a_1, a_2, \dots)$. Then, for any given k, $I_{\alpha_n} \subseteq I_{a_1 \dots a_k}$ for all large n. Thus $x \in I_{a_1 \dots a_k}$ for every k, and hence $x \in I_\alpha$.

Since $I_\alpha \cap I_\beta = \varnothing$ if $\alpha \neq \beta$, it follows that if $\alpha_n \to \alpha$ and $\alpha_n \neq \alpha$ for all large n, then any limit point of a sequence $x_n \in I_{\alpha_n}$ must be an endpoint of I_α. (If I_α is a point, we consider this point to be an endpoint.) Hence the set X of all endpoints of all I_α is also closed. Since $I_{a_1 \dots a_k}$ was chosen of minimal length, f maps the two endpoints of $I_{a_1 \dots a_k}$ onto the two endpoints of $I_{a_2 \dots a_k}$. Therefore, if I_α and $I_{\sigma(\alpha)}$ are intervals, f maps the two endpoints of I_α onto the two endpoints of $I_{\sigma(\alpha)}$. It follows that $f(X) = X$.

Define a map h of X onto Σ by setting $h(x) = \alpha$ if $x \in I_\alpha$. Then each point of Σ is the image of at most two points of X (and at most countably many points of Σ are the image of *two* points of X). Moreover the map h is continuous. For let $\delta_k > 0$ be the least distance between any two of the 2^k intervals $I_{a_1 \dots a_k}$. If $x \in I_\alpha$, $y \in I_\beta$ and $|x - y| < \delta_k$, then $d(\alpha, \beta) < 2^{-k}$. Finally, since $f(I_\alpha) = I_{\sigma(\alpha)}$, we have $h \circ f(x) = \sigma \circ h(x)$ for every $x \in X$.

Conversely, suppose there exist a set X and map h with the properties stated in the proposition. Choose $\alpha \in \Sigma$ so that $\sigma^3(\alpha) = \alpha$, $\sigma(\alpha) \neq \alpha$ and then choose $x \in X$ so that $h(x) = \alpha$. Since $h \circ f^j(x) = \sigma^j(\alpha)$, it follows that f has a periodic point of period 3 or 6. Hence f^4 is strictly turbulent. □

COROLLARY 16 *A map f is chaotic if and only if there exists a compact set $X \subseteq I$ such that $f^m(X) = X$ for some $m > 0$, and a continuous map h of X onto Σ such that each point of Σ is the image of at most 2 points of X and*

$$h \circ f^m(x) = \sigma \circ h(x) \quad \text{for every } x \in X. \quad \square$$

According to the general definitions in Chapter I, the map h in Proposition 15 is a semi-conjugacy and the shift σ is a factor of the restriction of f to X. The virtue of Proposition 15 is that it enables us to derive properties of any strictly turbulent map f from properties of the universal map σ. In the following application we use the notion of limit set, which will be studied in Chapter IV. Here it is sufficient to know that $\omega(x,f)$ is the set of all limit points of the sequence $\{f^n(x)\}$.

PROPOSITION 17 *If f is strictly turbulent, then there exist uncountably many points $x \in I$ such that the limit set $\omega(x,f)$ is uncountably infinite and contains a periodic point of period m or 2m, for every positive integer m.*

Proof There exist uncountably many $\alpha \in \Sigma$ whose expansion $(a_1, a_2, \ldots)$ contains every finite sequence of 0's and 1's once, and hence infinitely often. For example we can take

$$\alpha = (0,1,*,0,0,0,1,1,0,1,1,*,0,0,0,\ldots),$$

where all sequences of length 1 appear and then an arbitrary term, next all sequences of length 2 appear and then an arbitrary term, and so on. Given any such α, for every $\beta \in \Sigma$ there is a sequence of positive integers $n_1 < n_2 < \ldots$ such that $\sigma^{n_k}(\alpha) \to \beta$ as $k \to \infty$. If we choose $x_\alpha \in X$ so that $h(x_\alpha) = \alpha$, then for every $\beta \in \Sigma$ there exists a point $y_\beta \in \omega(x_\alpha, f)$ such that $h(y_\beta) = \beta$. It is now easily seen that we can take $x = x_\alpha$. Thus there are indeed uncountably many points x with the required properties. $\square$

COROLLARY 18 *If f is chaotic, then there exist uncountably many points $x \in I$ such that the limit set $\omega(x,f)$ is uncountably infinite and contains periodic points with periods divisible by every positive integer.* $\square$

Evans *et al.* [62] set up symbolic dynamics, in a modified form, for maps which are turbulent, but not necessarily strictly turbulent. They show, in particular, that Proposition 17 continues to hold if the word 'strictly' is omitted from its statement.

4 PERTURBED MAPS

It is of interest not only to study the behaviour of an individual map f but also to relate this behaviour to that of 'neighbouring' maps. This notion will now be made precise.

Let $C(I, I)$ denote the set of all continuous maps of I into itself. If the interval I is compact, then $C(I, I)$ becomes a metric space if we define the distance between two elements f,g by

$$d(f,g) = \sup_{x\in I} |f(x) - g(x)| .$$

If the interval I is not compact, let $J_1 \subset J_2 \subset \ldots$ be an increasing sequence of compact subintervals with union I. Then $C(I, I)$ becomes a metric space if we define the distance between two elements f,g by

$$d(f,g) = \Sigma_{n=1}^{\infty} d_n(f,g) / 2^n [1+d_n(f,g)] ,$$

where $d_n(f,g) = \sup |f(x) - g(x)|$, the supremum being taken over all $x \in J_n \setminus J_{n-1}$, with $J_0 = \emptyset$.

If $f \in C(I, I)$ then a *neighbourhood* of f in $C(I, I)$ is a subset **N** of $C(I, I)$ which contains all $g \in C(I, I)$ with $d(f,g) < \varepsilon$, for some $\varepsilon > 0$. Otherwise expressed, a set $\mathbf{N} \subseteq C(I, I)$ is a neighbourhood of f if there exists a compact interval $J \subseteq I$ and a number $\delta > 0$ such that **N** contains all continuous maps $g: I \to I$ with $|f(x) - g(x)| < \delta$ for every $x \in J$.

The usual topological notions can now be defined. For example, a subset **O** of $C(I, I)$ is *open* if, for every $f \in \mathbf{O}$, there exists a neighbourhood **N** of f which is contained in **O**. Again, if **E** is any subset of $C(I, I)$, its *interior* int **E** is the largest open subset contained in **E**. We can also define convergence. A sequence $f_n \in C(I, I)$ converges to $f \in C(I, I)$ if $d(f_n, f) \to 0$ as $n \to \infty$, i.e. if $f_n(x) \to f(x)$ uniformly on every compact subinterval of I.

The following theorem is due to Block [24].

THEOREM 19 *For any positive integers m, n*

$$\mathbb{P}_n \subset \text{int}\, \mathbb{P}_m \quad \text{if } n \prec m.$$

Proof Let $f: I \to I$ be a continuous map with a periodic point of period $n > 1$. We wish to show that there exists a neighbourhood **N** of f in $C(I, I)$ such that any $g \in \mathbf{N}$ has a periodic point of period m, for all m satisfying $n \prec m$. The proof will be carried out in a number of stages.

(i) n is odd.

By Šarkovskii's theorem we may assume that f has no periodic point of period ℓ, where ℓ is odd and $1 < \ell < n$. Furthermore, by Proposition I.8, we may assume that f has a periodic point x of period n such that

$$f^{n-1}(x) < f^{n-3}(x) < ... < f^2(x) < x < f(x) < ... < f^{n-2}(x).$$

Hence

$$f^{n+2}(x) < x < f(x).$$

There exists a neighbourhood **N** of f such that the last inequalities continue to hold when f is replaced by any $g \in$ **N**. Therefore, by Theorem 9 and Šarkovskii's theorem, g has a periodic point of period n+2.

(ii) $n = 4$.

By (i) and Šarkovskii's theorem we may assume that f does not have a periodic point of period 3. If $x_1 < x_2 < x_3 < x_4$ is an orbit of period 4 then, by Theorem 9, f interchanges the pairs x_1, x_2 and x_3, x_4. Thus $[x_1, x_2] \subseteq f[x_3, x_4]$ and $[x_3, x_4] \subseteq f[x_1, x_2]$. Hence $[x_1, x_2] = f(J)$, for some compact interval $J \subseteq [x_3, x_4]$. If y is that element of the pair x_1, x_2 for which $f(y) = x_4$ then $y = f(z)$ for some $z \in J$. Similarly, if y' is that element of the pair x_1, x_2 for which $f(y') = x_3$ then $y' = f(z')$ for some $z' \in J$. Since x_3 and x_4 do not have period 2, it follows that

$$f^2(z) > z,\ f^2(z') < z',\ f(x) < x_3 \quad \text{for all } x \in J.$$

There exists a neighbourhood **N** of f such that these inequalities continue to hold when f is replaced by any $g \in$ **N**. Therefore g has a periodic point of period 2.

(iii) $n = 2^d$, where $d > 2$.

If we set $s = 2^{d-2}$ then f^s has a periodic point of period 4. By (ii) there exists a neighbourhood **N** of f such that, for any $g \in$ **N**, g^s has a periodic point of period 2 and hence g has a periodic point of period $2s = 2^{d-1}$.

(iv) $n = 2^d q$, where $q > 1$ is odd and $d \geq 1$.

If we set $p = 2^d$ then f^p has a periodic point of period q. By (i) there exists a neighbourhood **N** of f such that, for any $g \in$ **N**, g^p has a periodic point of period $q + 2$ and hence g has a periodic point of period $2^e(q+2)$, where $0 \leq e \leq d$.

The theorem now follows immediately from the results already established and Šarkovskii's theorem. □

COROLLARY 20 *The set* $\mathbb{K}$ *of all chaotic maps is an open subset of* $C(I, I)$. □

There is an analogous result to Theorem 19 for turbulent maps.

THEOREM 21 *For any integer* $d \geq 0$,

$$\mathbb{T}_{2^d} \subset \text{int}\, \mathbb{T}_{3.2^d} .$$

Proof We give the proof for $d = 0$, the general case following by Lemma I.10. Let $f: I \to I$ be a continuous map which is turbulent. By Theorem 14, we need only show that there is a neighbourhood $\mathbf{N}$ of f in $C(I, I)$ such that any $g \in \mathbf{N}$ has a periodic point of period 3.

Since f is turbulent, there exist points a,b,c satisfying the conclusions of Lemma 2. Then either

$$f^3(c) < c < f(c)$$

or the same with the inequalities reversed. We may assume without loss of generality that the displayed inequalities hold. Then there exists a neighbourhood $\mathbf{N}$ of f such that the displayed inequalities continue to hold when f is replaced by any $g \in \mathbf{N}$. Then g has a periodic point of period 3, by Theorem 9. □

If $f: I \to I$ is an arbitrary continuous map of a compact interval into itself, then any C-neighbourhood of f contains strictly turbulent maps. To see this, let z be a fixed point of f and let y,w be interior points of I close to z with $y < w$. Choose points a,b,c,d so that

$$y < a < b < c < d < w.$$

Then the continuous map $g: I \to I$ defined by

$$g(x) = f(x) \quad \text{for } x \leq y \text{ or } x \geq w,$$
$$g(a) = a, \;\; g(b) = d, \;\; g(c) = d, \;\; g(d) = a,$$

and elsewhere by linearity, is strictly turbulent.

It follows that, for any positive integer m, $\mathbb{T}_m$ is not a closed subset of $C(I, I)$. It may be remarked also that $\mathbb{S}_{2^d}$ is not contained in the interior of $\mathbb{T}_{2^d}$. This can be shown, for $d = 0$, in the following way.

If f is the piecewise linear map defined by

$$f(0) = 0,\ f(1) = 3 = f(2),\ f(3) = 0,$$

then f is strictly turbulent. On the other hand, if g_ε is the piecewise linear map defined by

$$g_\varepsilon(0) = 0,\ g_\varepsilon(1) = 3 - \varepsilon = g_\varepsilon(2),\ g_\varepsilon(3) = 0,$$

where $0 < \varepsilon < 1$, then g_ε is not turbulent. Since any neighbourhood of f contains a map g_ε for ε sufficiently small, it follows that $\mathbb{S}_1$ is not contained in the interior of $\mathbb{T}_1$.

An interesting question remains open in connection with the turbulence stratification. Let $\mathbb{P}_{2^\infty}$ denote the set of all continuous maps $f: I \to I$ which have periodic points of period 2^d for all $d \geq 0$ and of no other periods. It may be asked whether $\mathbb{P}_{2^\infty}$ is the boundary of $\bigcup_{d\geq 0} \mathbb{P}_{2^d}$. In other words, if $f \in \mathbb{P}_{2^\infty}$, does any neighbourhood $\mathbf{N}$ of f in $C(I, I)$ contain not only chaotic maps but also maps whose periodic points have only finitely many periods?

5 SMOOTH MAPS

At this point we depart from our usual practice and establish some results which hold only for continuously differentiable maps. Let $C^1(I, I)$ denote the set of all continuously differentiable maps of I into itself. Then $C^1(I, I)$ becomes a metric space if we replace $\sup |f(x) - g(x)|$ by

$$\sup |f(x) - g(x)| + \sup |f'(x) - g'(x)|$$

in the definition of the metric on $C(I, I)$. This metric induces a corresponding topology. Furthermore, $f_n \to f$ in $C^1(I, I)$ if and only if $f_n(x) \to f(x)$ and $f_n'(x) \to f'(x)$, uniformly on every compact subinterval of I.

PROPOSITION 22 *If the interval I is compact then, for any positive integer m, $\mathbb{T}_m^1 := \mathbb{T}_m \cap C^1(I, I)$ is a closed subset of $C^1(I, I)$.*

Proof Suppose $f \in \mathbb{T}_m^1$ and set $g = f^m$. Then, by Lemma 2, there exist points $a,b,c \in I$ such that $g(b) = g(a) = a$, $g(c) = b$ and either $a < c < b$ or $b < c < a$. In either case it follows from the mean-value theorem that the interval $\langle a,b \rangle$ contains a point y such that $g'(y) = 0$ and a point z such that $g'(z) > 1$.

Suppose $f_n \in \mathbb{T}_m^1$ and $f_n \to f$ in $C^1(I, I)$. Without loss of generality we may assume that there exist points $a_n < c_n < b_n$ such that, if $g_n = f_n^m$, then $g_n(b_n) = g_n(a_n) = a_n$, $g_n(c_n) = b_n$.

Moreover, by restriction to a subsequence we may assume that $a_n \to a$, $b_n \to b$, $c_n \to c$. Then $a \le c \le b$ and $g(b) = g(a) = a$, $g(c) = b$. It follows that also $f \in \mathbb{T}^1_m$, unless $a = c = b$. But the latter case cannot occur, since it would imply both $g'(a) = 0$ and $g'(a) \ge 1$, by the first part of the proof. □

We next derive an analogous result to Proposition 22 for periodic points.

LEMMA 23 *Let $f: I \to I$ be a differentiable map of an interval into itself, and let $x_1 < \ldots < x_m$ be a periodic orbit of f of period $m > 2$. Then there exist points $y,z \in (x_1, x_m)$ such that $f'(y) \le -1, f'(z) > 0$.*

Proof For some $j > 1$ we have $f(x_{j-1}) > x_{j-1}$, $f(x_j) < x_j$. Then $f(x_{j-1}) \ge x_j$, $f(x_j) \le x_{j-1}$ and hence, by the mean-value theorem, $f'(y) \le -1$ for some $y \in (x_{j-1}, x_j)$.

On the other hand, for some k with $1 < k < m$ we have either $f(x_k) = x_1$ or $f(x_k) = x_m$. If $f(x_k) = x_1$ then $f'(z) > 0$ for some $z \in (x_k, x_{k+1})$, and if $f(x_k) = x_m$ then $f'(z) > 0$ for some $z \in (x_{k-1}, x_k)$. □

The following result is due to Block and Hart [38].

PROPOSITION 24 *Let I be a compact interval, m a positive integer and, for every $n \ge 1$, let $f_n \in C^1(I, I)$ have a periodic point x_n of period m. If $f_n \to f$ in $C^1(I, I)$ and $x_n \to x$ as $n \to \infty$, then x is a periodic point of f of period m if m is odd, and of period m or $m/2$ if m is even.*

Proof By continuity it follows at once that x is a fixed point of f^m. This establishes the result for $m = 1,2$. We now show that if $m > 2$ then x is not a fixed point of f.

Assume, on the contrary, that $f(x) = x$. Let y_n be the least point and z_n the greatest point in the orbit of x_n. By restricting attention to a subsequence we may suppose that $y_n \to y$, $z_n \to z$ as $n \to \infty$. There are positive integers j,k independent of n such that, for infinitely many n,

$$f_n^j(x_n) = y_n,\ f_n^k(x_n) = z_n .$$

Then, by continuity,

$$f^j(x) = y,\ f^k(x) = z.$$

Since x is a fixed point, this implies $y = z = x$. It follows from Lemma 23 that $f'(x) \le -1$ and $f'(x) \ge 0$, which is a contradiction.

This establishes the result for $m = 3$. Suppose $m > 3$. Evidently x is a periodic point of f with period h, where h divides m. Moreover x_n is a periodic point of $f_n{}^h$ with period m/h, and x is a fixed point of f^h. Hence, by what we have already proved, $m/h = 1$ or 2. □

COROLLARY 25 *If the interval I is compact then, for any positive integer m, $\mathbb{P}^1_m := \mathbb{P}_m \cap C^1(I, I)$ is a closed subset of $C^1(I, I)$ if m is not a power of 2, and has its closure contained in $\mathbb{P}^1_{m/2}$ if m is a power of 2. Hence $\bigcap_{k\geq 0} \mathbb{P}^1_{2^k}$ is a closed subset of $C^1(I, I)$.*

Proof This follows at once from Proposition 24 and Šarkovskii's theorem. □

If $m > 1$ is a power of 2, then $\mathbb{P}^1_m$ is *not* a closed subset of $C^1(I, I)$. The following example, provided by a referee, illustrates this for $m = 2$.

Take $I = [-1, 1]$ and $f_\alpha(x) = x^3 - \alpha x$. Then f_α maps I into itself, at least for $0 < \alpha < 3/2$. Evidently 0 is a fixed point of f_α for every α. Moreover, f_1 has no other periodic points, since $|f_\alpha(x)| < |x|$ if $\alpha = 1$ and $x \neq 0$. On the other hand, $(\alpha-1)^{1/2}$ and $-(\alpha-1)^{1/2}$ form a periodic orbit of period 2 for $1 < \alpha < 3/2$.

We are going to study now maps whose set of periodic points is closed. We consider first the case of an arbitrary continuous map of an interval into itself.

THEOREM 26 *Suppose the set of all periodic points of f is compact. Then f is not chaotic. Moreover, if x_n is a periodic point of f, if ξ_n is any point in the orbit of x_n, and if $x_n \to x$, $\xi_n \to \xi$ as $n \to \infty$, then ξ is in the orbit of x.*

Proof Assume, on the contrary, that f is chaotic. Then there exist disjoint compact subintervals J,K and a positive integer m such that

$$J \cup K \subseteq f^m(J) \cap f^m(K).$$

By Lemma I.4, for any $n > 0$ there exists a point $x_n \in J$ such that $f^{mn}(x_n) = x_n$ and $f^{jm}(x_n) \in K$ for $0 < j < n$. If y is a limit point of the sequence $\{x_n\}$ then $y \in J$ and $f^{jm}(y) \in K$ for every $j > 0$. Hence y is not periodic, which is a contradiction.

Thus every periodic point of f has period a power of 2. It follows from Theorem 9 that, for any $m > 0$, every periodic orbit of f^m is alternating. Let y_n be the least point in the orbit of x_n. By restriction to a subsequence we may assume that $y_n \to y$. Then y is periodic, with period m say. Hence $f^m(y) = y$, $f^m(y_n) \to y$, and y_n is the least point in a periodic orbit of f^m. If z_n is the greatest point in this orbit then $y_n \leq f^m(z_n) \leq f^m(y_n) \leq z_n$ and hence $f^m(z_n) \to y$.

But if $z_n \to z$ then z is periodic and $f^m(z) = y$, which implies $z = y$. Hence if η_n is any point in the f^m–orbit of y_n, then $\eta_n \to y$. Since $x_n = f^{k_n}(y_n)$, where $k_n = j_n m + r_n$ and $0 \le r_n < m$, it follows that $r_n = r$ is constant for all large n and $x = f^r(y)$. Similarly $\xi = f^s(y)$, and hence ξ is in the f-orbit of x. □

The first assertion of Theorem 26 was proved by Šarkovskii [104] and the second by Fedorenko and Šarkovskii [63].

If $f \notin P_{2^d}$ for some $d > 0$ then, by Šarkovskii's theorem, the periodic points of f are just the fixed points of $f^{2^{d-1}}$, and hence the set of all periodic points is closed. On the other hand, we have seen in Example I.14 that a continuous map with a compact set of periodic points may have orbits of period 2^d for *every* $d \ge 0$. It will now be shown, following Block and Hart [38], that this is impossible for continuously differentiable maps. Indeed we will show that it is also impossible for the practically important class of piecewise monotone maps, the definition of which is given below. The proof for continuously differentiable maps will make use of the following lemma.

LEMMA 27 *Let $f: I \to I$ be a differentiable map of an interval into itself, and let $x_1 < \ldots < x_n$ be a periodic orbit of f of even period $n > 2$. If this orbit is alternating, then there exist points $y, z \in (x_1, f(x_1))$ such that $f'(y) \le -1, f'(z) > 0$.*

Proof Let $m = n/2$. Since the orbit is alternating, f interchanges $\{x_1, \ldots, x_m\}$ and $\{x_{m+1}, \ldots, x_n\}$. In particular $f(x_m) \ge x_{m+1}, f(x_{m+1}) \le x_m$, and hence $f'(y) \le -1$ for some $y \in (x_m, x_{m+1}) \subseteq (x_1, f(x_1))$.

On the other hand $x_n = f(x_k)$, where $1 \le k \le m$. If $k = 1$ then, by Lemma 23, $f'(z) > 0$ for some $z \in (x_1, f(x_1))$. If $k > 1$ then, since $f(x_{k-1}) < f(x_k)$, $f'(z) > 0$ for some $z \in (x_{k-1}, x_k) \subseteq (x_1, f(x_1))$. □

A continuous map $f: I \to I$ is said to be *piecewise monotone* if I is the union of finitely many intervals on each of which f is either strictly increasing, strictly decreasing, or constant. A point $t \in I$ is said to be a *turning-point* of a piecewise monotone map if in every open interval containing t the map f is neither strictly increasing, strictly decreasing, nor constant. Thus a piecewise monotone map has at most finitely many turning-points. It follows from the definition that if f and g are piecewise monotone maps of an interval I to itself, then also $g \circ f$ is piecewise monotone. Hence all the iterates of a piecewise monotone map are again piecewise monotone.

The reader should be warned that some authors define a map $f: I \to I$ to be piecewise monotone if I is the union of finitely many intervals on each of which f is either non-decreasing or non-increasing.

PROPOSITION 28 *Let I be a compact interval and suppose either $f \in C^1(I, I)$ or f is piecewise monotone and $f \in C(I, I)$. If the set of all periodic points of f is closed, then there exists a positive integer d such that the set of periods of all periodic points is precisely* $\{1,2, \dots, 2^{d-1}\}$.

Proof By Theorem 26, every periodic point has period a power of 2 and by Theorem 9, for any $m > 0$, every periodic orbit of f^m is alternating.

Assume, on the contrary, that f has periodic points x_n of period $m_n = 2^{e_n}$, where $e_n \to \infty$ as $n \to \infty$. We may suppose that x_n is the least point in its orbit and that $x_n \to x$ as $n \to \infty$. Then x is a periodic point, of period $m = 2^e$ say. If we set $g = f^m$, then x is a fixed point of g and x_n is the least point in a periodic orbit of g, of period $m_n' = 2^{e_n - e}$ for all large n. Moreover $g(x_n) \to g(x) = x$.

If $f \in C^1(I, I)$ then also $g \in C^1(I, I)$. Hence it follows from Lemma 27 that $g'(x) \leq -1$ and $g'(x) \geq 0$, which is a contradiction.

If f is piecewise monotone then also g is piecewise monotone. Since the whole g^2-orbit of x_n lies between x_n and $g(x_n)$, the greatest point y_n in the g-orbit of x_n has the form $y_n = g(z_n)$, where $z_n \to x$. Hence also $y_n \to x$. Since g must have at least one turning-point in the interval $[x_n, y_n]$, it has exactly one turning-point in this interval for all large n. But the same argument applies to the g^2-orbits of x_n and $g(x_n)$. Thus g^2 has two turning-points in the interval $[x_n, y_n]$. Since x_n and y_n both converge to x, this gives a contradiction. □

Under the hypotheses of Proposition 28, if x_n is any periodic point of f, of period m_n, and if $x_n \to x$, where x has period m, then $m_n = m$ or $2m$ for all large n. To see this, suppose first that $f \in C^1(I, I)$. Then m_n can only take the values $1,2, \dots, 2^{d-1}$ and if $m_n = h$ for infinitely many n then $m = h$ or $h/2$, by Proposition 24.

Suppose next that f is piecewise monotone and $f \in C(I, I)$. Assume the conclusion is false, and let $g = f^m$. Then $g(x) = x$ and by restricting attention to a subsequence we may assume that the period of x_n under g is at least 4 for each n. Let y_n and z_n denote respectively the least and greatest points in the g-orbit of x_n. By again considering a subsequence, we may assume that

$y_n \to y$ and $z_n \to z$. Since $x_n \to x$, it follows from Theorem 26 that $y = z = x$. This yields a contradiction by a similar argument to that used at the end of the proof of Proposition 28.

III
Unstable Manifolds and Homoclinic Points

1 UNSTABLE MANIFOLDS OF FIXED POINTS

In the theory of smooth diffeomorphisms the notions of stable and unstable manifold play a significant role. It is remarkable that also for arbitrary continuous maps of an interval *unstable* manifolds can be defined and possess a number of basic properties.

Let $f: I \to I$ be a continuous map of an interval into itself and let z be a periodic point of f. The *unstable manifold* of z is defined to be the set

$$W(z,f) = \bigcap_{\varepsilon>0} \bigcup_{m\geq 0} f^m(z-\varepsilon, z+\varepsilon). \tag{1}$$

Thus $x \in W(z,f)$ if and only if $x = f^{m_k}(y_k)$ for some sequence of points $y_k \to z$ and some sequence of non-negative integers m_k. If x does not belong to the orbit of z this implies $m_k \to \infty$.

It is useful also to consider the *left and right unstable manifolds* of z defined by

$$\begin{aligned} W(z,f,L) &= \bigcap_{\varepsilon>0} \bigcup_{m\geq 0} f^m(z-\varepsilon, z], \\ W(z,f,R) &= \bigcap_{\varepsilon>0} \bigcup_{m\geq 0} f^m[z, z+\varepsilon). \end{aligned} \tag{2}$$

Thus $x \in W(z,f,L)$, resp. $W(z,f,R)$, if and only if $x = f^{m_k}(y_k)$ for some sequence of non-negative integers m_k and some sequence of points $y_k \to z$ with $y_k \leq z$, resp. $y_k \geq z$. If x is not in the orbit of z this again implies $m_k \to \infty$. Evidently

$$W(z,f) = W(z,f,L) \cup W(z,f,R). \tag{3}$$

We consider first the case in which the periodic point z is actually a *fixed* point.

PROPOSITION 1 *Suppose $f(z) = z$. Then each of $W(z,f)$, $W(z,f,L)$ and $W(z,f,R)$ is either an interval containing z or consists of the single point z.*

Proof It will be sufficient to give the proof for $W(z,f)$. Since z is a fixed point of f, $f^m(z - \varepsilon, z + \varepsilon)$ is a connected set containing z. It follows from (1) that $W(z,f)$ is also a connected set containing z. But a connected subset of the real line which contains more than one point is an interval. □

The left and right unstable manifolds of a fixed point are related in the following way.

PROPOSITION 2 *Suppose $f(z) = z$.*
If $W(z,f,L)$ contains a point $> z$, then $W(z,f,R) \subseteq W(z,f,L)$.
If $W(z,f,L)$ contains no point $< z$, then $W(z,f,L) = \{z\}$ or $W(z,f,R)$.
If $W(z,f,R)$ contains a point $< z$, then $W(z,f,L) \subseteq W(z,f,R)$.
If $W(z,f,R)$ contains no point $> z$, then $W(z,f,R) = \{z\}$ or $W(z,f,L)$.

Proof Only the first two assertions will be proved. Suppose $W(z,f,L)$ contains a point $c > z$. Then $[z,c] \subseteq W(z,f,L)$. Moreover it follows at once from (2) that $f(W(z,f,L)) \subseteq W(z,f,L)$. If $x \in W(z,f,R)$ then $x = f^{m_k}(y_k)$, where $y_k \to z$ and $y_k \geq z$. Thus $y_k \in W(z,f,L)$ for large k, and hence also $x \in W(z,f,L)$. This shows that $W(z,f,R) \subseteq W(z,f,L)$.

Now suppose, in addition, that $W(z,f,L)$ contains no point $< z$. Take any d such that $z < d < c$ and then choose $a < z$ so that $f(x) < d$ for $x \in (a,z)$. Since $a \notin W(z,f,L)$, there exists $b \in (a,z)$ such that $f^n(x) \neq a$ for all $n \geq 0$ if $x \in (b,z)$. We have $c = f^n(y)$ for some $y \in (b,z)$ and some $n > 0$. Let m be the greatest positive integer $\leq n$ such that $f^k(y) \in (a,z)$ for $0 \leq k < m$. If $f^m(y) < z$ then $f^m(y) \leq a$ and hence $f^m(y') = a$ for some $y' \in (b,z)$, which is a contradiction. Therefore $z < f^m(y) < d$. Since d may be arbitrarily close to z, it follows that $c \in W(z,f,R)$. Thus $W(z,f,L) \subseteq W(z,f,R)$. □

A subset S of the underlying interval I is said to be *invariant*, or *f-invariant*, if $f(S) \subseteq S$. It is said to be *strongly invariant* if $f(S) = S$.

In the proof of Proposition 2 we used the trivial fact that the left and right unstable manifolds are invariant. We now use Proposition 2 to show that they are actually strongly invariant.

PROPOSITION 3 *Suppose $f(z) = z$. Then $f(W(z,f)) = W(z,f)$ and*

$$f(W(z,f,L)) = W(z,f,L),\ f(W(z,f,R)) = W(z,f,R)\ .$$

Proof We give the proof for $W(z,f,L)$. The proof for $W(z,f,R)$ is analogous, and the result for $W(z,f)$ then follows from (3).

Evidently we may assume $W(z,f,L) \neq \{z\}$. Suppose first that $W(z,f,L)$ contains a left neighbourhood of z. If $c \in W(z,f,L)$ and $c \neq z$ then $c = f^m(y)$, where $m > 0$ and $y \in W(z,f,L)$. Then $d = f^{m-1}(y) \in W(z,f,L)$, since $W(z,f,L)$ is invariant, and $f(d) = c$. Suppose next that $W(z,f,L)$ has left endpoint z. Then $W(z,f,L) = W(z,f,R)$, by Proposition 2, and the strong invariance of $W(z,f,L)$ follows similarly from the invariance of $W(z,f,R)$. □

The unstable manifolds of a fixed point may in many cases be determined by combining their strong invariance with the following simple result.

PROPOSITION 4 *Suppose $f(z) = z$.*
If $f(x) < x$ for $a \leq x < z$, then $[a,z] \subseteq W(z,f,L)$.
If $f(x) > x$ for $z < x \leq b$, then $[z,b] \subseteq W(z,f,R)$.

Proof We prove only the first assertion. For any x_0 such that $a \leq x_0 < z$ there is an x_1 such that $x_0 < x_1 < z$ and $f(x_1) = x_0$. Thus there is an increasing sequence $\{x_k\}$, bounded above by z, such that $f(x_{k+1}) = x_k$ for every k. Morover $x_k \to z$ as $k \to \infty$, since the sequence $\{x_k\}$ converges and its limit is a fixed point of f. Consequently $[a,z] \subseteq W(z,f,L)$. □

If z is a fixed point of f, then it is also a fixed point of each of its iterates. The relation between the corresponding unstable manifolds is described in the next two propositions. We first treat two-sided unstable manifolds, since the result in this case is simpler.

PROPOSITION 5 *If $f(z) = z$, then $W(z,f^n) = W(z,f)$ for every $n > 1$.*

Proof It follows at once from the definition that $W(z,f^n) \subseteq W(z,f)$. Suppose $c \in W(z,f)$. Then $c = f^{m_k}(y_k)$, where $y_k \to z$. For some r such that $0 \leq r < n$ we must have $m_k \equiv r \pmod n$ for infinitely many k. Since $x_k = f^r(y_k) \to z$, it follows that $c \in W(z,f^n)$. Thus $W(z,f) \subseteq W(z,f^n)$. □

PROPOSITION 6 *If $f(z) = z$, then exactly one of the following holds:*

(i) *$W(z,f^n, S) = W(z,f,S)$ for $S = L$ and R and every $n > 1$;*

(ii) *$W(z, f^2, L)$ and $W(z,f^2, R)$ are intervals with z as right and left endpoint respectively,*

$$W(z,f) = W(z, f^2, L) \cup W(z,f^2, R),$$
$$f(W(z,f^2, L)) = W(z, f^2, R), \quad f(W(z, f^2, R)) = W(z, f^2, L),$$

and $W(z, f^n, S) = W(z,f)$ or $W(z, f^2, S)$, for $S = L$ and R, according as n is odd or even.

Moreover (ii) *holds if and only if z is an interior point of $W(z,f)$ and f exchanges sides at z, i.e. for every $x \in W(z,f)$, $f(x) \geq z$ if $x \leq z$ and $f(x) \leq z$ if $x \geq z$.*

Proof It is evident that (i) and (ii) cannot both hold and that $W(z, f^2, S) \subseteq W(z, f, S)$ for $S = L$ and R and every $n > 1$.

We show first that $W(z, f^n, L) = W(z, f, L)$ for every $n > 1$ if the following condition is satisfied:

(A) for any $\varepsilon > 0$ there exists $\delta > 0$ such that $f[z-\varepsilon, z] \supseteq [z-\delta, z]$.

Suppose $c \in W(z,f,L)$. Then $c = f^{m_k}(y_k)$, where $y_k \to z$ and $y_k \leq z$. For some r such that $0 \leq r < n$ we have $m_k \equiv r \pmod n$ for infinitely many k. But (A) implies that $y_k = f^{n-r}(x_k)$, where $x_k \to z$ and $x_k \leq z$ for all large k. It follows that $c \in W(z,f^n, L)$.

Similarly, $W(z, f^n, R) = W(z, f, R)$ for every $n > 1$ if the following condition is satisfied:

(B) for any $\varepsilon > 0$ there exists $\delta > 0$ such that $f[z, z+\varepsilon] \supseteq [z, z+\delta]$.

Consequently (i) holds if both (A) and (B) are satisfied. We show next that (i) holds also if (B) is satisfied, but not (A). Since (A) is not satisfied, there exists an $\varepsilon_0 > 0$ such that $f(x) \geq z$ for every $x \in [z-\varepsilon_0, z]$. If for some $\varepsilon > 0$ we have $f(x) = z$ for every $x \in [z-\varepsilon, z]$, then $W(z, f^n, L) = \{z\}$ for every $n \geq 1$ and hence, since (B) is satisfied, (i) holds. Thus we may suppose that for each small $\varepsilon > 0$ there exists $\delta > 0$ such that $f[z-\varepsilon, z] = [z, z+\delta]$, where $\delta = \delta(\varepsilon) \to 0$ as $\varepsilon \to 0$. It follows at once that $W(z, f, L) \subseteq W(z, f, R)$. On the other hand, if $c \in W(z, f^n, R)$ then $c = f^{nm_k}(y_k)$, where $y_k \to z$ and $y_k \geq z$. But (B) implies that, for all large k, $y_k = f^{n-1}(x_k)$, where $x_k \to z$ and $x_k \geq z$, and $x_k = f(w_k)$, where $w_k \to z$ and $w_k \leq z$. Hence $c \in W(z, f^n, L)$ and $W(z, f^n, R) \subseteq W(z, f^n, L)$. Since

$$W(z, f^n, L) \subseteq W(z, f, L) \subseteq W(z, f, R) = W(z, f^n, R),$$

it follows that (i) holds.

Similarly (i) holds if (A) is satisfied, but not (B). Suppose now that (i) does not hold. Then, by what we have proved, the following condition is satisfied:

(C) for each small $\varepsilon > 0$ there exist $\delta > 0$ and $\eta > 0$ such that $f[z-\varepsilon, z] = [z, z+\delta]$ and $f[z, z+\varepsilon] = [z-\eta, z]$.

This implies that

$$W(z, f, L) = W(z, f, R) = W(z,f)$$

and

(D) $$W(z,f) = W(z, f^2, L) \cup W(z, f^2, R).$$

Evidently also

(E) $$f(W(z, f^2, L)) \subseteq W(z, f^2, R), \; f(W(z, f^2, R)) \subseteq W(z, f^2, L).$$

We will show that $W(z, f^n, L) = W(z,f)$ for any odd n. If $c \in W(z,f)$ and $c \neq z$ then $c = f^{m_k}(y_k)$, where $m_k \to \infty$, $y_k \to z$ and $y_k < z$. For some r such that $0 \leq r < n$ we have $m_k \equiv r$ (mod n) for infinitely many k. If r is odd then $c \in W(z, f^n, L)$, since $x_k = f^r(y_k) \to z$ and $x_k < z$. If r is even then $n + r$ is odd and we reach the same conclusion. Similarly $W(z, f^n, R) = W(z,f)$ for any odd n. On the other hand $W(z, f^n, S) = W(z, f^2, S)$ for any even n, since (C) implies that f^2 satisfies both (A) and (B).

If $W(z, f^2, L) = W(z,f)$ then it follows from (E) and the strong invariance of $W(z,f)$ that also $W(z, f^2, R) = W(z,f)$. But then (i) holds, by the results of the previous paragraph, which is contrary to hypothesis. Hence $W(z, f^2, L)$ and $W(z, f^2, R)$ are properly contained in $W(z,f)$, and consequently they are both intervals, by (D). Moreover, by Proposition 2, $W(z, f^2, L)$ has right endpoint z and $W(z, f^2, R)$ has left endpoint z. Since $W(z, f^2, L) \cap W(z, f^2, R) = \{z\}$, it follows from (D) and the strong invariance of $W(z,f)$ that there is actually equality in (E). This completes the proof that (ii) holds if (i) does not.

Thus if (ii) holds then z is an interior point of $W(z,f)$. Furthermore, f exchanges sides at z. For assume, on the contrary, that $f(x) > z$ for some $x \in W(z,f)$ with $x > z$. We may choose x so that $f(x)$ is arbitrarily close to z and, by (C), $f^2(x) < z$. However, this contradicts the fact that z is the left endpoint of $W(z, f^2, R)$, since $x \in W(z, f^2, R)$ and hence $f^2(x) \in W(z, f^2, R)$.

Conversely, if z is an interior point of $W(z,f)$ and f exchanges sides at z, then $W(z, f^2, L)$ has right endpoint z, $W(z, f^2, R)$ has left endpoint z, and

$$W(z, f, S) = W(z, f) \neq W(z, f^2, S).$$

Thus (i) does not hold. □

LEMMA 7 *Suppose* $f(z) = z$ *and* $y \in W(z,f,S)$, *where* $S = L$ *or* R.
If $y < z$, *then* $y = f(x)$ *for some* $x \in W(z,f,S)$ *with* $x > y$.
If $y > z$, *then* $y = f(x)$ *for some* $x \in W(z,f,S)$ *with* $x < y$.

Proof Without loss of generality, suppose $y < z$. We will assume that $f(x) > y$ for every $x \in W(z,f,S)$ with $x > y$ and derive a contradiction. If $S = L$ then $y = f^n(x)$ for some $n > 0$ and some $x \in (y,z)$. Then $x \in W(z,f,L)$ and hence $f^k(x) \in W(z,f,L), f^k(x) > y$ for every $k \geq 0$, which gives a contradiction for $k = n$. If $S = R$ then, by Proposition 2, either $W(z,f,R) = W(z,f,L)$ or $W(z,f,R)$ contains a right neighbourhood of z, which gives a contradiction similarly. □

A sequence $\{x_k\}$ of real numbers will be said to be *bimonotonic* if for every $m \geq 0$, either $x_k > x_m$ for all $k > m$, or $x_k = x_m$ for all $k > m$, or $x_k < x_m$ for all $k > m$.

A sequence $\{x_k\}$ is bimonotonic if and only if, for some $c \in [-\infty,\infty]$, the terms $x_k < c$ form an increasing sequence, the terms $x_k > c$ form a decreasing sequence, and $x_k = c$ implies $x_{k+1} = c$. The notion of bimonotonicity will be used not only in the statement of the following result, but also in Chapter VI.

PROPOSITION 8 *Suppose* $f(z) = z$. *If* $x_0 \in W(z,f,S)$, *where* $S = L$ *or* R, *then there exists a bimonotonic sequence* $\{x_k\}$, *with* $x_k \in W(z,f,S)$, *such that* $f(x_k) = x_{k-1}$ *for every* $k \geq 1$ *and* $x_k \to z$ *as* $k \to \infty$.

Proof Since the result is trivial for $x_0 = z$, we may suppose $x_0 \neq z$ and, for definiteness, $x_0 < z$. By Lemma 7, $x_0 = f(y)$ for some $y \in W(z,f,S)$ with $y > x_0$.

We now begin the construction of the sequence $\{x_k\}$. There are two possibilities. First, if $f(x) \neq x_0$ for $x_0 < x \leq z$ then $f(x) = x_0$ for some $x \in W(z,f,S)$ with $x > z$ and we take x_1 to be the least. Then $f(y) = x_1$ for some $y \in (x_0, x_1)$, since $f(x) > x_0$ for every $x \in (x_0, x_1)$ and since $x_0 \in W(z,f,S)$ implies that (x_0, x_1) is not f-invariant. We now start again with x_1 in place of x_0.

Secondly, if $f(y) = x_0$ for some $y \in (x_0, z)$, let x'_1 be the greatest. Then $f(y) = x'_1$ for some $y \in (x'_1, z)$. Let x'_2 be the greatest. Continuing in this way, we define an increasing sequence $\{x'_k\}$ such that $f(x'_k) = x'_{k-1}$ for every k and $x'_k \to w$ as $k \to \infty$, where $w \leq z$ is a fixed point of f and $f(x) \geq w$ for $w \leq x \leq z$. If $w = z$ we can take $x_k = x'_k$ for every k. Suppose

$w < z$. Since $x_0 \in W(z,f,S)$, we have $f(x) < w$ for some $x \in W(z,f,S)$ with $x > w$ and hence with $x > z$. Thus there is a greatest $w' > z$ such that $f(x) \geq w$ for $z \leq x \leq w'$. Then $f(w') = w$ and $f(x)$ takes values less than w in any right neighbourhood of w'. Again since $x_0 \in W(z,f,S)$, for some $y \in (w, w')$ we have $f(y) \notin [w, w']$ and hence $f(y) > w'$. Choose any small $\delta_1 > 0$ so that $f(x) < z$ for $w' \leq x < w' + \delta_1$. For some large $n_1 > 0$ and some $u \in (w, w')$ we have $v = f(u) \in (w', w' + \delta_1)$ and $f(v) = x'_{n_1} \in (w-\delta_1, w)$. If v' is the nearest point to w' in $(w', w' + \delta_1)$ such that $f(v') = x'_{n_1}$, then $v' = f(u')$ for some $u' \in (w, w')$. We now take $x_k = x'_k$ for $1 \leq k \leq n_1$ and $x_{n_1+1} = v'$, and start again with x_{n_1+1} in place of x_0.

The infinite sequence $\{x_k\}$ constructed in this way is bimonotonic, since the terms $x_k < z$ form an increasing sequence and the terms $x_k > z$ form a decreasing sequence. The previous argument shows that $x_k \to z$ as $k \to \infty$ if all terms from some point on lie on the same side of z. In every other case the terms $x_p < z$ increase to a limit $\alpha \leq z$ and the terms $x_q > z$ decrease to a limit $\beta \geq z$. Moreover $\beta = f(\alpha)$ and $\alpha = f(\beta)$, since x_k lies on one side of z and x_{k+1} on the other for infinitely many k. We claim that $\alpha \leq f(x) \leq \beta$ for $\alpha \leq x \leq \beta$. To see this, we suppose that in the construction we have infinitely many points w_l to the left of z and infinitely many points w_r to the right of z, where w_l and w_r are of the same type as w in the preceding paragraph. (The argument in other cases is similar.) Then any left neighbourhood of α contains a point w_l and any right neighbourhood of β contains a point w_r. Since $f(x) \geq w_l$ for $w_l \leq x \leq w'_l$ and $f(x) \leq w_r$ for $w'_r \leq x \leq w_r$, the claim follows. Since $x_0 \in W(z,f,S)$, we now conclude that $\alpha = z = \beta$. Thus again $x_k \to z$. □

2 UNSTABLE MANIFOLDS OF PERIODIC POINTS

We now turn to the study of the unstable manifolds of *periodic* points.

PROPOSITION 9 *If $f^n(z) = z$ for some $n > 1$, then*

$$f(W(z,f^n)) = W(f(z), f^n). \tag{4}$$

Proof It follows at once from the definition (1) that $f(W(z,f^n)) \subseteq W(f(z), f^n)$. Replacing z by $f^k(z)$ $(1 \leq k < n)$, we obtain in this way

$$f^n(W(f(z), f^n)) \subseteq f(W(f^n(z), f^n)) = f(W(z,f^n)).$$

But $f^n(W(f(z), f^n)) = W(f(z), f^n)$, by Proposition 3. □

PROPOSITION 10 *If $f^n(z) = z$ for some $n > 1$, then*

$$W(z,f) = \bigcup_{j=0}^{n-1} W(f^j(z), f^n). \tag{5}$$

Proof It follows from Proposition 9 that the right side of (5) is contained in the left, since $W(z,f^n) \subseteq W(z,f)$ and $W(z,f)$ is f-invariant. Suppose on the other hand that $c \in W(z,f)$. Then $c = f^{m_k}(y_k)$, where $y_k \to z$. For some r such that $0 \le r < n$ we have $m_k \equiv r \pmod n$ for infinitely many k. Since $x_k = f^r(y_k) \to f^r(z)$, it follows that $c \in W(f^r(z), f^n)$. Thus the left side of (5) is contained in the right. □

It follows from (4) and (5) that the unstable manifold $W(z,f)$ is strongly invariant, as in the case of a fixed point.

We consider next the results corresponding to Propositions 9 and 10 for one-sided unstable manifolds.

PROPOSITION 11 *If $f^n(z) = z$ for some $n > 1$, then*

$$W(z,f,S) = \bigcup_{j=0}^{n-1} f^j(W(z,f^n,S)) \tag{6}$$

and

$$f(W(z,f,S)) = W(z,f,S),$$

where $S = L$ or R.

Proof The right side of (6) is contained in the left, since $W(z,f^n,S) \subseteq W(z,f,S)$ and $W(z,f,S)$ is f-invariant. We will show that the left side of (6) is also contained in the right. This is certainly true if $W(z,f^n,S) = W(z,f^n)$, since the right side of (6) is then $W(z,f)$, by Propositions 9 and 10. Thus we may assume that $W(z,f^n,S) \ne W(z,f^n)$.

Suppose, for definiteness, that $S = L$ and put $H_\varepsilon = \bigcup_{m\ge 0} f^{mn}(z-\varepsilon, z]$. Then

$$W(z,f^n,L) = \bigcap_{\varepsilon>0} H_\varepsilon,$$

$$W(z,f,L) = \bigcap_{\varepsilon>0} \{H_\varepsilon \cup f(H_\varepsilon) \cup \dots \cup f^{n-1}(H_\varepsilon)\}.$$

Thus if $x \in W(z,f,L)$ then $x = f^{j+mn}(y)$ for some $m \ge 0$, some j such that $0 \le j < n$, and some y in any prescribed left neighbourhood of z. If $W(z,f^n,L)$ contains a left neighbourhood of z we can take $y \in W(z,f^n,L)$ and then $x = f^j(w)$, where $w = f^{mn}(y) \in W(z,f^n,L)$. It only remains to consider the case where $W(z,f^n,L)$ has left endpoint z. But then $W(z,f^n,L) = \{z\}$ by Proposition 2, since $W(z,f^n,L) \ne W(z,f^n)$. Thus $\bigcap_{\varepsilon>0} H_\varepsilon = \{z\}$. Since f is continuous, we can choose $\varepsilon > 0$ so small that $f^j(H_\varepsilon)$ lies in any prescribed neighbourhood of $f^j(z)$ for $0 \le j < n$. It follows that $W(z,f,L)$ contains only points in the orbit of z. But this contradicts our assumption that $W(z,f^n,S) \neq W(z,f^n)$.

This proves the first assertion of the proposition, and the second assertion now follows from Proposition 3. □

PROPOSITION 12 *Suppose $f^n(z) = z$ for some $n > 0$ and $W(z,f^n,S) \neq \{z\}$, where $S \in \{L, R\}$.*

If $W(z,f^n,S) \neq W(z,f^n)$ then, for any $j \ge 0$, $f^j(W(z,f^n,S))$ is an S_j-neighbourhood of $f^j(z)$, where $S_j \in \{L,R\}$ and $S_0 = S$. Moreover

$$f^j(W(z,f^n,S)) = W(f^j(z),f^n,S_j) \tag{7}$$

for this choice of S_j and no other.

If $W(z,f^n,S) = W(z,f^n)$ then, for any $j \ge 0$, there is at least one $S_j \in \{L,R\}$ for which (7) *holds.*

Proof Put $z_j = f^j(z)$ and $W_j = f^j(W(z,f^n,S))$. Then $W_j \neq \{z_j\}$ for every $j > 0$. Indeed if equality held for some j we could obtain a contradiction by applying f^ℓ, where $\ell + j$ is a multiple of n. Thus W_j is an interval containing z_j. Moreover it is (strongly) invariant under f^n.

Suppose W_j contains a T-neighbourhood of z_j, where $T \in \{L,R\}$. If $x \in W(z_j,f^n,T)$ then $x = f^{nmk}(x_k)$, where $x_k \to z_j$ and x_k is on the T-side of z_j. Thus $x_k \in W_j$ for large k, and hence also $x \in W_j$. Thus $W(z_j,f^n,T) \subseteq W_j$.

Evidently $W_j \subseteq W(z_j,f^n)$. Assume first that the inclusion is strict. Then, by what we have just proved, W_j contains no T'-neighbourhood of z_j, where $T' \in \{L,R\}$ and $T' \neq T$. Hence W_j *is* a T-neighbourhood of z_j. If $W_0 = W(z,f^n,S)$ contained an S'-neighbourhood of z then, by Proposition 2, $W_0 = W(z,f^n)$, which gives a contradiction on applying f^j. Hence W_0 is an S-neighbourhood of z. If $y \in W_0$ then $y = f^{npk}(y_k)$, where $y_k \to z$ and y_k is on the S-side of z. Thus $y_k \in W_0$ for large k, and hence $f^j(y_k)$ is on the T-side of z_j. Since $f^j(y_k) \to z_j$, it follows that $f^j(y) \in W(z_j,f^n,T)$. Thus $W_j \subseteq W(z_j,f^n,T)$. Since the reverse inclusion has already been proved, this shows that (7) holds for $S_j = T$. On account of our assumption it cannot hold also for $S_j = T'$.

Assume next that $W_j = W(z_j,f^n)$ for some, and hence every, $j \ge 0$. If $W(z_j,f^n,T) = W(z_j,f^n)$ for some $T \in \{L,R\}$, then (7) holds with $S_j = T$. Otherwise we have

$$\{z_j\} \subset W(z_j, f^n, T) \subset W(z_j, f^n)$$

for $T = L$ and R. We will show that this leads to a contradiction. Again choose ℓ so that $\ell + j$ is a multiple of n. If $f^\ell(W(z_j, f^n, T))$ contained an S-neighbourhood of z for some T then, by the first part of the proof, $W_0 \subseteq f^\ell(W(z_j, f^n, T))$ and hence $W_j \subseteq W(z_j, f^n, T)$, which is a contradiction. It follows that $f^\ell(W(z_j, f^n)) = W(z, f^n)$ is an S'-neighbourhood of z. Then $W(z, f^n, S) = W(z, f^n, S')$, by Proposition 2, and $f^\ell(W(z_j, f^n, T))$ is an S'-neighbourhood of z for some T. But this gives a contradiction in the same way as before. □

In the relations (5) and (6) the terms on the right side need not be disjoint. The connected components of the left sides will now be determined.

PROPOSITION 13 *Suppose $f^n(z) = z$ for some $n > 1$, and for any $j \geq 0$ let $W_j = f^j(W(z, f^n, S))$, where $S = L$ or R. Then $f^i(z) \in \operatorname{int} W_k$ only if $W_i = W_k$.*

Moreover, if r is the least positive integer such that $W_r = W_0$, exactly one of the following holds:

(i) *r is odd or even and the connected components of $W(z,f,S)$ are $W_0, W_1, \dots, W_{r-1}$;*

(ii) *$r = 2s$ is even and the connected components of $W(z,f,S)$ are*

$$W_0 \cup W_s, W_1 \cup W_{s+1}, \dots, W_{s-1} \cup W_{2s-1}.$$

The same conclusion holds if we define $W_j = W(f^j(z), f^n)$ and replace $W(z,f,S)$ by $W(z,f)$.

Proof We give the proof for $W(z,f,S)$ only, the proof for $W(z,f)$ being the same. If $W_0 = \{z\}$ then $W_j = \{f^j(z)\}$, r is the period of the periodic point z and (i) holds. Thus we may suppose that $W_0 \neq \{z\}$. It follows directly from the definition of r that $W_i = W_k$ if and only if $i \equiv k \pmod r$. In particular, each W_j coincides with exactly one of $W_0, W_1, \dots, W_{r-1}$ and $W(z,f,S)$ is their union, by Proposition 11.

We now show that $W_i \subseteq W_k$ only if $W_i = W_k$. We may suppose $k = 0$, since the general case follows from this special case. But if $W_i \subseteq W_0$ then, by induction, $W_{pi} \subseteq W_i$ for every $p > 0$. Taking $p = n$, we obtain $W_0 \subseteq W_i$.

In particular, $W_i = W_k$ if $f^i(z)$ is an interior point of W_k, since then $W_i = W(f^i(z), f^n, S_i) \subseteq W_k$.

It follows also that the sets W_j may be totally ordered by writing $W_i \leq W_k$ if $W_i = W_k$ or if W_i contains a point to the left of W_k. We write $W_i < W_k$ if $W_i \leq W_k$ and $W_i \neq W_k$. That is,

$W_i < W_k$ if W_i contains a point to the left of W_k or, equivalently, if W_k contains a point to the right of W_i.

We claim that *any connected component of* $W(z,f,S)$ *contains at most two distinct sets* W_j. Assume, on the contrary, that a component of $W(z,f,S)$ contains at least three distinct sets W_j. Let $\mathbf{W}$ denote the set of W_j in this component. Also, let W_i denote the least element of $\mathbf{W}$ in the defined ordering, let W_j denote the least element of $\mathbf{W}\setminus\{W_i\}$, and let W_k denote the least element of $\mathbf{W}\setminus\{W_i,W_j\}$.

Then $W_i < W_j < W_k$. Moreover $W_i \cup W_j$ is connected and $W_j \cup W_k$ is connected. Since f^m maps W_j onto W_i for some m, and since both $f^m(W_i) \cup f^m(W_j)$ and $f^m(W_j) \cup f^m(W_k)$ are connected, it follows that $W_i \cup W_l$ is connected for some $l \neq j$. Hence, by the definition of W_k, $W_i \cup W_k$ is connected.

Let p be the right endpoint of W_i. If $p \notin W_i$ then $p \in W_j \cap W_k$. Moreover $W_i \cap W_j \neq \emptyset$, since $W_j < W_k$ implies that W_j contains a point to the left of p. If $p \in W_i$ then both W_j and W_k contain an open interval with left endpoint p. Moreover $p \in W_j$, since $W_j < W_k$. Thus, in every case,

$$W_i \cap W_j \neq \emptyset, \quad W_j \cap W_k \neq \emptyset .$$

If we define s by $i - j \equiv s \pmod r$ and $0 < s < r$, then $f^s(W_j) = W_i$ and hence $f^s(W_i) \cap W_i \neq \emptyset, f^s(W_k) \cap W_i \neq \emptyset$. Therefore $W_i \leq f^s(W_i)$ and $W_i \leq f^s(W_k)$. Let W_ℓ be the smaller and W_m the greater of the pair $f^s(W_i)$, $f^s(W_k)$. Then $W_\ell \subseteq W_i \cup W_m$ and hence $w = f^\ell(z)$ lies in W_i or W_m. Since W_i, W_ℓ and W_m are distinct, w can only be the right endpoint of W_i or the left endpoint of W_m, and so it must be both. Furthermore w must be an interior point of W_ℓ. Then $W_j = W_\ell$, since w must also be an interior point of W_j. Moreover $W_k = W_m$, since w cannot be an interior point of W_k. Since $W_k \neq f^s(W_k)$, it follows that $W_k = f^s(W_i)$ and $W_j = f^s(W_k)$. But then

$$f^s(W_i \cap W_j) = W_k \cap W_i = \{w\},$$

and similarly $f^{2s}(W_j \cap W_k) = \{w\}$. Since $f^{2s}(W_j)$ is an interval, this is a contradiction. This establishes our claim.

Thus if $W_i \cup W_j$ is connected, where $W_i \neq W_j$, then $W_i \cup W_j$ is a component of $W(z,f,S)$. If we define s in the same way as before, then $f^s(W_i) \cup W_i$ is connected and hence $f^s(W_i) = W_j$. Thus $f^{2s}(W_j) = W_j$ and hence r divides $2s$. Since $s < r$, this implies $r = 2s$. The rest of the proposition follows immediately. □

Finally we consider some properties of the closure of the unstable manifold.

PROPOSITION 14 *If z is a periodic point of f, then*

$$\overline{W(z,f)} = \bigcap_{\varepsilon>0} \overline{\bigcup_{m\geq 0} f^m(z-\varepsilon, z+\varepsilon)} ,$$

$$\overline{W(z,f,L)} = \bigcap_{\varepsilon>0} \overline{\bigcup_{m\geq 0} f^m(z-\varepsilon, z]} , \tag{8}$$

$$\overline{W(z,f,R)} = \bigcap_{\varepsilon>0} \overline{\bigcup_{m\geq 0} f^m[z, z+\varepsilon)} .$$

Proof Only the second relation (8) will be proved. The left side of (8) is contained in the right, since the right side is a closed set which contains $W(z,f,L)$. It remains to show that the right side of (8) is contained in the left.

Let z have period $n \geq 1$. If we put $H_\varepsilon = \bigcup_{m\geq 0} f^{mn}(z-\varepsilon, z]$ and

$$G_\varepsilon = H_\varepsilon \cup f(H_\varepsilon) \cup ... \cup f^{n-1}(H_\varepsilon)$$

then $W(z,f,L) = \bigcap_{\varepsilon>0} G_\varepsilon$ and the right side of (8) is just $\bigcap_{\varepsilon>0} \overline{G}_\varepsilon$. Suppose $x \notin \overline{W(z,f,L)}$, and let K be a compact interval containing x in its interior such that $K \cap W(z,f,L) = \emptyset$. Then, by Proposition 11,

$$K \cap f^j(W(z,f^n,L)) = \emptyset \text{ for } 0 \leq j < n.$$

Hence we can choose $\varepsilon > 0$ so that none of the sets $H_\varepsilon, f(H_\varepsilon), ..., f^{n-1}(H_\varepsilon)$ contains either endpoint of K. Since K contains no point in the orbit of z and $f^j(H_\varepsilon)$ is a connected set containing $f^j(z)$, it follows that $K \cap f^j(H_\varepsilon) = \emptyset$ for $0 \leq j < n$. Thus $K \cap G_\varepsilon = \emptyset$. Hence $x \notin \overline{G}_\varepsilon$ and x does not belong to the right side of (8). □

PROPOSITION 15 *Suppose the interval I is compact. If z is a periodic point of f and* $x \in \bar{J} \setminus J$, *where* $J = W(z,f)$, $W(z,f,L)$ *or* $W(z,f,R)$, *then x is periodic.*

Proof Since I is compact, the set $f(\bar{J})$ is closed. Since $f(J) = J$, it follows that $f(\bar{J}) = \bar{J}$. If $x \in \bar{J} \setminus J$ then $x = f(y)$ for some $y \in \bar{J}$. Moreover $y \notin J$, since $x \notin J$. Thus $\bar{J} \setminus J \subseteq f(\bar{J} \setminus J)$. But the set $\bar{J} \setminus J$ is finite, since J is a finite union of points or intervals. Therefore f maps $\bar{J} \setminus J$ one-to-one onto itself. Consequently every point in $\bar{J} \setminus J$ is periodic. □

3 HOMOCLINIC POINTS

For diffeomorphisms the notion of homoclinic point was introduced by Poincaré to describe a point belonging to both the stable and unstable manifolds of a hyperbolic periodic point. For maps of an interval Block has defined homoclinic points in the following way.

Let $f: I \to I$ be a continuous map of an interval into itself. A point y is *homoclinic* if there exists a point $z \neq y$ such that $f^n(z) = z$ for some $n > 0$, $y \in W(z,f^n)$ and $f^{kn}(y) = z$ for some $k > 0$. It follows that y is not in the f-orbit of z.

Since z is required to lie in the f^n-orbit of y, this specialises somewhat Poincaré's original idea. Nevertheless, it will be shown in Chapter VI that if there exists a homoclinic point in Poincaré's sense, then there exists also a homoclinic point in Block's sense.

The significance of homoclinic points is that maps which possess them behave wildly. In fact we will show that there is a close connection between turbulence and the existence of homoclinic points.

PROPOSITION 16 *If $f^n(z) = z$ for some $n > 0$ and there exists a point $y \neq z$ such that $y \in W(z,f^n)$ and $f^{kn}(y) = z$ for some $k > 0$, then f^{2n} is turbulent.*

Proof We may suppose $n = 1$ by replacing f by f^n. We may also suppose $k = 1$, since the unstable manifold is invariant. Thus

$$f(z) = z = f(y).$$

Without loss of generality we further suppose $y < z$.

Evidently we may assume that f is not turbulent. Then $f(x) \neq y$ for every $x \in (y,z)$. On the other hand, by Lemma 7, $y = f(w)$ for some least $w > z$. Then $f(x) > y$ for all $x \in (y,w)$. Similarly $f(x) \neq w$ for every $x \in (z,w)$. However, $w = f(v)$ for some $v \in (y,z)$, since $y \in W(z,f)$ implies that the interval (y,w) is not invariant. Then

$$[y,z] \subseteq f^2[y,v] \cap f^2[v,z].$$

Thus f^2 is turbulent, as we wished to prove. □

Proposition 16 says that $H(n) \subseteq T_{2n}$, where $H(n)$ denotes the set of all continuous maps $f: I \to I$ with a homoclinic point for which the associated periodic point has period n.

We now introduce two special types of homoclinic point, one orientation-preserving and the other orientation-reversing. Although there exist homoclinic points which are of neither type, any map with a homoclinic point must have a homoclinic point of one of these two types. However, the real justification for concentrating attention on these two types lies in the simple and precise results which are obtained.

Let $\mathbb{H}_n$ denote the set of all continuous maps $f: I \to I$ for which there exist points a,b,c such that

$$f^n(b) = f^n(a) = a, \; f^n(c) = b, \tag{9}$$

a has period n, and either

$$\begin{gathered} a < c < b, \\ f^n(x) > a \;\text{ for }\; a < x < b, \\ x < f^n(x) < b \;\text{ for }\; a < x < c, \end{gathered} \tag{10}$$

or the same with all inequalities in (10) reversed. Then c is a homoclinic point, since $c \in W(a, f^n)$ by Proposition 4, and

$$\mathbb{H}_n \subseteq \mathbb{T}_n ,$$

since $\langle a,b\rangle \subseteq f^n\langle a,c\rangle \cap f^n\langle c,b\rangle$. Moreover, by Lemma II.2 we actually have

$$\mathbb{H}_1 = \mathbb{T}_1 .$$

Similarly, let $\mathbb{H}_n^{\#}$ denote the set of all continuous maps $f: I \to I$ for which there exist points $x_0, \dots, x_4$ such that

$$f^n(x_0) = x_0 , \; f^n(x_k) = x_{k-1} \;\; (1 \le k \le 4), \tag{11}$$

x_0 has period n, and either

$$\begin{gathered} x_1 < x_3 < x_0 < x_4 < x_2 , \\ f^n(x) < x_0 < f^{2n}(x) \;\text{ for }\; x_0 < x < x_2 , \\ x < f^{2n}(x) < x_2 \;\text{ for }\; x_0 < x < x_4 , \end{gathered} \tag{12}$$

or the same with all inequalities in (12) reversed. Then x_4 is a homoclinic point and

$$\mathbb{H}_n^{\#} \subseteq \mathbb{T}_{2n} ,$$

since $\langle x_0, x_2\rangle \subseteq f^{2n}\langle x_0, x_4\rangle \cap f^{2n}\langle x_4, x_2\rangle$. It is readily seen that the relations (11) and (12) imply

$$\begin{aligned} x_1 &< f^n(x) \text{ for } x_0 < x < x_2 , \\ x_3 &< f^n(x) \text{ for } x_0 < x < x_4 , \\ x_0 &< f^n(x) \text{ for } x_1 < x < x_0 , \\ x_2 &> f^n(x) \text{ for } x_3 < x < x_0 , \end{aligned} \tag{13}$$

regardless of whether x_0 has period exactly n.

We are going to study the sets $\mathbb{H}_n$, $\mathbb{H}_n^{\#}$ when n is a power of 2. It may be shown that when n is not a power of 2 we have simply $\mathbb{T}_n = \mathbb{H}_n$.

PROPOSITION 17 *For any* $s \geq 0$,

$$\mathbb{H}_{2^s} \subseteq \mathbb{H}_{2^{s+2}} .$$

Proof Let $f \in \mathbb{H}_1$ and, without loss of generality, let a,b,c be points satisfying the relations (9) and (10) with $n = 1$.

Since $f[a,c] \supseteq [a,b]$, there is a greatest point $d \in (a,c)$ such that $f(d) = c$. Since $f^2[d,c] \supseteq [a,b] \supseteq [d,c]$, f^4 has a fixed point in $[d,c]$. If α is the least, then $d < \alpha < c$. Assume $f^2(\alpha) = \alpha$. Then $f^2(d,\alpha) \supseteq (\alpha,b)$ and hence $f^2(e) = c$ for some $e \in (d,\alpha)$. Then

$$f^3[d,e] \supseteq [a,b] \supseteq [a,c]$$

and hence

$$f^4[d,e] \supseteq [a,b] \supseteq [d,e].$$

Thus f^4 has a fixed point in $[d,e]$, which contradicts the definition of α. We conclude that $f^2(\alpha) \neq \alpha$, and so α has period 4.

Since $f^4(d) = a < d$, we have $f^4(x) < x$ for $d \leq x < \alpha$. On the other hand, since

$$f^2(a,d) \supseteq (a,b) \supseteq (a,d),$$

we have $f^4(a,d) \supseteq (a,b)$. Thus $\alpha = f^4(\beta)$ for some greatest $\beta \in (a,d)$. Similarly $\beta = f^4(\gamma)$ for some greatest $\gamma \in (d,\alpha)$. It now follows readily that $f \in \mathbb{H}_4$.

Thus the proposition holds for $s = 0$. Assume that $s > 0$ and it holds for all smaller values of s. If $f \in \mathbb{H}_{2^s}$, then $g = f^2 \in \mathbb{H}_{2^{s-1}}$. Hence $g \in \mathbb{H}_{2^{s+1}}$, by the induction hypothesis, and $f \in \mathbb{H}_{2^{s+2}}$, by Lemma I.10. □

To establish the analogous property of $\mathbb{H}_n^{\#}$ we make use of the following preliminary result.

LEMMA 18 *Suppose there exist points* $a < c < b$ *such that*

$$f(a) = a = f^2(b),\ f^2(c) = b,$$
$$f^2(x) > a \text{ for } a < x < b,$$
$$x < f^2(x) < b \text{ for } a < x < c.$$

Then $f \in \mathbb{H}_1^\#$ *if* $f(\xi) < a$ *for some* $\xi \in (a,b)$, *and* $f \in \mathbb{H}_1$ *otherwise.*

Proof The conditions evidently imply that $f(x) \neq a$ for $a < x < b$. Suppose first that $f(x) < a$ for every $x \in (a,b)$. In particular, $f(c) < a$ and $f(b) \leq a$. If $x \in (f(c), a)$ then $x = f(y)$ for some $y \in (a,c)$ and hence $f(x) = f^2(y)$ satisfies $a < f(x) < b$. Consequently $f^2(x) < a$ if $x \in (f(c), a)$. Since $f^3(b) = a$, it follows that we must have either $f(b) < f(c)$ or $f(b) = a$.

Consider first the case $f(b) < f(c)$. If we put

$$x_0 = a,\ x_1 = f(b),\ x_2 = b,\ x_3 = f(c),\ x_4 = c,$$

then the relations (11) and (12) are satisfied with $n = 1$. Thus $f \in \mathbb{H}_1^\#$.

Consider next the case $f(b) = a$. For some greatest $d \in (f(c), a)$ we must have $f(d) = c$. If $x \in (d,a)$ then $z = f(x) \in (a,c)$. Since $f^2(x) = x$ would imply $f^2(z) = z$ and $f^2(x) = f(c)$ would imply $f^2(z) = b$, it follows that $f(c) < f^2(x) < x$ for $d < x < a$. If we now put

$$x_0 = a,\ x_1 = b,\ x_2 = f(c),\ x_3 = c,\ x_4 = d,$$

then the relations (11) and (12), with the inequalities reversed, are satisfied with $n = 1$. Thus again $f \in \mathbb{H}_1^\#$.

Suppose next that $f(x) > a$ for every $x \in (a,b)$. Since $f^3(b) = a$ the hypotheses of the lemma imply that $f(b) \notin (a,b]$. If $f(b) > b$ then

$$[a, f(b)] \subseteq f[a,b] \cap f[b, f(b)]$$

and hence $f \in \mathbb{T}_1 = \mathbb{H}_1$. The only remaining possibility is $f(b) = a$. Since $f(c) > a$ we must have either $f(c) > b$ or $a < f(c) < b$. But if $f(c) > b$ then

$$[a,b] \subseteq f[a,c] \cap f[c,b],$$

and if $a < f(c) < b$ then

$$[a,b] \subseteq f[a,f(c)] \cap f[f(c), b].$$

Thus $f \in \mathbb{H}_1$ in every case. □

We can now derive the main result of this section. It establishes a precise relation between the set T_n and the sets H_n and $H_n^{\#}$ when n is a power of 2.

PROPOSITION 19 $T_{2^s} = H^{\#}_{2^{s-1}} \cup H_{2^s}$ *for any* $s > 0$.

Proof The right side is certainly contained in the left, as we have already seen. Thus we need only show that the left side is contained in the right.

We show first that $T_1 \subseteq H_1^{\#} \cup H_2$. Let $f \in T_1$. By Lemma II.2 we may assume that there exist points a,b,c which satisfy the relations (9) and (10) with $n = 1$. Since $f(c) > c$ and $f(b) < b$ we have $f(x) = x$ for some $x \in (c,b)$. Let ξ be the least fixed point of f^2 in (c,b). Then $f^2(x) < x < \xi$ for $c \le x < \xi$. For some $d \in (a,c)$ we have $f(d) = c$ and thus $f^2(d) = b$. Hence $f^2(\eta) = \xi$ for some greatest $\eta \in (a,c)$. Then $f^2(x) < \xi$ for $\eta < x \le c$. Similarly $f^2(\zeta) = \eta$ for some greatest $\zeta \in (c,\xi)$. Then $f^2(x) > \eta$ for $\zeta < x < \xi$. If $f(\xi) \neq \xi$ it follows at once that $f \in H_2$. If $f(\xi) = \xi$ then, since $f(c) > \xi$, it follows from Lemma 18 (with the inequalities reversed) that $f \in H_1^{\#}$. This establishes our claim.

Suppose next that $f \in T_2$. By Lemma II.2 we may assume that there exist points a, b, c which satisfy the relations (9) and (10) with $n = 2$. If $f(a) \neq a$ then a has period 2 and hence $f \in H_2$. If $f(a) = a$ then $f \in H_1 \cup H_1^{\#}$, by Lemma 18, and hence $f \in H_1^{\#} \cup H_2$, by the previous part of the proof. This proves the theorem for $s = 1$.

Suppose finally that $f \in T_{2^{s+1}}$ for some $s > 0$. Then $g = f^{2^s} \in T_2$ and hence $g \in H_1^{\#} \cup H_2$. If $g \in H_2$ then $f \in H_{2^{s+1}}$. If $g \in H_1^{\#}$ we may assume that relations (11) and (12) hold with $n = 2^s$. If $f^{n/2}(x_0) \neq x_0$ then $f \in H^{\#}_{2^s}$. Thus to complete the proof we will assume $f^{n/2}(x_0) = x_0$ and deduce a contradiction. By (12) we now have $f^{n/2}(x) \neq x_0$ for $x_0 < x < x_2$. If $f^{n/2}(x) > x_0$ for $x_0 < x < x_2$ then $f^n(x) > x_0$ for $x > x_0$ and x close to x_0, which contradicts (12). Hence $f^{n/2}(x) < x_0$ for $x_0 < x < x_2$. Similarly from (13) we obtain $f^{n/2}(x) > x_0$ for $x_1 < x < x_0$. But then we again obtain $f^n(x) > x_0$ for $x > x_0$ and x close to x_0.
□

Proposition 19 immediately implies the following counterpart to Proposition 17.

COROLLARY 20 $H_1 \subseteq H_1^{\#} \cup H_2$ *and* $H^{\#}_{2^{s-1}} \cup H_{2^s} \subseteq H^{\#}_{2^s} \cup H_{2^{s+1}}$ *for any* $s > 0$. □

The following examples show that the preceding results are sharp:

If f: $[0,4] \to [0,4]$ is the piecewise linear map defined by

(i) $f(0) = 0, f(2) = 4, f(9/4) = 19/8, f(15/4) = 7/4, f(4) = 0$,
then $f \in H_1 \cap H_2$ but $f \notin H_1^{\#} \cup H_2^{\#}$;

(ii) $f(0) = 4, f(2) = 2, f(3) = 0, f(4) = 2$,
then $f \in H_1^{\#} \cap H_2^{\#}$ but $f \notin H_1 \cup H_2 \cup H_4$;

(iii) $f(0) = 4, f(2) = 2, f(3) = 0, f(15/4) = 3/4, f(4) = 2$,
then $f \in H_1^{\#} \cap H_4$ but $f \notin H_1 \cup H_2 \cup H_2^{\#}$;

(iv) $f(0) = 0, f(2) = 4, f(5/2) = 7/4, f(7/2) = 3/2, f(4) = 0$,
then $f \in H_1 \cap H_1^{\#}$ but $f \notin H(2)$;

(v) $f(0) = 5/2, f(1) = 4, f(3) = 0, f(4) = 1/2$,
then $f \in H_2^{\#}$ but $f \notin H_1 \cup H_1^{\#} \cup H_2 \cup H_4$.

By combining Proposition 19 with Proposition 16 we obtain at once

PROPOSITION 21 *A continuous map $f: I \to I$ is chaotic if and only if it has a homoclinic point.*
□

We conclude this section with some results which provide additional information about the unstable manifolds of a periodic point, when this point does *not* possess a homoclinic point.

LEMMA 22 *Suppose $f(z) = z$. If f is not strictly turbulent, then exactly one of the following alternatives holds:*

(i) $W(z,f,L) = W(z,f,R)$,

(ii) $W(z,f) \neq \{z\}$ *and*

$$W(z,f,L) = W(z,f) \cap \{x: x \le z\},$$
$$W(z,f,R) = W(z,f) \cap \{x: x \ge z\}.$$

Proof We show first that if $x < z < f(x)$, then $f(y) > x$ for all $y \in [x,z]$. Assume, on the contrary, that $f(y) = x$ for some $y \in (x,z)$. Choose $x' \in (x,y)$ so close to x that $f(x') > z$ and then choose $y' \in (y,z)$ so close to y that $f(y') < x'$. Then

$$[x',z] \subseteq f[x',y] \cap f[y',z],$$

contrary to the hypothesis that f is not strictly turbulent. Similarly if $f(x) < z < x$, then $f(y) < x$ for all $y \in [z,x]$.

Suppose $x \in W(z,f)$ and $x \neq z$. Then, by Proposition 8, there exists a bimonotonic sequence $\{x_k\}$ with $x_0 = x, f(x_k) = x_{k-1}$ for $k \geq 1$ and $x_k \to z$ as $k \to \infty$.

Assume first that $x < z < f(x)$. Then by the previous part of the proof we must have

$$x_0 < x_2 < x_4 < \ldots < z < \ldots x_5 < x_3 < x_1 .$$

Hence $[x_0 , x_1] \subseteq W(z,f,L) \cap W(z,f,R)$, which implies $W(z,f,L) = W(z,f,R)$ by Proposition 2. Similarly $W(z,f,L) = W(z,f,R)$ if $f(x) < z < x$.

Suppose now that $W(z,f,L) \neq W(z,f,R)$. If $W(z,f,R)$ contains a point $y < z$ then, by Proposition 2, $W(z,f,R)$ contains some point $> z$. Hence $y = f^n(x)$ for some positive integer n and some $x \in W(z,f,R)$ with $x > z$. It follows that for some $x' \in W(z,f,R)$ with $x' > z$ we have $f(x') < z$, contrary to the previous part of the proof. We conclude that $W(z,f,R)$ contains no point $< z$, and similarly $W(z,f,L)$ contains no point $> z$. □

LEMMA 23 *Let z_1 , z_2 be fixed points of f, with $z_1 < z_2$, such that $z_2 \in \overline{W(z_1,f,R)}$ and $z_1 \in \overline{W(z_2,f,L)}$. Then f is turbulent.*

Proof Suppose first that the interval (z_1, z_2) is invariant. For any $y \in (z_1, z_2)$ let $\mu(y)$ denote the minimum value of f on the interval $[y, z_2]$. Then $z_1 < \mu(y) < y$, since the interval (z_1, z_2) is invariant and since $z_1 \in \overline{W(z_2,f,L)}$ implies that the interval $[y, z_2]$ is not invariant. Since for the same reason $\mu(\mu(y)) < \mu(y)$, there exists a point $y' \in [\mu(y), y]$ with $f(y') < y'$.

Since $y \in W(z_1 ,f,R)$ there exists a point $x \in (z_1 , \mu(y))$ such that $f^k(x) = y$ for some $k > 0$. Moreover, we may suppose $f^j(x) \geq \mu(y)$ for $1 \leq j \leq k$, by replacing x by a suitable iterate. But $f(x') < x'$ for some $x' \in [\mu(x), x]$. Hence if a is the greatest fixed point of f less than x then $a > z_1$. Since the interval $[a, z_2]$ is not invariant we have $f(b) = a$ for some $b \in (a, z_2)$. Evidently $b \notin (a,x]$. Moreover $b \notin [y, z_2]$, since $f(b) < \mu(y)$. Thus $b \in (x,y)$. Since $f^k(x) = y$ we must have $f(c) = b$ for some $c \in (a,b)$. Thus f is turbulent.

Suppose next that the interval (z_1, z_2) is not invariant. Then $f(z) = z_1$ or z_2 for some $z \in (z_1, z_2)$. Without loss of generality, assume that $f(z) = z_2$. We may suppose that $f(x) > z$ for $z \leq x \leq z_2$, since otherwise f is turbulent. We may suppose further that $f(x) > x$ for $z_1 < x \leq z$. For if z' is the greatest fixed point of f in the interval $[z_1, z]$, then $f(x) > x$ for $z' < x \leq z$, hence $[z', z] \subseteq W(z',f,R)$ and $z_2 = f(z) \in W(z',f,R)$, so that z_1 can be replaced by z'.

Assume first that $f(x) \neq z_1$ for all $x > z_2$, and hence for all $x > z_1$. Then

$$w_1 := \min_{x \geq z} f(x) > z_1 .$$

Choose $w \in (z_1, z)$ so close to z_1 that $w < w_1$. Since $w \in W(z_2, f, L)$, we have $w = f^n(y)$ for some $y \in (z, z_2)$ and some $n > 0$. Let $m \leq n$ be the least positive integer such that $f^m(y) \leq w$. If $f^{m-1}(y) \in (z_1, z)$ then $f^{m-1}(y) < f^m(y)$, and if $f^{m-1}(y) \geq z$ then $f^m(y) \geq w_1$. In either case we have a contradiction.

Hence f takes the value z_1 at some point to the right of z_2. If z_3 is the nearest such point to z_2, then $f(z_3) = z_1$ and $f(x) > z_1$ for $z_1 < x < z_3$. Assume that f is not turbulent. Then $f(x) < z_3$ for $z_1 \leq x \leq z_3$, and hence the interval $[z_1, z_3]$ is invariant. Moreover, there exists a $\delta > 0$ such that $f(x) < z_3 - \delta$ for all $x \in [z_1, z_3]$. If $w \in (z_1, z)$ then $w = f^n(y)$ for some $y \in (z, z_2)$ and some $n > 0$. Let $m \leq n$ be the least positive integer such that $f^m(y) \leq w$. Then $m > 1$ and $f^{m-1}(y) > z_2$. By choosing w sufficiently close to z_1 we can ensure that $f^{m-1}(y) > z_3 - \delta$. But this is a contradiction. □

PROPOSITION 24 *Let $z_1 < \ldots < z_n$ be an orbit of f of period $n > 1$. If f^n is not turbulent and $i \neq j$, then $z_i \notin \overline{W(z_j, f^n)}$.*

Proof Put $g = f^n$ and assume that $z_i \in \overline{W(z_j, g)}$ for some pair i,j with $i \neq j$. Then, by Proposition 9, for every j we have $z_i \in \overline{W(z_j, g)}$ for some $i \neq j$ and for every i we have $z_i \in \overline{W(z_j, g)}$ for some $j \neq i$. In particular, this implies $z_2 \in \overline{W(z_1, g)}$ and $z_{n-1} \in \overline{W(z_n, g)}$.

Since g is not turbulent we must actually have $z_2 \in \overline{W(z_1, g, R)}$ and $z_{n-1} \in \overline{W(z_n, g, L)}$, by Lemma 22. Then $z_1 \notin \overline{W(z_2, g, L)}$ by Lemma 23, and hence $z_1 \notin \overline{W(z_2, g)}$ by Lemma 22. Consequently $z_3 \in \overline{W(z_2, g)}$, and actually $z_3 \in \overline{W(z_2, g, R)}$.

Proceeding in this manner we obtain $z_{k+1} \in \overline{W(z_k, g, R)}$ for $1 \leq k < n$. Since $z_{n-1} \in \overline{W(z_n, g, L)}$, this contradicts Lemma 23. □

NOTES

The unstable manifolds of a periodic point were defined for continuous maps of an interval by Block [20], who also proved Propositions 1, 3, 10 and 15. Proposition 4 is the essential content of Lemma 4 in Block [21]. Propositions 5, 9 and 14 (in the two-sided case) are given by Xiong [125]. Nitecki [96] states less precise forms of Propositions 6 and 12, and outlines the proof of Proposition 13. Coven and Nitecki [58] also have a version of Proposition 12. It should be noted that in the last two references the unstable manifolds are the closures of those defined here. Proposition 8 was suggested by Fedorenko and Šarkovskii [63].

Block's definition of homoclinic points appeared in Block [21]. The definitions of the sets $\mathbb{H}_n$ and $\mathbb{H}_n^*$, and the proofs of Propositions 17 and 19, are given in Block and Coppel [26]. It is also proved there that $\mathbb{T}_n = \mathbb{H}_n$ if n is not a power of 2. Proposition 21 was proved by Block [21]. The proof given here shows that chaotic maps have many homoclinic points of a very specific type. Weaker forms of Proposition 24 were first proved by Xiong [125] and Nitecki [95].

IV
Topological Dynamics

Topological dynamics began with the work of Poincaré and G.D. Birkhoff, and there are today sizeable books on the subject. Traditionally it studies qualitative properties of homeomorphisms of a compact metric space. However, we will be concerned with arbitrary endomorphisms of a compact interval. Thus the space will be more special, but the map more general. A number of results will actually be valid for arbitrary endomorphisms of a compact metric space, and these results will be marked with a dagger (†).

1 LIMIT SETS

Let I be a compact interval and $f: I \to I$ a continuous map of this interval into itself. Since we regard I as the underlying topological space, we will always use the relative topology. Thus a subset G of I is open if, for any $x \in G$, all points of I near x are also in G. For example, $[0,1/2)$ is an open subset of $I = [0,1]$.

As in Chapter I, we define the *trajectory* of a point $x \in I$ to be the sequence

$$\gamma(x) = \gamma(x,f) = \{f^n(x)\}_{n \geq 0}.$$

Distinct points $f^j(x), f^k(x)$ $(j \neq k)$ in the trajectory of x yield the same point in the orbit of x if $f^j(x) = f^k(x)$. The following lemma describes precisely what happens in this case. For later use we consider the trajectory of an interval, as well as that of a point.

†LEMMA 1 *Let H be a connected subset of I and let $E = \bigcup_{k \geq 0} f^k(H)$. Then either the connected components of E are the sets $f^k(H)$ $(k \geq 0)$, or there exist integers $m \geq 0$ and $p > 0$ such that the connected components of E are $f^k(H)$ $(0 \leq k < m)$ and $E_j := \bigcup_{k \geq 0} f^{m+j+kp}(H)$ $(0 \leq j < p)$.*

Proof Let S denote the set of non-negative integers m such that $f^m(H)$ and $f^{m+i}(H)$ are in the same component of E for some positive integer i. If $S = \emptyset$, then the components of E are the sets $f^k(H)$ ($k \geq 0$). Hence we suppose $S \neq \emptyset$.

Let m denote the smallest element of S. Then the sets $f^k(H)$ ($0 \leq k < m$) are components of E. It suffices to determine the components of $E' = \bigcup_{k \geq m} f^k(H)$.

Let p denote the smallest positive integer such that $f^m(H)$ and $f^{m+p}(H)$ are in the same component of E'. Since f maps a component of E' into a component of E', this implies that

(#) $f^{m+i}(H)$ and $f^{m+i+p}(H)$ are in the same component of E', for each non-negative integer i.

Hence, for any fixed j with $0 \leq j < p$, the set $E_j = \bigcup_{k \geq 0} f^{m+j+kp}(H)$ is contained in a component of E'. In particular, E' has r components, where $1 \leq r \leq p$.

We claim that, for any $i \geq m$, the sets $f^i(H), \ldots, f^{i+r-1}(H)$ are contained in distinct components of E'. Assume on the contrary that $f^j(H)$ and $f^l(H)$ are contained in the same component of E', where $i \leq j < l \leq i+r-1$. Then all the sets $f^k(H)$ with $k \geq j$ are contained in at most $l-j$ components of E'. Since $l-j < r$ and since each component of E' contains $f^k(H)$ for infinitely many k, by (#), this is a contradiction. This establishes our claim.

In particular, $f^m(H), f^{m+1}(H), \ldots, f^{m+r-1}(H)$ are contained in distinct components of E', and so are $f^{m+1}(H), \ldots, f^{m+r-1}(H), f^{m+r}(H)$. It follows that $f^m(H)$ and $f^{m+r}(H)$ are contained in the same component of E'. By the definition of p, this implies that $p = r$.

Since $E' = E_0 \cup \ldots \cup E_{p-1}$ and each E_j is contained in a component of E', each E_j is a component of E'. □

Thus if $f^j(x) = f^k(x)$ for some pair j, k with $j \neq k$ then the trajectory $\gamma(x)$ consists of m distinct points, followed by the points of an orbit of period p repeated periodically. In this case the point x is said to have a *finite orbit* or to be *eventually periodic*. We do not exclude the possibility that $m = 0$, i.e. that x itself is periodic.

We define the *limit set* of a point $x \in I$ to be the set

$$\omega(x) = \omega(x,f) = \bigcap_{m \geq 0} \overline{\bigcup_{n \geq m} f^n(x)} .$$

Equivalently, $y \in \omega(x)$ if and only if y is a limit point of the trajectory $\gamma(x)$, i.e. $f^{n_k}(x) \to y$ for some sequence of integers $n_k \to \infty$.

The basic properties of limit sets will now be established. For any positive integer m we have

$$\omega(x,f) = \bigcup_{j=0}^{m-1} \omega(f^j(x), f^m) . \tag{1}$$

Indeed the right side is obviously contained in the left, and if $f^{n_k}(x) \to y$ then for some j with $0 \le j < m$ we must have $n_k \equiv j \pmod m$ for infinitely many k. Evidently also

$$f(\omega(x, f^m)) = \omega(f(x), f^m). \tag{2}$$

†LEMMA 2 *For any $x \in I$, the limit set $\omega(x)$ is non-empty, closed and strongly invariant.*

Proof It follows at once from the definition that $\omega(x)$ is non-empty and closed, since a closed subset of I is compact and the intersection of a decreasing sequence of non-empty compact sets is again a non-empty compact set. Moreover we clearly have $f(\omega(x)) \subseteq \omega(x)$. If $y \in \omega(x)$ and $f^{n_k}(x) \to y$, then by restricting attention to a subsequence we may assume that $f^{n_k-1}(x) \to z$. Then $y = f(z)$, which shows that we actually have $f(\omega(x)) = \omega(x)$. □

Limit sets also possess the following general property.

†LEMMA 3 *If $L = \omega(x)$ is a limit set and if F is any non-empty proper closed subset of L, then*

$$F \cap \overline{f(L \setminus F)} \neq \emptyset.$$

Proof Assume, on the contrary, that F and $\overline{f(L \setminus F)}$ are disjoint. Then there exist open sets G_1, G_2 such that $L \setminus F \subset G_1$, $F \subset G_2$ and $\overline{G}_2$ is disjoint from $f(\overline{G}_1)$. For all large n, $f^n(x)$ belongs to either G_1 or G_2 and it belongs to each of them for infinitely many n. Hence there is an infinite sequence $n_1 < n_2 < \ldots$ such that $f^{n_k}(x) \in G_1$ and $f^{n_k+1}(x) \in G_2$. If y is a limit point of the sequence $f^{n_k}(x)$, then $y \in \overline{G}_1$ and $f(y) \in \overline{G}_2$, which is a contradiction. □

Lemma 3 admits the alternative formulation: if $L = \omega(x)$ is a limit set and if G is a non-empty subset of L, which is open in L and such that $f(\overline{G}) \subseteq G$, then $G = L$.

A point $x \in I$ is said to be *asymptotically periodic* if there exists a periodic point z such that

$$d[f^n(x), f^n(z)] \to 0 \text{ as } n \to \infty,$$

where $d(x_1, x_2)$ denotes the distance between the points x_1 and x_2. The limit set $\omega(x)$ is then just the orbit of z.

†LEMMA 4 *A limit set $\omega(x)$ contains only finitely many points if and only if x is asymptotically periodic.*

If $\omega(x)$ contains infinitely many points, then no isolated point of $\omega(x)$ is periodic.

Proof Suppose first that $L = \omega(x)$ is finite. Then f permutes the elements of L, since $f(L) = L$, and hence L contains a periodic orbit.

Suppose next that L is finite or infinite and contains a periodic orbit $C \neq L$. It follows from Lemma 3 that $F = L\backslash C$ is not closed. Hence every neighbourhood of C contains points of $L\backslash C$. Thus L must be infinite. This already proves that a finite limit set is necessarily a periodic orbit.

Suppose finally that L contains a periodic orbit C, of period m, and that some point $y \in C$ is isolated in L. Then y is a fixed point of $g = f^m$. Put $x_j = f^j(x)$ and let $\omega_j = \omega(x_j, g)$ denote the limit set of x_j under g. Then, by (1), $y \in \omega_j$ for some j such that $0 \leq j < m$. Since y is isolated in ω_j, we must have $\{y\} = \omega_j$ by what we proved in the previous paragraph. Thus y is the unique limit point of the sequence $f^{km+j}(x)$ and hence $f^{km+j}(x) \rightarrow y$ as $k \rightarrow \infty$. If we put $z = f^{m-j}(y)$, it follows that $f^{km+i}(x) \rightarrow f^i(z)$ as $k \rightarrow \infty$ for $0 \leq i < m$. Hence $d[f^n(x), f^n(z)] \rightarrow 0$ as $n \rightarrow \infty$. Thus x is asymptotically periodic and $L = C$ is finite. □

LEMMA 5 *If a limit set $L = \omega(x)$ contains an interval, then L is the union of finitely many disjoint closed intervals $J_1, ..., J_p$ such that $f(J_k) = J_{k+1}$ $(1 \leq k < p)$ and $f(J_p) = J_1$.*

Proof Let J_1 be a maximal closed subinterval of L. Since J_1 contains more than one point of the trajectory $\gamma(x)$, we have $f^k(J_1) \cap J_1 \neq \emptyset$ for some $k > 0$. It follows from Lemma 1 that there exists an integer $p > 0$ such that the closed connected sets $J_k = f^{k-1}(J_1)$ $(1 \leq k \leq p)$ are pairwise disjoint and $f(J_p) \subseteq J_1$. Moreover $L = \bigcup_{k=1}^{p} J_k$, since if $f^m(x) \in J_1$ then $f^n(x) \in \bigcup_{k=1}^{p} J_k$ for every $n > m$. Since L is strongly invariant, it follows that $f(J_p) = J_1$ and each J_k is an interval. □

Let

$$\Lambda = \Lambda(f) = \bigcup_{x \in I} \omega(x)$$

denote the set of all limit points of all trajectories. It follows immediately from Lemma 2 that

$$f(\Lambda) = \Lambda.$$

Moreover, by (1), for any positive integer m

$$\Lambda(f) = \Lambda(f^m).$$

The following remarkable result, due to Šarkovskii, provides a sufficient condition for a point to belong to Λ.

PROPOSITION 6 *A point $c \in I$ lies in Λ if every open interval with left endpoint c contains at least two points of some trajectory.*

The same conclusion holds if 'left' is replaced by 'right'.

Proof The proof will be conducted in a number of stages.

(i) We show first that c is a limit point of some trajectory if for each $\varepsilon > 0$ there exist $\delta > 0$ and $n > 0$ such that $[c,c+\delta] \subseteq f^n[c,c+\varepsilon]$.

The assumption implies that there exists a decreasing sequence $\varepsilon_k \to 0$ and a sequence of positive integers n_k such that $H_k = [c,c+\varepsilon_k]$ satisfies $H_{k+1} \subseteq f^{n_k}(H_k)$ $(k \geq 1)$. Let F_1 be a compact subinterval of H_1 such that $H_2 = f^{n_1}(F_1)$ and, for $k > 1$, let F_k be a compact subinterval of F_{k-1} such that $H_{k+1} = f^{m_k}(F_k)$, where $m_k = n_1+\ldots+n_k$. Then for any $x \in \bigcap_{k>0} F_k$ we have $f^{m_k}(x) \to c$ as $k \to \infty$.

(ii) Thus we may suppose that for some interval $H_0 = [c,c+\varepsilon_0]$ we have either $c \notin f^n(H_0)$ or $H_0 \cap f^n(H_0) = \{c\}$ for every $n > 0$. Since the proposition is trivial if $c \in \omega(c)$, we may suppose also that $f^n(c) \notin H_0$ for every $n > 0$, by decreasing ε_0 if necessary. By hypothesis, if $G \subset H_0$ is any open interval with left endpoint c then $G \cap f^k(G) \neq \emptyset$ for some $k > 0$. Moreover, by the choice of H_0, if $G \cap f^k(G) \neq \emptyset$ then c lies to the left of $f^k(G)$ and $f^k(c) > c+\varepsilon_0$. It follows that if $\tilde{G} \subset H_0$ is any other open interval with left endpoint c then $\tilde{G} \cap f^k(G) \neq \emptyset$ for infinitely many k.

(iii) We now show that for any open interval $G \subset H_0$ with left endpoint c there exist positive integers m, n such that $f^m(c)$ is an interior point of $f^n(G)$.

If $f^k(c)$ is not an interior point of $f^k(G)$ then it is an endpoint. Thus we may assume that $f^k(c)$ is an endpoint of $f^k(G)$ for every $k > 0$.

Suppose $G \cap f^m(G) \neq \emptyset$, $G \cap f^n(G) \neq \emptyset$, and $m \neq n$. We may choose the notation so that $f^m(c) \leq f^n(c)$. Then $f^m(G) \subseteq G \cup f^n(G)$. If $f^m(c) \neq f^n(c)$ then $f^m(c)$ is an interior point of $f^n(G)$. Thus we may assume that all intervals $f^k(G)$ for which $G \cap f^k(G) \neq \emptyset$ have a common right endpoint $b = f^k(c)$.

Let $m < n$ be two positive values of k for which $G \cap f^k(G) \neq \emptyset$. The set $S = f^m(G) \cup f^{m+1}(G) \cup \ldots \cup f^{n-1}(G)$ does not intersect some open interval $\tilde{G}$ with left endpoint c. If $f^n(G) \subseteq f^m(G)$ then $f(S) \subseteq S$, which is impossible since $\tilde{G} \cap f^k(S) \neq \emptyset$ for some $k > 0$. Hence $f^m(G) \subseteq f^n(G)$. Thus if we put $d = n - m$ then

$$f^m(G) \subseteq f^{m+d}(G) \subseteq f^{m+2d}(G) \subseteq \dots .$$

Hence $G \cap f^{m+jd}(G) \neq \emptyset$ and $b = f^{m+jd}(c)$ is the right endpoint of the interval $f^{m+jd}(G)$ for every $j \geq 0$. Now choose m, n so that $d = n - m$ has its least value. Then $G \cap f^{m+k}(G) \neq \emptyset$ if and only if $k \geq 0$ is a multiple of d. It follows that for any open interval $\tilde{G} \subseteq G$ with left endpoint c we can find $j \geq 0$ such that $\tilde{G} \cap f^{m+jd}(G) \neq \emptyset$. Thus if we put $E = \bigcup_{j \geq 0} f^{m+jd}(G)$, then $f^d(E) = E$ and $\bar{E} = [c,b]$. This implies that $c = f^d(y)$ for some $y \in (c,b)$. But this is a contradiction, since $c \notin f^k(G)$ if $G \cap f^k(G) \neq \emptyset$. This establishes the claim that $f^m(c)$ is an interior point of $f^n(G)$ for some $m, n > 0$.

(iv) We show next that if $f^m(c)$ is an interior point of $f^n(G)$ for some $m, n > 0$ then there exists a compact interval $H \subset G$ with the property that for any compact interval $E \subset G$ we have $E \subset f^k(H)$ for some $k > 0$.

Let $F \subset f^n(G)$ be a compact interval containing $f^m(c)$ in its interior and take $H \subset G$ to be a compact interval such that $f^n(H) = F$. Choose a small open interval $\tilde{G} \subset G \backslash E$ with left endpoint c so that $f^m(\tilde{G}) \subset F$. Then for any $j > m$ such that $\tilde{G} \cap f^j(\tilde{G}) \neq \emptyset$ we have $E \subset f^j(\tilde{G})$, since $\tilde{G}$ lies to the left of E and $f^j(c)$ to the right of E. Hence $E \subset f^{j-m}(F) = f^{j-m+n}(H)$.

(v) Finally, let $G_k = (c, c + \varepsilon_k)$, where the decreasing sequence $\varepsilon_k \to 0$. For each $k \geq 1$ let $H_k \subseteq G_k$ be a compact interval with the property described in (iv). Then we can find a positive integer n_k such that $H_{k+1} \subset f^{n_k}(H_k)$. Let $F_1 \subseteq H_1$ be a compact interval such that $H_2 = f^{n_1}(F_1)$ and, for $k > 1$, let $F_k \subseteq F_{k-1}$ be a compact interval such that $H_{k+1} = f^{m_k}(F_k)$, where $m_k = n_1 + \dots + n_k$. Then for any $x \in \bigcap F_k$ we have $f^{m_k}(x) \to c$ as $k \to \infty$. □

It is worth noting that the proof of Proposition 6 gives not only $f^{m_k}(x) \to c$ as $k \to \infty$, but also $f^{m_k}(x) > c$ for all k if the first condition is satisfied, and $f^{m_k}(x) < c$ for all k if the second condition is satisfied. The following example shows that one or other of the conditions of Proposition 6 is not necessary for the point c to belong to Λ.

EXAMPLE 7 Let $I = [-1,1]$ and let $f: I \to I$ be the piecewise linear function defined by

$$f(-1) = f(0) = 0, \; f(1) = -1.$$

Then $c = 0$ is a limit point, but no open interval with left or right endpoint c contains two points of the same trajectory.

In fact it is easily seen that if a point $c \in \Lambda$ does not satisfy one or other of the conditions of Proposition 6 then it must be periodic. The next example shows that we cannot replace 'open' by 'compact' in the statement of Proposition 6.

EXAMPLE 8 Let $I = [0,7]$ and let $f: I \to I$ be the piecewise linear function defined by

$$f(0) = 3,\ f(1) = f(2) = 4,\ f(3) = 7,$$
$$f(4) = 4,\ f(5) = 1,\ f(7) = 3\,.$$

In any compact interval with right endpoint 1 there is a point y such that $f^m(y) = 1$ for some $m > 0$, since f maps a left neighbourhood of 1 onto a left neighbourhood of 4 and $1 \in W(4,f,L)$. On the other hand $1 \notin \Lambda$, since $f(x) \geq 1$ for all $x \in I$ and $f^n(x) = 4$ for all $n > 0$ if $x \in [1,2]$.

Before deriving a necessary and sufficient condition for a point to belong to Λ we prove a simple, but useful, lemma.

LEMMA 9 *Let J be a subinterval of I which contains no periodic point of f. If $x \in J, f^m(x) \in J$ for some $m > 0$ and $y \in J, f^n(y) \in J$ for some $n > 0$, then*

$$y < f^n(y) \text{ if } x < f^m(x) \quad and \quad y > f^n(y) \text{ if } x > f^m(x)\,.$$

Proof Without loss of generality, suppose $x < f^m(x)$. If we put $g = f^m$ then the interval $[x,g(x)]$ contains no periodic point of g. If $g^k(x) > x$ for some $k \geq 1$ then $g^{k+1}(x) > g(x)$, since g^k does not have a fixed point in the interval $[x,g(x)]$. It follows by induction that $g^k(x) > x$ for every $k \geq 1$. In particular, $f^{mn}(x) > x$. If we had $f^n(y) < y$ then in the same way we would obtain $f^{mn}(y) < y$. But then $f^{mn}(z) = z$ for some z between x and y, which is a contradiction. □

COROLLARY 10 *If J is a subinterval of I which contains no periodic point of f then, for any $x \in I$, the points of the trajectory $\gamma(x)$ which lie in J form a strictly monotonic (finite or infinite) sequence.* □

We will say that an interval J without periodic points such that $J \cap f^k(J) \neq \varnothing$ for some $k > 0$ is of *increasing type* or *decreasing type* according as the points of all trajectories in J form an increasing or decreasing sequence.

PROPOSITION 11 *A point $c \in I$ lies in Λ if and only if every open interval containing c contains at least 3 points of some trajectory.*

Proof The necessity of the condition follows at once from the definition of a limit point. Suppose then that the condition is satisfied. If G is any open interval containing c we put

$$G_- = \{x \in G: x < c\}, \ G_+ = \{x \in G: x > c\},$$

so that $G = G_- \cup \{c\} \cup G_+$. By Proposition 6 we may suppose that there exists a G such that

$$G_- \cap f^k(G_-) = \varnothing, \ G_+ \cap f^k(G_+) = \varnothing \quad \text{for all } k > 0.$$

Moreover, since the result is trivial if c is periodic, by restricting G we may suppose that $f^k(c) \notin G$ for all $k > 0$.

By hypothesis for some $x \in G$ we have also $f^m(x) \in G$ and $f^n(x) \in G$, where $0 < m < n$. Then $x \neq c$ and we suppose, for definiteness, that $x \in G_+$. Since $f^m(x) \neq c$ we must have $f^m(x) \in G_-$. But then, by Corollary 10, $f^n(x) \in G_-$, which is a contradiction. □

COROLLARY 12 *The set Λ is closed.* □

COROLLARY 13 *For any open set $G \supset \Lambda$ there exists a positive integer $m = m(G)$ such that at most m points of any trajectory $\gamma(x)$ lie outside G.*

Proof We observe that each $y \in I \setminus G$ lies in an open interval H_y which contains at most 2 points of any trajectory. The compact set $I \setminus G$ is covered by a finite number p of the intervals H_y. Hence any trajectory has at most $2p$ points in $I \setminus G$. □

In the statement of Corollary 13 the set Λ cannot be replaced by any proper closed subset, since if $y \in \Lambda$ any open set containing y contains infinitely many points of some trajectory. Corollary 13 sharpens, for endomorphisms of a compact interval, a theorem of G.D. Birkhoff for homeomorphisms of a compact metric space in which Λ is replaced by the (in general) larger nonwandering set Ω defined below.

2 RECURRENT AND NONWANDERING POINTS

We now introduce some further concepts. Let $f: I \to I$ be a continuous map of a compact interval I into itself. A point $x \in I$ is said to be *recurrent* if $x \in \omega(x)$ and *nonwandering* if every open set containing x contains at least two points of some trajectory. Equivalently, x is recurrent if $f^{n_k}(x) \to x$ for some sequence of integers $n_k \to \infty$, and nonwandering if $f^{n_k}(x_k) \to x$ for some sequence of points $x_k \to x$ and some sequence of integers $n_k \to \infty$.

The preceding definitions remain valid if I is replaced by any compact metric space X. It should be noted also that some authors use 'Poisson stable' in place of our 'recurrent', and then use 'recurrent' in a stronger sense, which will be considered in Chapter V.

Let P, R and Ω denote respectively the set of all periodic, recurrent and nonwandering points. When we wish to emphasise the dependence on f we will write, for example, P(f) instead of P. It follows at once from the definitions that

$$f(\mathrm{P}) = \mathrm{P} \subseteq \mathrm{R} = f(\mathrm{R})$$

and

$$\mathrm{R} \subseteq \Lambda \subseteq f(\Omega) \subseteq \Omega = \overline{\Omega}.$$

Examples show that each of the inclusions here can be strict, even for maps of an interval.

LEMMA 14 *Let J be an open subinterval which contains no periodic point of f. Then*

(i) *J contains at most one point of any limit set $\omega(x)$,*
(ii) *J contains no recurrent point,*
(iii) *if $x \in J$ is nonwandering, then no other point of its trajectory lies in J.*

Proof The first two assertions follow at once from Corollary 10. To prove the third, suppose we had $f^m(x) \in J$ for some $m > 0$. Since $f^m(x) \neq x$, there is an open interval G containing x such that $G \subseteq J$, $f^m(G) \subseteq J$ and $G \cap f^m(G) = \emptyset$. Since $x \in \Omega$ we can choose $y \in G$ and $n > m$ so that also $f^n(y) \in G$. Then $f^m(y)$ does not lie between y and $f^n(y)$, which contradicts Lemma 9. □

The closure $\overline{\mathrm{R}}$ of the set of all recurrent points is known as the *centre* of f. Birkhoff showed that for homeomorphisms of a compact metric space the centre could be characterized as the nonwandering set of the nonwandering set of the nonwandering set ..., continued by transfinite induction until one gets nothing smaller. The following result shows that for endomorphisms of a compact interval one need make at most one step past Ω itself, and that in this case the centre can also be characterized as the closure of the set of all periodic points.

PROPOSITION 15 *Any non-isolated point of* Ω *lies in* Λ, *and any non-isolated point of a limit set* $\omega(x)$ *lies in* $\overline{\mathrm{P}}$. *Moreover*

$$\overline{\mathrm{P}} = \overline{\mathrm{R}} = \Omega(f\,|\Omega),$$

i.e., $\overline{\mathrm{R}}$ *is the nonwandering set of* f, *regarded as a map of the compact invariant set* Ω.

Proof The first assertion follows at once from Proposition 6, and the second from Lemma 14(i). The relation $\overline{\mathrm{P}} = \overline{\mathrm{R}}$ follows similarly from Lemma 14(ii) and the obvious inclusion $\overline{\mathrm{P}} \subseteq \overline{\mathrm{R}}$. The inclusion $\overline{\mathrm{P}} \subseteq \Omega(f\,|\Omega)$ is also obvious. Suppose on the other hand that $y \in \Omega(f\,|\Omega)$ and let G be any open interval containing y. Then $G \cap \Omega$ has non-empty intersection with $f^m(G \cap \Omega)$ for some $m > 0$. Hence, by Lemma 14(iii), G must contain a periodic point. Therefore $\Omega(f\,|\Omega) \subseteq \overline{\mathrm{P}}$. □

By definition, a point x lies in the nonwandering set Ω if there exist points $x_k \to x$ and integers $n_k \to \infty$ such that $f^{n_k}(x_k) \to x$. The following result shows that for endomorphisms of an interval the definition can be sharpened.

PROPOSITION 16 *A point* x *is nonwandering if and only if there exist points* $x_k \to x$ *and integers* $n_k \to \infty$ *such that* $f^{n_k}(x_k) = x$.

Proof Let $x \in \Omega$. We may assume that x is not periodic, since if x has period m we can take $x_k = x$ and $n_k = km$. We show first that for any open interval G containing x we have $x \in f^n(G)$ for some $n > 0$.

Put $E = \bigcup_{k\geq 0} f^k(G)$. Since $G \cap f^k(G) \neq \emptyset$ for some $k > 0$, the set E has only finitely many connected components, by Lemma 1. Assume $x \notin f^m(\bar{E})$ for some $m > 0$. Then there exists an open interval $H \subset G$ containing x such that $H \cap f^m(\bar{E}) = \emptyset$. Since $f^m(\bar{E})$ is f-invariant, this implies that $H \cap f^k(H) = \emptyset$ for all $k \geq m$, which contradicts $x \in \Omega(f)$. We conclude that $x \in f^m(\bar{E})$ for every $m > 0$. Since $\bar{E}\setminus E$ is a finite set and x is not periodic, it follows that $x \in f^m(E)$ for some $m > 0$, and hence $x \in f^n(G)$ for some $n > 0$.

Thus there exist points $x_k \to x$ and integers n_k such that $f^{n_k}(x_k) = x$. Moreover, since x is not periodic, we must have $n_k \to \infty$. □

COROLLARY 17 *If an endpoint of I is nonwandering, then it lies in $\overline{P}$.*

Proof Let $I = [a,b]$. If $a \notin \overline{P}$ there is an interval $J = [a, a+\varepsilon]$ such that $J \cap P = \emptyset$. Since $f^n(a) > a$, it follows that $f^n(x) > x$ for every $x \in J$ and all $n \geq 1$. Hence, by Proposition 16, $a \notin \Omega$. □

Further information about the nonwandering set is provided by the following results.

PROPOSITION 18 *The set Ω of nonwandering points is contained in the closure of the set of eventually periodic points.*

Proof If $x \in \Omega(f)$ then for any open interval G containing x there is a positive integer m such that $G \cap f^m(G) \neq \emptyset$. We will prove the proposition by showing that some point of G is eventually periodic.

It follows from Lemma 1 that there exists an integer $p > 0$ such that $E = \bigcup_{k \geq 0} f^{kp}(G)$ is an interval and $f^j(G) \cap E = \emptyset$ if j is not divisible by p. Since $x \in \Omega(f)$, it follows that also $x \in \Omega(f^p)$. Evidently p divides m and $f^p(E) \subseteq E$. If E contains no fixed point of f^p, then either $f^p(y) > y$ for all $y \in E$ or $f^p(y) < y$ for all $y \in E$. Suppose, for definiteness, that the first alternative holds. Since $f^p(x) > x$, we can choose a small open interval $H \subset G$ containing x so that $f^p(H)$ is disjoint from H and lies to the right of H. Then for any $y \in H$ and any $k \geq 1$, $f^{kp}(y)$ lies to the right of H. But this contradicts $x \in \Omega(f^p)$. We conclude that some point of E is a fixed point of f^p, and hence some point of G is eventually periodic. □

PROPOSITION 19 *Let $x \in \Omega \setminus \overline{P}$. Then there exists a $\delta > 0$ such that, for any $\varepsilon \in (0,\delta)$ we have*

$$J \cap f^n(J_1) = \emptyset \quad \text{for all } n > 0,$$

where $J = [x-\varepsilon, x+\varepsilon]$ and J_1 denotes exactly one of $[x, x+\varepsilon]$, $[x-\varepsilon, x]$.

If $x \in \Omega \setminus \Lambda$ we can choose δ so that in addition

$$J \cap f^n(J_2) \subseteq J_1 \text{ for all } n > 0,$$

where $J_2 = [x-\varepsilon, x]$ or $[x, x+\varepsilon]$ according as $J_1 = [x, x+\varepsilon]$ or $[x-\varepsilon, x]$.

Proof Choose $\varepsilon_0 > 0$ so that for $J_0 = (x-\varepsilon_0, x+\varepsilon_0)$ we have $J_0 \cap P = \emptyset$. Then, by Lemma 14, $f^n(x) \notin J_0$ for every $n > 0$. On the other hand, $J_0 \cap f^n(J_0) \neq \emptyset$ for infinitely many $n > 0$.

Suppose the interval J_0 is of increasing type, in the terminology introduced following Corollary 10. Then, for any $\varepsilon \in (0, \varepsilon_0)$,

$$[x-\varepsilon, x] \cap f^n[x, x+\varepsilon] = \emptyset \quad \text{for all } n > 0.$$

We claim that, for some $\varepsilon_1 \in (0, \varepsilon_0)$,

$$[x, x+\varepsilon_1] \cap f^n[x, x+\varepsilon_1] = \emptyset \quad \text{for all } n > 0.$$

Assume on the contrary that, for any $\varepsilon \in (0, \varepsilon_0)$,

$$(x, x+\varepsilon) \cap f^n(x, x+\varepsilon) \neq \emptyset \quad \text{for some } n > 0.$$

Then, by Proposition 6 and the remark following its proof, there exists a point $y \in I$ and a sequence of integers $n_k \to \infty$ such that $f^{n_k}(y) > x$ for all k and $f^{n_k}(y) \to x$ as $k \to \infty$. But this contradicts the hypothesis that J_0 is of increasing type.

By Proposition 6 again, if $x \notin \Lambda$ then for some $\varepsilon_2 > 0$

$$(x-\varepsilon_2, x) \cap f^n(x-\varepsilon_2, x) = \emptyset \quad \text{for all } n > 0.$$

Hence for any $\varepsilon < \min(\varepsilon_1, \varepsilon_2)$ the requirements of the proposition are met with $J_1 = [x, x+\varepsilon]$ and $J_2 = [x-\varepsilon, x]$. The argument when J_0 is of decreasing type is the same, but in this case $J_1 = [x-\varepsilon, x]$ and $J_2 = [x, x+\varepsilon]$. □

COROLLARY 20 *The set $\Omega(f) \setminus \overline{P}(f)$ is at most countable and is nowhere dense in I.*

Proof By Proposition 19, each point of $\Omega \setminus \overline{P}$ is isolated on one side in Ω. □

It may be noted that Nitecki [96] gives an example of a map f for which $\Omega(f) \setminus \overline{P}(f)$ is infinite, although f is non-chaotic and *unimodal*, i.e. there exists a point $c \in I$ such that f is strictly monotonic to the left of c and strictly monotonic in the opposite sense to the right of c.

The following example illustrates the behaviour described in Proposition 19.

EXAMPLE 21 Let $I = [0,1]$ and let $f: I \to I$ be the piecewise linear function defined by

$$f(0) = 0, \ f(1/4) = 3/4, \ f(1/2) = 0 = f(3/4), \ f(1) = 1/4.$$

Then $3/4 \in \Omega$, but $3/4 \notin \Lambda$ and in fact $3/4 \notin f(\Omega)$.

Our next result shows that for piecewise monotone maps we can replace Λ by $\overline{P}$ in the second statement of Proposition 19, since the two are actually equal. It appears to be an open problem whether $\Lambda = \overline{P}$ also for continuously differentiable maps.

PROPOSITION 22 *If f is piecewise monotone, then $\Lambda(f) = \overline{P}(f)$.*

Proof We will assume that there exists a point $x \in \Lambda \setminus \overline{P}$ and derive a contradiction. Let J be the component of $I \setminus \overline{P}$ which contains x. We may suppose that J is of increasing type. Let c denote the left endpoint of J.

Let T denote the set of turning-points t of f such that $f^k(t) \in (c,x)$ for some positive integer k. For each $t \in T$ let $j(t)$ denote the least positive integer k for which $f^k(t) \in (c,x)$. Let z denote the largest element of the set $\{f^{j(t)}(t) : t \in T\}$ if $T \neq \emptyset$, and let z denote any point of (c,x) if $T = \emptyset$.

Since $x \in \Omega$, there is a point $w \in (z,x)$ and a positive integer n such that $f^n(w) = x$. Since $x \in \Lambda$, there is a point $y \in (w,x)$ and an integer $m > n$ such that $y < f^m(y) < x$. We have $f^m(x) > x$ and $f^m(w) > w$, because $f^m(y) > y$ and there are no periodic points in J. In fact $f^m(w) > x$, since either $f^m(w) = f^{m-n}(x) \in J$ or $f^m(w)$ is to the right of J.

Since $f^m(y) < x$, there is a point $v \in (w,x)$ at which f^m assumes its minimum value on the closed interval $[w,x]$. If there is more than one such point we take v to be the nearest to x. Then $v < f^m(v) < x$. From the definition of v, $f^i(v)$ is a turning-point of f for some i with $0 \leq i < m$. Moreover $f^i(v) \in T$, since $f^m(v) \in (c,x)$. Let $t_0 = f^i(v)$ and $j = j(t_0)$. Then

$$f^{j+i}(v) = f^j(t_0) \leq z < v.$$

Since v and $f^{j+i}(v)$ are in J, and J is of increasing type, this is a contradiction. □

The following result is also related to Proposition 19.

LEMMA 23 *Suppose $x \in \Omega$ and the orbit of x is infinite. If for some $d > x$ the interval $J = [x,d)$ contains no periodic points and is of increasing type, then the connected sets $\overline{J}, f(\overline{J})$, $f^2(\overline{J})$, ... are pairwise disjoint and $\omega(y) = \omega(x)$ for every $y \in J$.*

If for some $c < x$ the interval $J = (c,x]$ contains no periodic points and is of decreasing type, then the same conclusions hold.

Proof Without loss of generality we restrict attention to the first case. If the connected sets $f^n(\bar{J})$ $(n \geq 0)$ are not pairwise disjoint then, by Lemma 1, the set $E = \bigcup_{n\geq 0} f^n(\bar{J})$ has only finitely many connected components. Thus $E \setminus \text{int}\, E$ is a finite set. Since $f(E) \subseteq E$ and $x \in E$ has an infinite orbit it follows that $f^m(x) \in \text{int}\, E$ for some m. But, since $x \in \Omega$, there exist points $x_k \to x$ and integers $n_k \to \infty$ such that $f^{n_k}(x_k) = x$. For all large k we have $f^m(x_k) \in E$ and $n_k > m$. Thus $f^m(x_k) = f^{q_k}(y_k)$ for some $y_k \in \bar{J}$ and some $q_k \geq 0$, and $x = f^{n_k - m + q_k}(y_k)$. Since x is not periodic it follows that $x = f^q(y)$ for some $y \in \text{int}\, J$ and some $q > 0$. But this contradicts the fact that J is of increasing type.

Thus the connected sets $f^n(\bar{J})$ $(n \geq 0)$ are pairwise disjoint. It follows that for any $y \in \bar{J}$ the orbit of y is also infinite and the distance between $f^n(y)$ and $f^n(x)$ tends to zero as $n \to \infty$. Hence $\omega(y) = \omega(x)$. □

COROLLARY 24 *Any open interval J such that $J \cap \mathrm{P} = \varnothing$ contains at most one nonwandering point which is not eventually periodic.* □

It follows at once from Lemma I.10 that, for any positive integer m,

$$\mathrm{P}(f) = \mathrm{P}(f^m).$$

It will now be shown that there is an analogous result for the set of recurrent points.

†LEMMA 25 *For any positive integer m,*

$$\mathrm{R}(f) = \mathrm{R}(f^m).$$

Proof It follows at once from the definition that if x is recurrent for f^m then it is also recurrent for f. Suppose, on the other hand, that x is recurrent for f. That is, $x \in \omega(x,f)$. Then, by (1), $x \in \omega(f^j(x), f^m)$ for some j with $0 \leq j < m$. It follows that $\omega(x,f^m) \subseteq \omega(f^j(x), f^m)$. Using (2), we obtain successively

$$\omega(f^j(x), f^m) \subseteq \omega(f^{2j}(x), f^m)$$

$$\cdots$$

$$\omega(f^{j(m-1)}(x), f^m) \subseteq \omega(f^{jm}(x), f^m) = \omega(x,f^m).$$

Hence we must actually have $\omega(x,f^m) = \omega(f^j(x), f^m)$ and $x \in \omega(x,f^m)$. □

The analogue of Lemma 25 for the nonwandering set is false, although obviously $\Omega(f^m) \subseteq \Omega(f)$. Thus in Example 8 we saw that $1 \in \Omega(f)$. However $1 \notin \Omega(f^2)$, since $f^{2n}[0,2] = [4,7]$ for every $n > 0$. (Note also that $5 \notin \Omega(f)$, and hence $f(\Omega) \neq \Omega$.) In this example the point 1 is eventually periodic. We now show that this is no accident.

PROPOSITION 26 *If $x \in \Omega(f)$ and x is not eventually periodic, then $x \in \Omega(f^n)$ for every $n \geq 1$.*

Proof We may suppose that $x \notin \overline{P}(f)$, since $\overline{P}(f) = \overline{P}(f^n) \subseteq \Omega(f^n)$. By Proposition 16 we have $x = f^{n_k}(x_k)$, where $x_k \to x$ and $n_k \to \infty$. Then $x_k \neq x$ for all k. Without loss of generality, suppose $x_k < x$ for all k.

Let S denote the set of all non-negative integers r such that for every compact interval H with right endpoint x there exists a positive integer $m \equiv r \pmod n$ such that $x \in f^m(H)$. Evidently if $r \in S$ and $r' \equiv r \pmod n$, then also $r' \in S$. We wish to show that $0 \in S$.

The set S is not empty, since there are only finitely many residue classes modulo n. For the same reason we can choose a compact interval H_0 with right endpoint x so that $x \in f^m(H_0)$ implies $m \in S$. Moreover, by shortening H_0 we may suppose that $H_0 \cap P = \emptyset$.

We show first that if $f^{jn}(x) \in \operatorname{int} f^{kn+s}(H_0)$ for some $j, k, s \geq 0$ then $r \in S$ implies $r+s \in S$. Let $H \subseteq H_0$ be a compact interval with right endpoint x such that $f^{jn}(H) \subseteq f^{kn+s}(H_0)$. For infinitely many $m \equiv r \pmod n$ there exists a point $y \in H$ such that $f^m(y) = x$, since $r \in S$ and $x \notin P$. If $m = \ell n + r$ with $\ell > j$ then $x = f^{\ell n+r}(y) \in f^{(\ell-j+k)n+r+s}(H_0)$, and hence $r + s \in S$.

We show next that for any $r \in S$ there exist $j, k \geq 0$ such that $f^{jn}(x) \in \operatorname{int} f^{kn+r}(H_0)$. Since $x \notin P$, we can choose $k_1 < k_2 < k_3$ so that $x \in H_i = f^{k_i n+r}(H_0)$ for $i = 1, 2, 3$. If $x \in \operatorname{int} H_i$ for some i we can take $j = 0, k = k_i$. Thus we may suppose that x is an endpoint of each H_i. Since $f^{k_i n+r}(y_i) = x > y_i$ for some $y_i \in H_0$, we must have $f^{k_i n+r}(x) > x$. Hence x is the left endpoint of each H_i. If we put $x_i = f^{(k_i-k_1)n}(x)$, then $x_i \in H_i$. Moreover the three points x_i are distinct, since x is not eventually periodic. Since $x_1 = x$ is the left endpoint of both H_2 and H_3, we must have either $x_1 < x_2 < x_3$ or $x_1 < x_3 < x_2$. In the first case $x_2 \in \operatorname{int} H_3$ and we can take $j = k_2 - k_1$, $k = k_3$. In the second case $x_3 \in \operatorname{int} H_2$ and we can take $j = k_3 - k_1$, $k = k_2$.

Combining the results of the last two paragraphs, we see that $r \in S$ implies $nr \in S$ and hence $0 \in S$. □

It is still possible to say something when $x \in \Omega(f)$ is eventually periodic, but the results in this case are more involved.

LEMMA 27 *Suppose $x \in \Omega(f)$ and the trajectory of x contains a periodic point of period $n \geq 1$. Then for any neighbourhood G of x there exists a positive integer N such that $x \in f^{2jn+N}(G)$ for all $j \geq 0$.*

Proof Let $z = f^p(x)$ have period n. We may suppose that x does not lie in the orbit of z, since otherwise the result is trivial. By Proposition 16 we have $x = f^{n_k}(x_k)$, where $x_k \to x$ and $n_k \to \infty$. Then $y_k = f^p(x_k) \to z$ and $x = f^{n_k-p}(y_k)$. Thus $y_k \neq z$ for all k. Without loss of generality, suppose $y_k < z$ for all k.

There exists a left neighbourhood H of z such that $H \subseteq f^p(G)$ and $x \notin f^k(H)$ for $0 \leq k < 2n$. Let $m \geq 2n$ be the least positive integer such that $x \in f^m(H)$. We cannot have $f^n(H) \subseteq H$, since then $H \cup f(H) \cup \ldots \cup f^{n-1}(H)$ would be f-invariant and so would contain x. If $f^n(H) \supseteq H$ then $x \in f^{jn+m}(H)$ for all $j \geq 0$. Otherwise $f^n(H) = H' \cup H''$, where H'' is a right neighbourhood of z and $H' \subset H$. Then $x = f^{m-n}(f^n(y))$, where $y \in H$ and hence $f^n(y) \in H''$. We cannot have $f^{2n}(H) \subseteq H \cup f^n(H)$, since then $H \cup f(H) \ldots \cup f^{2n-1}(H)$ would be f-invariant and so would contain x. Hence either $f^n(H'') \supseteq H''$ or $f^n(H'') \supseteq H$. If $f^n(H'') \supseteq H''$ then $x \in f^{jn+m}(H)$ for all $j \geq 0$. If $f^n(H'') \supseteq H$ then $x \in f^{2jn+m}(H)$ for all $j \geq 0$. In every case, $x \in f^{2jn+m+p}(G)$ for all $j \geq 0$. □

LEMMA 28 *Suppose $x \in \Omega(f) \setminus \overline{P}(f)$, and the trajectory of x contains a periodic point of period $n > 2$. Then either $x \in \Omega(f^n)$ or n is even and $x \in \Omega(f^{n/2})$.*

Proof Let $y = f^q(x)$ have period n. Since $x \in \Omega(f)$ we have $x = f^{n_k}(x_k)$, where $x_k \to x$ and $n_k \to \infty$. Then $y_k = f^q(x_k) \to y$ and $x = f^{n_k-q}(y_k)$. For some $p \geq q$ we must have $n_k \equiv p \pmod{n}$ for infinitely many k. Then $z_k = f^p(x_k) \to f^p(x) = z$. Moreover $f^{n_k-p}(z_k) = x$ and we may suppose that all z_k lie on the S-side of z. Then z has period n, $x \in W(z,f^n,S)$ and, for every neighbourhood H of x, $f^p(H)$ contains an S-neighbourhood of z.

Put $W_j = f^j(W(z,f^n,S))$, so that $x \in W_0$. Assume $x \in \operatorname{int} W_0$. Take $H = \langle a,b \rangle$ to be any small closed neighbourhood of x contained in W_0, where the endpoint b lies between x and z and the endpoint a lies on the opposite side of x to z. Then $a = f^{in}(c)$ for some $i > 0$, where $c \in f^p(H)$, and thus $a \in f^{in+p}(H)$. Since $z = f^{in+p}(x)$, it follows that $H \subseteq f^{in+p}(H)$. Hence H contains a periodic point. Thus $x \in \overline{P}(f)$, which is contrary to hypothesis. We conclude that x is an endpoint of W_0.

There are unique integers s,t with $0 \leq s < n$ and $t > 0$ such that $p + s = tn$. Suppose first that $x \in W_s$. That is, $x = f^s(w)$, where $w = f^{nl_k}(w_k)$, $w_k \to z$ and w_k is on the S-side of z. Then $w_k = f^p(\xi_k)$, where $\xi_k \to x$, and hence $x = f^{n(t+l_k)}(\xi_k) \in \Omega(f^n)$.

Suppose next that $x \notin W_s$. This implies, in particular, that $s > 0$ and $W_s \neq W_0$. On the other hand, $f^s(z) = f^{tn}(x) \in W_0$, since W_0 is f^n-invariant. Since W_0 and W_s are distinct, and $f^s(z)$ lies in their intersection, it follows from Proposition III.13 that $W_0 \cup W_s$ is a connected component of $W(z,f,S)$ and $W_{2s} = W_0$. Thus W_0 contains the f^s-orbit of z. Moreover $f^s(z)$ is the endpoint $\neq x$ of W_0, since $f^s(z) \in \operatorname{int} W_0$ would imply $W_s = W_0$ by Proposition III.13.

Assume $f^{2s}(z) \neq z$. Then there exists a least integer $h > 2$ such that $f^{hs}(z) = z$. It follows that $W_{ks} = W_0$ for $1 < k \leq h$, since $f^{ks}(z) \neq f^s(z)$ and hence $f^{ks}(z) \in \operatorname{int} W_0$. Therefore $W_s = f^s(W_{(h-1)s}) = W_{hs} = W_0$, which is a contradiction. We conclude that $f^{2s}(z) = z$ and $n = 2s$. Hence s divides p and $x \in \Omega(f^s)$. □

We can now derive without difficulty

PROPOSITION 29 *If $x \in \Omega(f)$ and x is eventually periodic, then $x \in \Omega(f^n)$ for every odd $n > 1$.*

Proof As in the proof of Proposition 26, we may suppose that $x \notin \overline{P}(f)$. By Lemma 28 there exists a positive integer m such that for $g = f^m$ we have $x \in \Omega(g) \setminus \overline{P}(g)$ and the g-orbit of x contains a fixed point of g^2. Then, by Lemma 27, for any neighbourhood G of x there exists a positive integer N such that $x \in g^{4j+N}(G)$ for all $j \geq 0$. But for any odd n there exist integers p, q such that $np - 4q = N$. Then $j = q + nk \geq 0$ for all large k and $4j + N$ is a multiple of n. Hence $x \in \Omega(g^n)$ and, *a fortiori*, $x \in \Omega(f^n)$. □

By combining Propositions 26 and 29 we obtain at once

PROPOSITION 30 *For any odd positive integer m,*

$$\Omega(f) = \Omega(f^m). \quad \square$$

It follows that if $n = 2^d q$, where q is odd, then $\Omega(f^n) = \Omega(f^{2^d})$. Thus all possible sets $\Omega(f^n)$ occur in the sequence

$$\Omega(f) \supseteq \Omega(f^2) \supseteq \Omega(f^4) \supseteq \dots .$$

Moreover, every succession of equalities and strict containments is possible here, as may be shown by using the constructions of Examples I.13 and I.14.

We show finally that this is no longer true if we place restrictions on the map f. We will make use of the following preliminary result.

LEMMA 31 *If $x \in \Omega(f)$ and the limit set $\omega(x,f)$ contains a periodic point z, then $x \in W(z,f)$.*

Proof We may suppose that $x \neq z$. Let G be an open interval which contains z but not x. Since $z \in \omega(x)$, there exists a least positive integer p such that $f^p(x) \in G$. Let G_0 be an open interval containing x such that $f^p(G_0) \subseteq G$. Then $x \in f^k(G_0)$ for some $k > p$, by Proposition 16, and hence

$$x \in f^{k-p}(f^p(G_0)) \subseteq f^{k-p}(G).$$

Since G was arbitrary, it follows that $x \in W(z,f)$. □

PROPOSITION 32 *If f has no periodic point of period $2q$, for every odd integer $q > 1$, then*

$$\Omega(f) = \Omega(f^2).$$

Proof Assume, on the contrary, that there exists a point $x \in \Omega(f) \setminus \Omega(f^2)$. Then x is eventually periodic, but not periodic, by Proposition 26. Let z be a periodic point in the orbit of x and let its period be $n = 2^d p$, where $p \geq 1$ is odd and $d \geq 0$. If we put $g = f^p$ then $x \in \Omega(g) \setminus \Omega(g^2)$, by Proposition 30, and the g–orbit of x contains a point of period 2^d. In fact we must have $d = 0$ or 1, since $d > 1$ would imply $x \in \Omega(g^{2^{d-1}}) \subseteq \Omega(g^2)$, by Lemma 28. Moreover g also has no periodic point of period $2q$, for every odd $q > 1$.

Thus we may now assume that $x \in \Omega(f) \setminus \Omega(f^2)$ and that the f-orbit of x contains a fixed point z of f^2. By Lemma 31 we have $x \in W(z,f)$. Since the hypothesis of the proposition implies that f^2 is not turbulent, z cannot be a fixed point of f, by Proposition III.16. Thus z has period 2 and $x \in W(u,g)$, where $g = f^2$ and $u = z$ or $f(z)$. Put $v = f(u)$, so that u and v are fixed points of g, and let k be the least positive integer such that $f^k(x) = u$.

Suppose first that k is odd. Then also $v = f^{2k}(x) \in W(u,g)$, and hence $u \in W(v,g)$. Without loss of generality, assume $u < v$. Then, by Lemma III.22, $v \in W(u,g,R)$ and $u \in W(v,g,L)$. It follows from Lemma III.23 that g is turbulent, which contradicts the hypothesis of the proposition.

Suppose next that k is even and, without loss of generality, assume $x < u$. Since $x \notin \Omega(g)$, we cannot have $W(u,g,L) = W(u,g,R)$. Therefore, by Lemma III.22, $x \in W(u,g,L)$ and $x \notin W(u,g,R)$. Again since $x \notin \Omega(g)$, if G is a sufficiently small open interval containing x then $f^k(\overline{G})$ is a closed interval with left endpoint u. Since $x \in \Omega(f)$, it follows that $x \in W(u,f,R)$. Since $x \notin W(u,g,R)$, this implies $x \in W(v,g)$. Since the least positive integer l such that $f^l(x) = v$ is odd, this yields a contradiction by what we have already proved. □

COROLLARY 33 *If f is non-chaotic, then $\Omega(f^n) = \Omega(f)$ for every $n \geq 1$.* □

We show finally that any isolated point of the centre is also isolated in the nonwandering set. The proof is based on the following lemma, which is of interest in itself.

LEMMA 34 *Let J be an open subinterval of I which contains no periodic point of f. Then no point of I belongs to the trajectories of three distinct nonwandering points in J.*

Proof Let x_1, x_2, x_3, where $x_1 < x_2 < x_3$, be three points in $\Omega(f) \cap J$. Assume on the contrary that there exist integers $m_i \geq 0$ $(i = 1,2,3)$ such that

$$f^{m1}(x_1) = f^{m2}(x_2) = f^{m3}(x_3) = y, \text{ say.}$$

Without loss of generality we may suppose that the interval J is of increasing type.

We show first that $J_3 = f^{m3}[x_2, x_3]$ is an interval. Assume on the contrary that $J_3 = \{y\}$. There exists a point u with $x_2 < u < x_3$ and an integer $n > m_3$ such that $f^n(u) = x_3$. Then $f^n(x_3) = x_3$, since $n > m_3$ implies that $f^n[x_2, x_3]$ is a single point. But this contradicts $J \cap P = \emptyset$.

We show next that y is an interior point of $J_2 \cup J_3$, where $J_2 = f^{m2}[x_1, x_2]$. There exist points $v_i \to x_2$ and integers $p_i \to \infty$ such that $f^{p_i}(v_i) = x_2$. We may suppose that $v_i \in (x_1, x_2)$, since J is of increasing type, and that $p_i > m_2$ for all i. If $f^{m2}(v_i) \in J_3$ for some i, then $f^{m2}(v_i) = f^{m3}(v)$ for some $v \in [x_2, x_3]$ and hence $x_2 = f^{p_i - m2 + m3}(v)$, contrary to the supposition that J is of increasing type. Consequently $f^{m2}(v_i) \notin J_3$ for all i. Since $f^{m2}(v_i) \to f^{m2}(x_2) = y$, it follows that y is an endpoint of J_3 and that y is an interior point of $J_2 \cup J_3$.

Finally, there exist points $w_i \to x_1$ and integers $q_i \to \infty$ such that $f^{q_i}(w_i) = x_1$. We may suppose that $w_i \in J$ and that $q_i > m_1$ for all i. The preceding argument, with x_2 replaced by x_1, shows that $f^{m1}(w_i) \notin J_3$ for all i. Replacing x_3 by x_2, we obtain similarly $f^{m1}(w_i) \notin J_2$ for all i. Since $f^{m1}(w_i) \to y$, and y is an interior point of $J_2 \cup J_3$, this is a contradiction. □

PROPOSITION 35 *Any isolated point of* P(*f*) *is also an isolated point of* $\Omega(f)$.

Proof Let x be an isolated periodic point and let n be its period. Let J be an open interval such that $J \cap P = \{x\}$ and let K be an open interval containing x such that $K \subset J, f^n(K) \subset J$, and $f^{2n}(K) \subset J$. If $y \in K \cap \Omega$, then $f^{kn}(y) \in J \cap \Omega$ for $k = 0,1,2$. This implies that $f^{2n}(y) = x$, by Lemma 14. It now follows from Lemma 34 that on each side of x there are at most two points of Ω in K. □

NOTES

Some books on topological dynamics are Gottschalk and Hedlund [65], Nemytskii and Stepanov [91], Sell [114] and Sibirskii [115].

Lemmas 2-4 are stated in Šarkovskii [105], and Lemmas 3 and 5 are proved in Šarkovskii [108]. Šarkovskii [106] has strengthened the second statement of Lemma 4 by showing that, for maps of an interval, if a limit set $L = \omega(x)$ is infinite then the non-periodic points of L are dense in L. Propositions 6 and 11, and Corollaries 12 and 13, appeared in Šarkovskii [109].

Agronsky *et al.* [2] have shown that Lemma 5 is 'best possible'. That is, if L is a non-empty closed subset of an interval I which either contains no interval or is the union of finitely many disjoint closed intervals, then $L = \omega(x,f)$ for some continuous map $f: I \to I$ and some $x \in I$.

The concept of recurrent point is due to Poincaré [99], who used the term 'Poisson stable' point, and that of nonwandering point is due to Birkhoff [19]. The results of Proposition 15 were announced by Šarkovskii [103] with a sketch of the proof. The relation $\overline{P} = \overline{R}$ was rediscovered by Coven and Hedlund [55], and Nitecki [94] also obtained the relation $\overline{P} = \Omega(f\,|\Omega)$ for piecewise monotone maps. Proposition 16 was first proved for piecewise monotone maps by Young [136] and then in general by Coven and Nitecki [58].

Proposition 18 is due to Block [20]. Nitecki [94] proved Proposition 19 for piecewise monotone maps, and also Proposition 22. Corollaries 20 and 24 were proved by Xiong [126], to whom Lemma 23 is also essentially due.

Lemma 25 was already proved for endomorphisms of a compact metric space by Erdös and Stone [61]. Propositions 26, 29, and 30 are due to Coven and Nitecki [58]. They also give a detailed proof of the claim at the end of the paragraph following Proposition 30.

Proposition 32 is stated without proof (but with a misprint) by Blokh [40]. Corollary 33 was first proved independently by Nitecki [95] and Zhou [137]. Lemma 34 and Proposition 35 are due to Xiong [131].

V
Topological Dynamics (continued)

In this chapter we study two further types of recurrence, one stronger and the other weaker than the type already considered in Chapter IV. The first derives its significance from the notion of minimal set, whereas the second derives its significance from the notion of asymptotically stable set.

We again denote by X a compact metric space and by $f: X \to X$ a continuous map of this space into itself. Of course we are especially interested in the case where $X = I$ is a compact interval. However, many of our results will be valid without this restriction and these results will again be marked with a dagger (†).

1 MINIMAL SETS AND STRONG RECURRENCE

A set $M \subseteq X$ is said to be a *minimal set* if it is non-empty, closed and invariant and if no proper subset has these three properties. Equivalently, a non-empty set M is minimal if $\overline{\gamma(x,f)} = M$ for every $x \in M$.

For example, a finite set is a minimal set if and only if it is a periodic orbit. It follows at once from the definition that two minimal sets either coincide or are disjoint.

It is easy to see that Lemma I.17 admits the following generalization: if $f: X \to X$ and $g: Y \to Y$ are topologically conjugate, and if $h: X \to Y$ is the corresponding conjugacy, then M is a minimal set for f if and only if $h(M)$ is a minimal set for g.

†LEMMA 1 *A non-empty set M is minimal if and only if* $\omega(x,f) = M$ *for every* $x \in M$.

Proof If M is a minimal set and $x \in M$ then $\omega(x)$ is a non-empty, closed and invariant subset, hence $\omega(x) = M$.

If $\omega(x) = M$ for every $x \in M$ then M is closed and invariant. Moreover if N is a closed invariant subset of M and $y \in N$ then $M = \omega(y) \subseteq N$, hence $N = M$ and M is a minimal set. □

†COROLLARY 2 *A minimal set is strongly invariant.* □

For the following result we give two proofs. The first holds for an arbitrary compact metric space, but the second is restricted to the case of a compact interval. However, the second proof is more elementary and more constructive.

†LEMMA 3 *Any non-empty closed invariant set F contains a minimal set.*

Proof (i) The collection of all non-empty, closed invariant subsets of F is partially ordered by inclusion. Therefore by Hausdorff's Maximality Theorem, which is proved in Rudin [100] for example, it contains a maximal totally ordered subcollection S. The intersection M of all elements of S is evidently closed and invariant. It is also non-empty, since X is compact. It now follows at once from the definition of S that M is a minimal set.

(ii) For any $x \in F$ let $\delta(x)$ denote the distance between the least and greatest points in $\overline{\gamma(x)}$. Let μ be the greatest lower bound of $\delta(x)$ for $x \in F$ and let (x_k) be a sequence of points of F such that $\delta(x_k) \to \mu$. By restriction to a subsequence we may suppose that $x_k \to y$, where $y \in F$. Then $\delta(y) = \mu$, since $\delta(y) \le \lim \delta(x_k)$.

Put $E = \overline{\gamma(y)}$. If $z \in E$ then evidently $\delta(z) = \mu$. Hence if m is the least point in E then $m \in \overline{\gamma(z)}$ for every $z \in E$. Thus $\overline{\gamma(m)} \subseteq \overline{\gamma(z)}$ for every $z \in E$, and in particular for every $z \in \overline{\gamma(m)}$. It follows that $\overline{\gamma(m)}$ is a minimal set. □

In the case where $X = I$ is a compact interval any finite set can be a minimal set, since there exists a continuous map for which the points of the set form a periodic orbit. We show next that, in the same case, by no means every closed infinite set can be a minimal set.

LEMMA 4 *Any infinite minimal set is a Cantor set, i.e. a closed subset of I which has no isolated point and which contains no interval.*

Proof Let M be an infinite minimal set. Then M contains no periodic points. Since every point of M is recurrent it follows that no point of M is isolated. It also follows from Lemma IV.5 that M contains no interval. □

It is not difficult to see that any Cantor set can be a minimal set. In fact, in Chapter VI we will give an example of a continuous self-map f_0 of the interval $I_0 = [0,1]$ for which the classical 'middle-third' Cantor set C_0 is a minimal set. On the other hand, as proved in Hocking and Young [69] for example, for any Cantor set $C \subseteq I$ there exists a homeomorphism $h: C \to C_0$. The complement of C in I is a union of countably many open intervals. Evidently C is a minimal set for the continuous self-map f of I defined by setting $f(x) = h^{-1} \circ f_0 \circ h(x)$ for $x \in C$ and by linearity on any component of the complement of C.

It may be noted that Lemma 4 does not hold for continuous maps of an arbitrary compact metric space, with the general definition of a Cantor set given in Chapter II. A simple counterexample, which will be considered in Chapter IX, is an irrational rotation of a circle.

A point $x \in X$ is said to be *strongly recurrent* if for every open set U containing x there exists a positive integer $N = N(U)$ such that if $f^m(x) \in U$, where $m \geq 0$, then $f^{m+k}(x) \in U$ for some k with $0 < k \leq N$.

Thus a strongly recurrent point is one which is recurrent with 'bounded return times'. A simple sufficient condition for a point x to be strongly recurrent is that it be *regularly recurrent*, i.e. for each open set U containing x there exists a positive integer $N = N(U)$ such that $f^{kN}(x) \in U$ for all $k > 0$.

The close connection between strong recurrence and minimal sets is brought out in the following result, which was first proved in 1912 by G.D. Birkhoff and is also discussed in Birkhoff [19]. He used a different, but equivalent, definition of strong recurrence and called it simply 'recurrence'. The term 'almost periodic' is sometimes also used with the same meaning (although this does not agree with its use in analysis).

†PROPOSITION 5 *If M is a minimal set, then any point $x \in M$ is strongly recurrent. Conversely, if x is strongly recurrent then its orbit closure $\overline{\gamma(x)}$ is a minimal set.*

Proof Let M be a minimal set and assume that $x \in M$ is not strongly recurrent. Then there exists an open neighbourhood U of x such that, for some increasing sequence (n_k) of positive integers, $f^{n_k}(x) \in U$ but $f^n(x) \notin U$ for $n = n_k+1,\ldots,n_k+k$. Moreover we may suppose that $f^{n_k}(x) \to y$, where $y \in \overline{U} \cap M$. Since $f^m(y) \in U$ for some positive integer m there exists an open neighbourhood V of y such that $f^m(V) \subseteq U$. But we can choose $k \geq m$ so that $f^{n_k}(x) \in V$. Then $f^{n_k+m}(x) \in U$, which is a contradiction.

Assume next that x is strongly recurrent but $M = \overline{\gamma(x)}$ is not a minimal set. Then there is a non-empty proper closed invariant subset L of M and $x \notin L$. Let U be an open neighbourhood of x such that $\bar{U} \cap L = \varnothing$. There exists a positive integer N such that $f^k(x) \in U$ for some positive integer k in every set of N consecutive positive integers. Since L is invariant there exists, by uniform continuity, an open set V containing L such that

$$U \cap f^j(V) = \varnothing \quad \text{for } 0 \leq j \leq N.$$

But $f^p(x) \in V$ for some positive integer p and $f^{p+q}(x) \in U$ for some positive integer $q \leq N$. Thus we have a contradiction. □

†COROLLARY 6 *A point $x \in X$ is strongly recurrent if and only if, for every point $y \in \omega(x,f)$, also $x \in \omega(y,f)$.* □

We denote by SR = SR(f) the set of all strongly recurrent points and by RR = RR(f) the set of all regularly recurrent points. It follows from Proposition 5 that SR is the union of all minimal sets and thus f(SR) = SR. It will now be shown that Lemma IV.25 continues to hold if recurrence is replaced by strong recurrence or regular recurrence.

†LEMMA 7 *For any positive integer m,*

$$\mathrm{SR}(f^m) = \mathrm{SR}(f) \ \textit{and} \ \mathrm{RR}(f^m) = \mathrm{RR}(f).$$

Proof It is obvious that $\mathrm{SR}(f^m) \subseteq \mathrm{SR}(f)$. Let $x \in \mathrm{SR}(f)$. Then $M = \overline{\gamma(x,f)}$ is an f-minimal set. Since M is a closed f^m-invariant set, it contains an f^m-minimal set L. If $y \in L$ then

$$\gamma(y,f) \subseteq L \cup f(L) \cup \ldots \cup f^{m-1}(L) \subseteq M.$$

Since $M = \overline{\gamma(y,f)}$, it follows that

$$M = L \cup f(L) \cup \ldots \cup f^{m-1}(L).$$

But it is easily seen that $f^k(L)$ is also an f^m-minimal set for each k. Since $x \in f^k(L)$ for some k with $0 \leq k < m$ it follows that x is f^m-strongly recurrent. The corresponding assertion for regular recurrence follows at once from the definition. □

For the remainder of this section we suppose that $X = I$ is a compact interval. It will be shown, following Block and Coven [27], that *if $x \in \Lambda(f) \setminus \overline{\mathrm{P}}(f)$ then $\omega(x,f)$ is an infinite minimal set.* This result is the key to the construction of a continuous map f for which $\Lambda(f) \neq \overline{\mathrm{P}}(f)$. Block and Coven give an example of such a map, which in addition is non-chaotic and such that $f|\Omega(f)$ is not one-to-one. They also give an example of a continuous map f with a point $x \in \Omega(f)\setminus\overline{P}(f)$ such that $\omega(x,f)$ is a Cantor set, but not a minimal set under the action of f.

Put

$$\tilde{\Lambda}(f) = \bigcap_{n\geq 0} f^n(\Omega(f)) .$$

Since $f(\Lambda) = \Lambda \subseteq \Omega$ we have $\Lambda \subseteq f^n(\Omega)$ for every $n \geq 0$ and hence $\Lambda \subseteq \tilde{\Lambda}$. We will show subsequently that $\Lambda = \tilde{\Lambda}$. Thus the notation $\tilde{\Lambda}$ is not really needed. For this reason the statements of the following two propositions refer to Λ, rather than to $\tilde{\Lambda}$. However, in the proofs we will make the *a priori* weaker hypothesis that $x \in \tilde{\Lambda}$ and then show that also $x \in \Lambda$. In this way the relation $\Lambda = \tilde{\Lambda}$ will be proved simultaneously.

PROPOSITION 8 *If x has a finite orbit and $x \in \Lambda(f)$, then $x \in \overline{P}(f)$.*

Proof We assume first that $x \in \tilde{\Lambda}(f)$ and prove that $x \in \Lambda(f)$. Evidently we may suppose that $x \notin P(f)$. Let n be the least positive integer such that $f^n(x) \in P(f)$ and put $z = f^n(x)$. Then $x \in W(z,f)$, by Lemma IV.31.

For each $k > 0$ there exists a point $y = y_k \in \Omega(f)$ such that $f^k(y) = x$. Then $y \in W(z,f)$ by Lemma IV.31 again. Since $W(z,f)$ has only finitely many boundary points, we can choose k so that also $y \in \operatorname{int} W(z,f)$.

Assume first that, for any neighbourhood $V \subseteq W(z,f)$ of y, $f^{k+n}(V)$ contains a neighbourhood of z. Since $y \in \Omega(f)$ we have $f^m(v) = y$ for some $v \in V$ and some $m > 0$. Since $v \in W(z,f)$ and $f^{k+n}(V)$ contains a neighbourhood of z, it follows that $f^{k+n+p}(w) = v$ for some $w \in V$ and some $p > 0$. Thus any neighbourhood of y contains three points of some trajectory, and hence $y \in \Lambda(f)$ by Proposition IV.11. Since $x = f^k(y)$, it follows that also $x \in \Lambda(f)$.

Assume next that there is a neighbourhood V of y such that $f^{k+n}(V)$ contains only a one-sided neighbourhood of z. Let this neighbourhood be on the side S, where $S = R$ or L. Then the same holds if V is replaced by any smaller neighbourhood of y. Since $y \in \Omega(f)$ it follows from Proposition IV.16 that $y \in W(z,f,S)$. If the previous assumption holds for no y then the present assumption holds for infinitely many y. Hence we may suppose also $y \in \operatorname{int} W(z,f,S)$. By the same argument as before, with $W(z,f)$ replaced by $W(z,f,S)$, we obtain $y \in \Lambda(f)$ and hence also $x \in \Lambda(f)$.

It remains to prove the proposition as stated. Since $\Lambda(f) = \Lambda(f^m)$ and $P(f) = P(f^m)$ we may assume that $z = f(x)$ is a fixed point of f. We may also assume that $x \neq z$, since otherwise there is nothing to prove. Without loss of generality suppose $x < z$.

Let $x \in \omega(y,f)$, where $y \in I$. If there exists an increasing sequence (n_k) of positive integers such that the sequence $f^{n_k}(y)$ is decreasing and converges to x then, since

$$f^{n_{k+1}-n_k}(f^{n_k}(y)) = f^{n_{k+1}}(y) < f^{n_k}(y)$$

and

$$f^{n_{k+1}-n_k}(x) = z > x,$$

there is a periodic point in $(x, f^{n_k}(y))$ and hence $x \in \overline{P}(f)$.

It only remains to consider the case where there exists an increasing sequence (n_k) of positive integers such that the sequence $f^{n_k}(y)$ is increasing and converges to x. There is a side S of z such that a subsequence of $f^{n_{k+1}}(y)$ converges to z from this side. Since $f^{n_{k+1}-n_k-1}(z_k) = x$ for some z_k between $f^{n_{k+1}}(y)$ and z we have $x \in W(z,f,S)$ and hence $[x,z] \subseteq W(z,f,S)$. If x is in the interior of $W(z,f,S)$ let N be a left neighbourhood of x contained in $W(z,f,S)$. Then $f(N)$ contains an S-neighbourhood of z and thus $f^j(N) \supseteq N$ for some positive integer j. Hence $x \in \overline{P}(f)$.

It now remains to consider the possibility that x is an endpoint of $W(z,f,S)$. Then, since $W(z,f,S)$ is invariant under f, it can contain no element of the orbit of y. Hence $S = R$, $W(z,f,R) = [x,z]$ and the interval $[x,z]$ contains no point in the orbit of y. Choose $w > z$ so that $f(t) > x$ for all $t \in [z,w]$. Since $w \notin W(z,f,R)$ there is a $\delta > 0$ such that $w \notin f^n[z,z+\delta]$ for all $n \geq 0$. Since $f^k(y) \in (z,z+\delta)$ for some positive integer k it follows that $f^n(y) < w$ for all $n \geq k$. Thus if $f^n(y) > x$ then actually $z < f^n(y) < w$ and hence $f^{n+1}(y) > x$. Therefore $f^n(y) > x$ for all $n \geq k$, which is a contradiction. □

PROPOSITION 9 *If x has an infinite orbit and $x \in \Lambda(f) \setminus \overline{P}(f)$, then $\omega(x,f)$ is an infinite minimal set. Moreover, if J is the component of $I \setminus \overline{P}(f)$ which contains x, then one endpoint of J is a regularly recurrent point of $\omega(x,f)$ and the other is either eventually periodic or an endpoint of I.*

Proof At first we suppose merely that x has an infinite orbit and $x \in \Omega(f) \setminus \overline{P}(f)$. Let J be the component of $I \setminus \overline{P}(f)$ which contains x, and without loss of generality suppose J is of increasing type. Also, let c and d be the left and right endpoints of J. Then, by Lemma IV.23, the connected sets $f^k[x,d]$ $(k \geq 0)$ are pairwise disjoint. Thus the orbit of d is also infinite and $\omega(d,f) = \omega(x,f)$.

Since J is of increasing type and $x \in \Omega(f)$ the orbit $\gamma(x,f)$ contains a point to the right of x. If z is the greatest lower bound of all such points then $z \geq d$. Moreover if $z = d$ then $d \in \omega(x,f)$.

Let W denote the set of all $w \in I$ such that, for every open interval V containing x, $w \in f^k(V)$ for some $k > 0$. Then $f(W) \subseteq W$ and $(c,x) \cap W = \emptyset$, since J is of increasing type. By Proposition IV.16, if $V \subseteq J$ is any open interval containing x then $f^k(v) = x$ for some $v \in V$ and some $k > 0$. Actually $v < x$, since J is of increasing type. Since $f^k(x) \geq z$ it follows that $[x,z] \subseteq f^k(V)$. Therefore, since V is arbitrary, $[x,z] \subseteq W$. Thus if K is the component of W which contains x then K is an interval with left endpoint x and $z \in K$.

From now on we suppose that actually $x \in \tilde{\Lambda}(f)$. Thus for each $n > 0$ there exists a point $w_n \in \Omega(f)$ such that $f^n(w_n) = x$. We will show that $w_n \in W$. Indeed if $w_n \notin W$ then for some open interval V containing x we have $w_n \notin \bigcup_{k \geq 0} f^k(V)$. Since there is a neighbourhood U of w_n with $f^n(U) \subseteq V$, this contradicts $w_n \in \Omega(f)$.

We show next that $z = d$. Assume on the contrary that $z > d$. Since $d \in \overline{P}(f)$, it follows that the interval K contains a periodic point p, of period n say. Thus if we put $g = f^n$ then K contains the fixed point p of g. Since $g(W) \subseteq W$, this implies $g(K) \subseteq K$. It follows that $g^j(x) \in \operatorname{int} K$ either for $j = 1$ or $j = 2$.

For some $w \in \Omega(f)$ we have $g(w) = x$. Moreover $w \in \Omega(g)$, by Proposition IV.26. Let $M \subseteq K$ be a neighbourhood of $g^j(x)$ and let $L \subseteq J$ be a neighbourhood of x such that $g^j(L) \subseteq M$. Then $g^k(L) \subseteq K$ for all $k \geq j$. If H is a neighbourhood of w such that $g(H) \subseteq L$ then $g^k(H) \subseteq K$ for all $k \geq j+1$. Since $w \in \Omega(g)$, it follows from Proposition IV.16 that $w \in K$.

Assume $w \in \operatorname{int} K$. Since $g(K) \subseteq K$, there is a neighbourhood U of w such that $g(U) \subseteq [x,d]$. Since $w \notin \bigcup_{k \geq 0} f^k[x,d]$, this contradicts $w \in \Omega(g)$. Hence w is the right endpoint of K.

Between d and p there must also be a periodic point p' of f, of period $n' > n$. Since $f^{n'}(w') = x$ for some $w' \in \Omega(f)$ it follows by the same reasoning that $w' = w$. Since $x \notin P(f)$, this is a contradiction.

We conclude that $z = d$, and hence also $d \in \omega(x,f)$. It will now be shown that for any open interval N with left endpoint d there is a positive integer k such that $f^k(N) \cap (c,x) \neq \emptyset$.

Assume on the contrary that the set $S = \bigcup_{k \geq 0} f^k(N)$ is disjoint from (c,x). Since N contains a periodic point, S has only finitely many components. For some $m > 0$ we have $f^m(x) \in N$. If w is sufficiently close to x then also $f^m(w) \in N$, and for some such w and some $n > m$ we have $f^n(w) = x$. Hence $x \in f^{n-m}(N)$. Thus $x \in S$ and x is the left endpoint of the component of S which contains it. If $w_k \in \Omega(f)$ and $f^k(w_k) = x$ then $w_k \in S$, by a similar argument. Since the points w_k are distinct it follows that $w_k \in \operatorname{int} S$ for some k. Since $f(S) \subseteq S$, there is a neighbourhood $U \subseteq S$ of this w_k such that $f^k(U) \subseteq [x,d]$. Since $w_k \notin \bigcup_{n \geq 0} f^n[x,d]$, this contradicts $w_k \in \Omega(f)$.

Now let $K \subseteq J$ be any open interval with right endpoint x. Then $x \in f^n(K)$ for some $n > 0$. Hence $f^n(x) > d$ and $f^n(K)$ contains an open interval N with left endpoint d. Therefore, by what we have just proved, some iterate $f^q(K)$ contains a point $y \in (c,x)$. Since J is of increasing type we must actually have $y \in K$. Thus K contains at least two points of some orbit. Since K was arbitrary, it follows from Proposition IV.6 that $x \in \Lambda(f)$.

Thus $x \in \omega(y,f)$ for some $y \in (c,x)$. Again let N be any open interval with left endpoint d. Since $d \in \omega(x,f)$ and the interval $[x,d]$ contains no point in the orbit of y we have $f^m(y) \in N$ for some $m > 0$. We also have

$$y < f^n(y) < f^{n+i}(y) < x$$

for some $n > m$ and some $i > 0$. Then $f^i[f^n(y),x]$ contains an open interval M with left endpoint d. Moreover we may choose M so that $f^m(y) \notin \bar{M}$. We claim that *if* $f^k(M) \cap M \neq \emptyset$ *for some* $k > 0$, *then* $f^k(d) \in N$.

Assume on the contrary that $f^k(d)$ lies to the right of N. Then the interval $f^k(M)$ contains $f^m(y)$. Hence $f^m(y) \in f^{k+i}[f^n(y),x]$ and $f^n(y) \in f^\ell[f^n(y),x]$ for $\ell = k+i+n-m$. Since J is of increasing type, this is a contradiction.

Assume on the other hand that $f^k(d) < d$. Then actually $f^k(d) \leq c$, since the intervals $f^j[x,d]$ are disjoint and J is of increasing type. Hence $y \in f^k(M)$ and $f^m(y) \in f^{k+m}(M)$, which leads to a contradiction in the same way as before.

This establishes our claim. Since M contains a periodic point it follows that d is strongly recurrent, and even regularly recurrent. Hence $\overline{\gamma(d,f)}$ is an infinite minimal set, by Proposition 5. But $\overline{\gamma(d,f)} = \omega(d,f) = \omega(x,f)$.

Finally, since $x \in \Lambda(f)$, it follows from Lemma IV.23 that the left endpoint c of J either has a finite orbit or is the left endpoint of I. □

In the proofs of Propositions 8 and 9 we have shown in addition that $\tilde{\Lambda} \subseteq \Lambda$. Thus we have now also proved

PROPOSITION 10 *For any continuous map* $f: I \to I$,

$$\Lambda(f) = \bigcap_{n \geq 0} f^n(\Omega(f)). \quad \square$$

Proposition 10 was announced by Blokh [40]. A proof has not previously been published. If f is piecewise monotone then $\Lambda(f) = \bar{P}(f)$, by Proposition IV.22, but possibly $\Omega(f) \neq \Lambda(f)$. A description of $\Omega(f)$, for piecewise monotone f, is given by the spectral decomposition theorem of Nitecki [96]. A description of $\Omega(f)$, for arbitrary continuous maps f, is given by a more general spectral decomposition theorem of Blokh (unpublished).

2 ASYMPTOTICALLY STABLE SETS

A non-empty closed set A is said to be *stable* if for each open set $U \supseteq A$ there exists an open set $V \supseteq A$ such that $\gamma(x) \subseteq U$ for every $x \in V$. It is said to be *asymptotically stable* if in addition there exists an open set $U_0 \supseteq A$ such that $\omega(x) \subseteq A$ for every $x \in U_0$.

When the set A consists of a single point these are exactly the original definitions due to Lyapunov of a stable, resp. asymptotically stable, equilibrium point. The second requirement in the definition of an asymptotically stable set is sometimes taken as the definition of an *attractor*. However, we will not use this term since several other definitions of attractor have appeared in the literature, one of them being equivalent to our definition of asymptotically stable set. It will be noted that the whole space X is an asymptotically stable set.

It follows at once from the definition that any stable set, and hence also any asymptotically stable set, is invariant. The union of two asymptotically stable sets is again asymptotically stable and so also is their intersection, provided it is not empty. Furthermore, if A is an asymptotically stable set and if C is a closed invariant set then $A \cap C$, when it is not empty, is an asymptotically stable set for the restriction of f to C.

†PROPOSITION 11 *If A is an asymptotically stable set then there exists an open set $U_0 \supseteq A$ with the property that, for each open set $U \supseteq A$, there is a positive integer* $N = N(U)$ *such that $f^n(\bar{U}_0) \subseteq U$ for all $n \geq N$.*

Proof There exists an open set $U_0 \supseteq A$ such that $\omega(x) \subseteq A$ for every $x \in \bar{U}_0$. Let U be any open set such that $A \subseteq U \subseteq U_0$ and let V be an open set such that $A \subseteq V$ and $\gamma(x) \subseteq U$ for every $x \in V$. Evidently $V \subseteq U$. For each $x \in \bar{U}_0$ there exists a positive integer $n = n(x)$ such that $f^n(x) \in V$, since $\omega(x) \subseteq A$. Hence, for some open set W_x with $x \in W_x$ we have $f^n(W_x) \subseteq V$. The sets W_x form an open cover of $\bar{U}_0$. Let $W_{x_1}, \ldots, W_{x_r}$ be a finite subcover and let $N = \max\{n(x_1), \ldots, n(x_r)\}$. Then, for each $x \in \bar{U}_0, f^n(x) \in V$ for some $n \leq N$ and hence $f^n(x) \in U$ for all $n \geq N$. □

Proposition 11 admits the following converse.

†PROPOSITION 12 *Let A be a closed invariant set and let* $U_0 \supseteq A$ *be an open set such that, for each open set* $U \supseteq A$, *there is a positive integer* $N = N(U)$ *with the property that* $f^n(\bar{U}_0) \subseteq U$ *for all* $n \geq N$. *Then A is asymptotically stable.*

Proof Evidently $\omega(x) \subseteq A$ for every $x \in \bar{U}_0$. Hence we need only show that A is stable. If this is not the case then there exists an open set $U \supseteq A$, points $x_k \in \bar{U}_0$ with $x_k \to x \in A$ and integers $n_k \geq 0$ such that $f^{n_k}(x_k) \notin U$. Then $n_k < N$ for all k, where $N = N(U)$. Thus for some $m < N$ we have $n_k = m$ for infinitely many k. But $f^m(x) \in A$, since A is invariant, and hence $f^m(x_k) \in U$ for all large k. Thus we have a contradiction. □

Propositions 11 and 12 provide an alternative definition of asymptotically stable sets. It will now be shown that asymptotically stable sets can also be characterized in another way.

†PROPOSITION 13 *If A is a closed invariant set and if there exists an open set* $V \supseteq A$ *such that*

(i) $f(\bar{V}) \subseteq V$,
(ii) $\bigcap_{n\geq 0} f^n(\bar{V}) \subseteq A$,

then A is asymptotically stable.

Proof By (i), the compact sets $f^n(\bar{V})$ form a decreasing sequence. Hence (ii) implies that the hypothesis of Proposition 12 is satisfied. □

†COROLLARY 14 *If U is a non-empty open set such that* $f(\bar{U}) \subseteq U$, *then the set* $A = \bigcap_{n\geq 0} f^n(\bar{U})$ *is asymptotically stable.* □

The converse of Proposition 13 is also valid, and in fact the following stronger statement holds.

†PROPOSITION 15 *If A is an asymptotically stable set, then there exists an open set* $W \supseteq A$ *such that*

(i) $f(\bar{W}) \subseteq W$,
(ii) $\bigcap_{n\geq 0} f^n(\bar{W}) \subseteq A$.

Moreover, for any open set $U \supseteq A$ *we can choose W so that* $\bar{W} \subseteq U$.

Proof Let U_0 be an open set containing A with the property guaranteed by Proposition 11. Then (ii) holds with W replaced by U_0.

Let U be any open set such that $A \subseteq U$ and $\gamma(x) \subseteq U_0$ for every $x \in \bar{U}$. Also let V be the set of all points x such that $\gamma(x) \subseteq U$. Then $A \subseteq V \subseteq U$, $f(V) \subseteq V$, V contains an open neighbourhood of A, and $\bigcap_{n \geq 0} f^n(\bar{V}) \subseteq A$. We will prove that V itself is open.

Choose a positive integer $N = N(U)$ as in Proposition 11. If $x \in V$ then $f^j(x) \in U$ for $0 \leq j \leq N$. Hence, for all y in a neighbourhood of x, $f^j(y) \in U$ for $0 \leq j \leq N$. Since $y \in U \subseteq U_0$ and $f^n(\bar{U}_0) \subseteq U$ for $n \geq N$, this implies that $\gamma(y) \subseteq U$. Thus V is open.

For some positive integer m we must have $f^m(\bar{V}) \subseteq V$. Let V_{m-1} be an open set containing $f^{m-1}(\bar{V})$ such that $f(\bar{V}_{m-1}) \subseteq V$ and define successively open sets $V_{m-2}, \ldots, V_1$, so that $f^j(\bar{V}) \subseteq V_j$ and $f(\bar{V}_j) \subseteq V_{j+1}$ for $1 \leq j < m-1$. If

$$W = V_1 \cup V_2 \cup \ldots \cup V_{m-1} \cup V$$

then W is open and contains A. Moreover (i) holds and for each $x \in \bar{W}$ we have $f^k(x) \in \bar{V}$ for some k such that $0 \leq k < m$. Hence if $x \in \bigcap_{n \geq 0} f^n(\bar{W})$ and p is a positive integer then $x \in f^k(\bar{V})$ for some $k \geq p$. Since the nested intersection $\bigcap_{n \geq 0} f^n(\bar{V})$ is contained in A, it follows that (ii) holds.

If in the preceding argument we replace U by an open set whose closure is contained in some given U we see that we can choose W so that also $\bar{W} \subseteq U$. □

There exists yet one more characterization of asymptotically stable sets. A *complete negative trajectory* of a point $x \in X$ is an infinite sequence (x_n) with $x_0 = x$ and $f(x_n) = x_{n-1}$ for all $n \geq 1$. Since the mapping f is an endomorphism rather than a homeomorphism, such a sequence need not exist and need not be uniquely determined if it does exist.

†PROPOSITION 16 *A closed invariant set A is asymptotically stable if and only if there exists an open set $U_0 \supseteq A$ such that $U_0 \backslash A$ contains no complete negative trajectory.*

Proof Suppose first that A is asymptotically stable. Then there exists an open set $U \supseteq A$ with $f(\bar{U}) \subseteq U$ and $\bigcap f^n(\bar{U}) \subseteq A$. If U contains a complete negative trajectory (x_n) then $x_0 = f^n(x_n) \in f^n(\bar{U})$ for every n and hence $x_0 \in A$.

Suppose conversely that there exists an open set $U_0 \supseteq A$ such that $U_0 \backslash A$ contains no complete negative trajectory. We show first that A is stable. If this is not the case there exists an open set $U \supseteq A$ with $\bar{U} \subseteq U_0$, a point $x \in A$ and a sequence $x_k \to x$ such that $f^m(x_k) \notin U$ for some $m \geq 0$ which depends on x_k. Evidently we may assume $x_k \in U$ for all k. Let n_k be the least positive integer such that $f^{n_k}(x_k) \notin U$. The sequence (x_k) contains a subsequence, which will still be denoted by (x_k), such that $f^{n_k-1}(x_k) \to y_0$, where $y_0 \in \bar{U}$ and $f(y_0) \notin U$. Therefore $y_0 \notin A$ and $n_k \to \infty$, since $n_k = m$ for infinitely many k would imply $y_0 = f^{m-1}(x) \in A$. The sequence (x_k) contains a subsequence, again denoted by (x_k), such that $f^{n_k-2}(x_k) \to y_1$, where $f(y_1) = y_0$ and $y_1 \in \bar{U}\setminus A$. Proceeding in this way, we construct an infinite sequence (y_n) such that $f(y_n) = y_{n-1}$ and $y_n \in \bar{U}\setminus A$ $(n \geq 1)$. But this is contrary to hypothesis.

Thus for any open set U with $A \subseteq \bar{U} \subseteq U_0$ there exists an open set V with $A \subseteq V \subseteq U$ and $\gamma(x) \subseteq U$ for every $x \in V$. To complete the proof we show that $\omega(x) \subseteq A$ for every $x \in V$. If this is not the case there exists an $x \in V\setminus A$ and an open set W with $A \subseteq W \subseteq V$ such that $f^n(x) \notin W$ for infinitely many n. Hence there exists an increasing sequence (n_k) of positive integers such that $f^{n_k}(x) \to y_0$, where $y_0 \in \bar{U}\setminus W$. Since $y_0 \in \omega(x)$ and $\omega(x)$ is strongly invariant, there exists a point $y_1 \in \omega(x)$ such that $f(y_1) = y_0$. Evidently $y_1 \in \bar{U}\setminus A$. Proceeding in this way, we again obtain a contradiction. □

To illustrate the utility of this last characterization we now prove

†PROPOSITION 17 *If A is an asymptotically stable set and if A_1 is a closed invariant subset of A which is asymptotically stable for the restriction of f to A, then A_1 is itself an asymptotically stable set.*

Proof Let U be any open set containing A and let $U_1 \supseteq A_1$ be a subset of A which is open in A with $f(\bar{U}_1) \subseteq U_1$ and $\bigcap f^n(\bar{U}_1) \subseteq A_1$. Then there exists an open set V with $A_1 \subseteq V \subseteq U$ such that $f(\bar{V}) \cap A \subseteq U_1$. If A_1 is not asymptotically stable then, by Proposition 16, there exists a complete negative trajectory (x_n) in $V\setminus A_1$. If $x_n \in A$ then $x_{n-1} = f(x_n)$ lies in $f(\bar{V}) \cap A \subseteq U_1$ and $x_0 = f^n(x_n) \in f^{n-1}(\bar{U}_1)$. Since $x_0 \notin A_1$ it follows that $x_n \notin A$ for all large n. Thus there exists a complete negative trajectory in $V\setminus A$. But, since A is asymptotically stable, this contradicts Proposition 16. □

The statements and proofs of the preceding results have evolved gradually, so that it is difficult to give precise references. However, we mention Bhatia *et al.* [18] for Propositions 11 and 12, Block and Franke [35] for Propositions 13 and 15, and Izman [72] for Proposition 16.

The next result, taken from Šarkovskii *et al.* [113], gives a simple characterization of an asymptotically stable fixed point when $X = I$ is a compact interval.

PROPOSITION 18 *A fixed point z is asymptotically stable if and only if it is contained in an open interval G such that, for every $x \in G$,*

$$f^2(x) >, =, \text{ or } < x \text{ according as } x <, =, \text{ or } > z.$$

Proof Suppose first that z is asymptotically stable. Then there exists an open interval G with $f(\overline{G}) \subseteq G$ and $\bigcap f^n(\overline{G}) = \{z\}$, since if an open set G has these properties then so also does the connected component of G which contains z. Evidently $f^2(x) \neq x$ for all $x \in G$ with $x \neq z$. Moreover we cannot have either $f^2(x) > x$ for all $x \in G$ with $x > z$, or $f^2(x) < x$ for all $x \in G$ with $x < z$. It follows that the condition is necessary.

Suppose, conversely, that there exists an open interval G containing z such that, for all $x \in G$, $f^2(x) > x$ or $< x$ according as $x < z$ or $> z$. Then $f(x) \neq x$ for all $x \in G$ with $x \neq z$ and hence $f(x) > x$ or $< x$ according as $x < z$ or $> z$. If H is a sufficiently small open interval containing z, then $\overline{H} \subseteq G$ and $f(\overline{H}) \subseteq G$.

Assume first that $f(x) \leq z$ for all $x \in H$ with $x < z$ and $f(x) \geq z$ for all $x \in H$ with $x > z$. Then for every closed interval $K \subseteq H$ with $z \in K$ we have $f(K) \subseteq K$. Moreover $f^n(x) \to z$ for every $x \in H$, since the monotonic sequence $\{f^n(x)\}$ can only converge to a fixed point. Thus the fixed point z is asymptotically stable.

It remains to consider the case where, for some $x_0 \in H$, either $x_0 < z < f(x_0)$ or $f(x_0) < z < x_0$. Assume for definiteness that $x_0 < z < f(x_0)$. If we put $b = f(x_0)$ then we can choose $a \in [x_0, z)$ so that $f(a) = b$ and $f(x) < b$ for $a < x \leq z$. Hence $f(x) > a$ for $z \leq x \leq b$, since $f(x) = a$ would imply $f^2(x) = b \geq x$. It follows that for $J = (a,b)$ we have $f(J) \subseteq J$, since $f(x) > x$ for $a < x < z$ and $f(x) < x$ for $z < x < b$. Moreover $K := \bigcap f^n(\overline{J})$ is either $\{z\}$ or a closed interval containing z with $f(K) = K$. To show that z is asymptotically stable it now only remains to exclude the second alternative.

Assume $K = [c,d]$, where $c < d$. Then $z > c$, since $f(x) < x$ for $z < x \leq d$, and $z < d$, since $f(x) > x$ for $c \leq x < z$. For some $x_2 \in (z,d]$ we have $f(x_2) = c$, and for some $x_1 \in [c,z)$ we have $f(x_1) = x_2$. Then $f^2(x_1) = c \leq x_1$, which is a contradiction. □

COROLLARY 19 *A fixed point z is asymptotically stable if and only if there exists an open interval G containing z such that $\omega(x) = \{z\}$ for every $x \in G$.*

Proof The necessity of the condition is obvious, since it is part of the definition of asymptotic stability. Suppose then that the condition is satisfied. Evidently this implies that z is the only periodic point in G. In particular, $f^2(x) \neq x$ if $x \in G$ and $x \neq z$. Put $G_- = \{x \in G: x < z\}$ and

$G_+ = \{x \in G: x > z\}$. If $f^2(x) < x$ for all $x \in G_-$ then, for every $k > 0, f^{2k}(x) < x$ for all $x \in G_-$, because the inequality holds if x is sufficiently close to z. Since this is contrary to hypothesis, we conclude that $f^2(x) > x$ for all $x \in G_-$. Similarly we can show that $f^2(x) < x$ for all $x \in G_+$. Hence, by Proposition 18, z is asymptotically stable. □

Thus, in the case of an *interval*, the second requirement in the definition of an asymptotically stable *fixed point* actually implies the first. On the other hand, this does not hold in the case of a circle. For example, let g: [0,1] → [0,1] be defined by $g(x) = x^2$, and let $f: S^1 \to S^1$ be the homeomorphism of the circle to itself obtained by identifying 0 and 1. Then f has a unique fixed point y, and $\omega(x) = \{y\}$ for every $x \in S^1$. However, y is not stable, since points close to y on one side of y move away from y, under iteration by f, before approaching y on the other side.

We now derive the standard stability criteria for fixed points and periodic orbits.

PROPOSITION 20 (i) *Let z be a fixed point and let f be differentiable at z. Then z is asymptotically stable if $|f'(z)| < 1$ and not stable if $|f'(z)| > 1$.*

(ii) *Let* $P = \{z_1 < \ldots < z_n\}$ *be a periodic orbit and let f be differentiable at every point of* P. *Then* P *is asymptotically stable if* $\rho < 1$ *and not stable if* $\rho > 1$, *where*

$$\rho = |f'(z_1) \ldots f'(z_n)| .$$

Proof Consider first the case of a fixed point z. If $|f'(z)| < \theta < 1$ then, by the definition of a derivative, there exists an open interval $G = (z-\delta, z+\delta)$, where $\delta > 0$, such that

$$|f(x) - z| \le \theta |x - z| \quad \text{for all } x \in G.$$

It follows from the definition that z is asymptotically stable. If $|f'(z)| > 1$ then in the same way there exists an open interval G containing z such that

$$|f(x) - z| \ge \mu |x - z| \quad \text{for all } x \in G,$$

where $\mu > 1$. It follows that z is not stable, by the definition of stability.

Consider next the case of a periodic orbit P. If $g = f^n$ then, by the chain rule for differentiation,

$$g'(z) = f'(z) f'[f(z)] \ldots f'[f^{n-1}(z)].$$

Hence the value of $\rho = |g'(z_i)|$ is independent of i. If $\rho < 1$ then, by the previous part of the proof, each point of P is an asymptotically stable fixed point of g. It follows easily that P is an asymptotically stable set of f. Similarly, if $\rho > 1$ then no point of P is a stable fixed point of g and P itself is not a stable set of f. □

A fixed point z for which $|f'(z)| = 1$ may be either asymptotically stable, stable but not asymptotically stable, or not stable. The three possibilities are exhibited by the fixed point $z = 0$ of the maps $f(x) = x - x^3$, $f(x) = x$, and $f(x) = x + x^3$ respectively. An analogous remark applies to periodic orbits.

If an asymptotically stable set A has the form $A = A_1 \cup A_2$, where A_1 and A_2 are disjoint closed invariant proper subsets, then A_1 and A_2 are themselves asymptotically stable sets, by Proposition 17. We will say that an asymptotically stable set – or, more generally, any closed invariant set – is *indecomposable* if it cannot be represented as the union of two disjoint closed invariant proper subsets.

The following result gives an overview of asymptotically stable sets for maps of an interval.

PROPOSITION 21 *An arbitrary asymptotically stable set is the union of finitely many disjoint indecomposable asymptotically stable sets.*

If A is an indecomposable asymptotically stable set, then so also is $B = \bigcap f^n(A)$. *Moreover, there exists a positive integer d such that B is the union of d disjoint connected closed sets (i.e., points or closed intervals), which are permuted cyclically by f and are asymptotically stable for* f^d. *For some open set* $V \supseteq A$ *with* $f(\overline{V}) \subseteq V$ *we have* $B = \bigcap f^n(\overline{V})$.

Proof Let A be an arbitrary asymptotically stable set and let V be an open set containing A such that $f(\overline{V}) \subseteq V$ and $\bigcap f^n(\overline{V}) \subseteq A$. The compact set A is contained in finitely many components of V, say $V_1, \ldots, V_m$. Moreover for each i with $1 \leq i \leq m$ there exists a unique $j = j(i)$ with $1 \leq j \leq m$ such that $f(\overline{V}_i) \subseteq V_j$.

It follows that for at least one i there is an integer $n > 0$ such that $f^n(\overline{V}_i) \subseteq V_i$. If s_1 is the least such integer n then, with a suitable choice of notation,

$$f(\overline{V}_1) \subseteq V_2 , f(\overline{V}_2) \subseteq V_3 , \ldots , f(\overline{V}_{s_1}) \subseteq V_1 .$$

If $s_1 < m$ and if, for some i with $s_1 < i \leq m$ and some $n > 0$, we have $f^n(\overline{V}_i) \subseteq V_i$ then this process can be repeated. In this way we arrange the components V_i into groups, so that

$$f(\overline{V}_{s_1+1}) \subseteq V_{s_1+2}, \ldots, f(\overline{V}_{s_2}) \subseteq V_{s_1+1},$$

$$\cdot \quad \cdot \quad \cdot \quad \cdot \quad \cdot$$

$$f(\overline{V}_{s_{h-1}+1}) \subseteq V_{s_{h-1}+2}, \ldots, f(\overline{V}_{s_h}) \subseteq V_{s_{h-1}+1},$$

whereas for each i with $s_h < i \leq m$ there is a positive integer $n \leq m$ such that $f^n(\overline{V}_i) \subseteq V_{s_k}$ for a unique $k = k(i)$ with $1 \leq k \leq h$.

Let A_k denote the set of all points $x \in A$ such that $f^n(x) \in V_{s_k}$ for some positive integer n. Then $A_1, \ldots, A_h$ are disjoint closed invariant sets and

$$A = A_1 \cup \ldots \cup A_h .$$

Hence $A_1, \ldots, A_h$ are themselves asymptotically stable. If we put $d_1 = s_1$ and $d_k = s_k - s_{k-1}$ $(1 < k \leq h)$, then $f^{d_k}(\overline{V}_{s_k}) \subseteq V_{s_k}$. Hence the connected closed set

$$C_k = \bigcap_{n>0} f^{nd_k}(\overline{V}_{s_k})$$

is asymptotically stable for f^{d_k}. Moreover $f^{d_k}(C_k) = C_k$ and $C_k, f(C_k), \ldots, f^{d_k-1}(C_k)$ are disjoint connected closed subsets of A_k. Thus

$$B_k = C_k \cup f(C_k) \cup \ldots \cup f^{d_k-1}(C_k)$$

is strongly invariant for f. Moreover B_k is asymptotically stable, since if we put

$$W_k = V_{s_{k-1}+1} \cup \ldots \cup V_{s_k}$$

then $f(\overline{W}_k) \subseteq W_k$ and $B_k = \bigcap f^n(\overline{W}_k)$. In fact B_k is an indecomposable asymptotically stable set, because C_k is connected. Since $B_k \subseteq A_k$ and $f^m(A_k) \subseteq W_k$, we have $B_k = \bigcap f^n(A_k)$. It follows that A_k is also an indecomposable asymptotically stable set. Thus if A is itself indecomposable we must have $h = 1$. □

It will be observed that the proof of Proposition 21 remains valid in any locally connected compact metric space, since connected components of open sets are then open. In the case of an interval Proposition 21 implies that if a minimal set is asymptotically stable it must be a periodic orbit.

3 CHAIN RECURRENCE

For any $x \in X$, let $Q(x) = Q(x,f)$ denote the intersection of all asymptotically stable sets which contain $\omega(x)$. Evidently $Q(x)$ is a closed invariant set and $\omega(x) \subseteq Q(x)$. It will now be shown that the sets $Q(x)$ enjoy a number of other general properties.

†LEMMA 22 *If* $y \in Q(x)$, *then* $Q(y) \subseteq Q(x)$.

Proof Any asymptotically stable set containing $\omega(x)$ contains y, and hence also $\omega(y)$ and $Q(y)$. □

†LEMMA 23 *The map* $x \to Q(x)$ *is upper semi-continuous; i.e., for any open set* U *containing* $Q(x)$ *there exists an open set* V *containing* x *such that* $Q(y) \subseteq U$ *for every* $y \in V$.

Proof Assume on the contrary that for some open set $U \supseteq Q(x)$ there exists a sequence $x_n \to x$ and points $y_n \in Q(x_n)$ such that $y_n \notin U$. By restriction to a subsequence we may suppose that $y_n \to y$, where $y \notin U$.

Let A be an asymptotically stable set containing $\omega(x)$ and let $V \supseteq A$ be an open set with $f(\overline{V}) \subseteq V$ and $\bigcap f^n(\overline{V}) \subseteq A$. Then $f^m(x) \in V$ for some $m \geq 0$. Hence $f^m(x_n) \in V$ for all $n \geq N$, say. Thus $Q(x_n) \subseteq A$ and $y_n \in A$ for all $n \geq N$. Hence $y \in A$. Since A was arbitrary it follows that $y \in Q(x)$, which is a contradiction. □

†LEMMA 24 *If* $y \in \overline{\gamma(x)}$, *then* $Q(y) = Q(x)$.

Proof It follows at once from the definition that $Q(f^n(x)) = Q(x)$, for any positive integer n. Suppose $y \in \omega(x)$, so that $f^{n_k}(x) \to y$ for some increasing sequence (n_k) of positive integers. Then $y \in Q(x)$ and hence $Q(y) \subseteq Q(x)$, by Lemma 22. On the other hand, by Lemma 23, for any open set $U \supseteq Q(y)$ we have $Q(x) = Q(f^{n_k}(x)) \subseteq U$ for all large k and hence $Q(x) \subseteq Q(y)$. □

†LEMMA 25 *For every open set* $U \supseteq Q(x)$ *there exists an asymptotically stable set* A *containing* $\omega(x)$ *with* $A \subseteq U$.

Proof Let **A** denote the collection of all asymptotically stable sets which contain $\omega(x)$ and assume, on the contrary, that for some open set $U \supseteq Q(x)$ every set $A \in$ **A** has a point in $X \setminus U$. Then, for each $A \in$ **A**, the intersection S_A of A with $X \setminus U$ is non-empty and closed. Moreover, since the intersection of finitely many elements of **A** is again an element of **A**, the closed sets S_A

have the finite intersection property. Therefore, since X is compact, there exists a point $y \in S_A$ for all $A \in \mathbf{A}$. Then $y \in Q(x) \cap X \setminus U$, which is a contradiction. □

†COROLLARY 26 *There exists a decreasing sequence of asymptotically stable sets* $A_1 \supseteq A_2 \supseteq \ldots$ *such that* $Q(x) = \bigcap A_n$. □

We consider next the dependence of the set $Q(x,f)$ on the map f. If X and Y are metric spaces, we denote by $C(X,Y)$ the set of all continuous maps from X into Y. If X is compact, as we assume, then $C(X,Y)$ becomes a metric space if we define the distance between two elements f,g by

$$d(f,g) = \sup_{x \in X} d_Y(f(x), g(x)),$$

where d_Y denotes the metric on Y. The usual topological notions can now be introduced, as for the special case considered in Chapter II. The statement of the following lemma is to be understood in this sense (with $Y = X$).

†LEMMA 27 *The map* $f \to Q(x,f)$ *is also upper semi-continuous; i.e., for any open set* U *containing* $Q(x,f)$ *there exists an open set* $\mathbf{V}$ *containing* f *such that* $Q(x,g) \subseteq U$ *for every* $g \in \mathbf{V}$.

Proof By Lemma 25 and Proposition 15, there exists an open set W such that $Q(x,f) \subseteq W \subseteq U$ and $f(\overline{W}) \subseteq W$. Then $f^n(x) \in W$ for some $n > 0$. If we take $\mathbf{V}$ to be an open set containing f such that $g(\overline{W}) \subseteq W$ and $g^n(x) \in W$ for every $g \in \mathbf{V}$, then $Q(x,g) \subseteq W$ for every $g \in \mathbf{V}$. □

†LEMMA 28 *For any* $x \in X$, $f(Q(x)) = Q(x)$.

Proof Since $Q(x)$ is invariant we need only show that $Q(x) \subseteq f(Q(x))$. Suppose $y \in Q(x)$. Let A be any asymptotically stable set containing $\omega(x)$ and let $V \supseteq A$ be an open set with $f(\overline{V}) \subseteq V$. Then $f^m(x) \in V$ for some $m > 0$. If $y \notin f(\overline{V})$ then there exists an open set W with $f(\overline{V}) \subseteq W \subseteq \overline{W} \subseteq V$ and $y \notin \overline{W}$. Then $f(\overline{W}) \subseteq W$ and $\bigcap f^n(\overline{W})$ is an asymptotically stable set which contains $\omega(x)$ but not y. But this contradicts $y \in Q(x)$.

Hence $y \in f(\overline{V})$. Since V can be chosen in any neighbourhood of A and the space X is compact, it follows that $y \in f(A)$. Since A can be chosen in any neighbourhood of $Q(x)$, by Lemma 25, it further follows that $y \in f(Q(x))$. □

†LEMMA 29 *If A is an asymptotically stable set containing* $\omega(x)$, *then* $Q(x,f\,|A) = Q(x)$.

Proof If $\tilde{A}$ is any asymptotically stable set containing $\omega(x)$ then $A \cap \tilde{A}$ is an asymptotically stable subset of A containing $\omega(x)$, and so it contains $Q(x,f\,|A)$. Hence $Q(x,f\,|A) \subseteq Q(x)$. On the other hand, any subset of A which contains $\omega(x)$ and is asymptotically stable for $f\,|A$ is also asymptotically stable for f, by Proposition 17, and so it contains $Q(x)$. Hence $Q(x) \subseteq Q(x,f\,|A)$. □

†LEMMA 30 *Let $A_1 \supseteq A_2 \supseteq \ldots$ be a decreasing sequence of closed invariant sets and let $A = \bigcap A_n$. Then, for any $x \in A$,*

$$\bigcap_{n>0} Q(x,f\,|A_n) \subseteq Q(x,f\,|A)\,.$$

Proof If $y \in Q(x,f\,|A_n)$ for every n then obviously $y \in A$. If $y \notin Q(x,f\,|A)$ then there exists an asymptotically stable subset C of A containing $\omega(x)$ but not y. Let U_1 be an open subset of A, containing C but not y, such that $f(\overline{U}_1) \subseteq U_1$.

There exists an open set U in X with $U \cap A = U_1$ and $f(\overline{U}) \cap A \subseteq U_1$. We claim that there exists a positive integer m such that $f(\overline{U \cap A_n}) \subseteq U \cap A_n$ for all $n \geq m$.

If this is not the case then for infinitely many n there exists a point $y_n \in \overline{U \cap A_n}$ such that $f(y_n) \notin U$. Without loss of generality we may suppose that $y_n \to y$ as $n \to \infty$. Then $y \in \overline{U} \cap A$ but $f(y) \notin U$, which is a contradiction.

Since $V_m = U \cap A_m$ is an open subset of A_m with $f(\overline{V}_m) \subseteq V_m$, the set $\bigcap_k f^k(\overline{V}_m)$ is an asymptotically stable subset of A_m. This set contains $\omega(x)$, since $U_1 \subseteq V_m$ and $\omega(x)$ is strongly invariant. However, it does not contain y, since $f(\overline{V}_m) \subseteq U$ and $y \notin U \cap A$. But this contradicts $y \in Q(x,f\,|A_m)$. □

†LEMMA 31 *For any $x \in X$,* $Q(x,f\,|Q(x)) = Q(x)$.

Proof By Corollary 26, there exists a sequence A_n of asymptotically stable sets with $A_1 \supseteq A_2 \supseteq \ldots$ and $Q(x) = \bigcap A_n$. By Lemma 29, for every n we have

$$Q(x,f\,|A_n) = Q(x).$$

Hence, by Lemma 30, $Q(x) \subseteq Q(x,f\,|Q(x))$. The reverse inclusion is trivial. □

†LEMMA 32 *For any* $x \in X$ *and any integer* $m > 1$,

$$Q(x,f) = \bigcup_{j=0}^{m-1} Q(f^j(x), f^m)$$

and

$$f(Q(x,f^m)) = Q(f(x), f^m).$$

Proof Let A be an asymptotically stable set containing $\omega(x,f)$. Then A is an asymptotically stable set also for f^m. Moreover it contains $\omega(f^j(x), f^m)$ and hence also $Q(f^j(x), f^m)$ for $0 \le j < m$. It follows that

$$\bigcup_{j=0}^{m-1} Q(f^j(x), f^m) \subseteq Q(x,f).$$

Let $\tilde{A}_1$ be an asymptotically stable set for f^m containing $\omega(x,f^m)$. Let $V_1 \supseteq \tilde{A}_1$ be an open set such that $f^m(\overline{V}_1) \subseteq V_1$ and $A_1 = \bigcap_k f^{km}(\overline{V}_1) \subseteq \tilde{A}_1$. Then A_1 is also an asymptotically stable set for f^m containing $\omega(x,f^m)$, since $f^{hm}(x) \in V_1$ for some $h > 0$. If we put $V_2 = f^{-1}(V_1)$ then V_2 is an open set and $f^m(\overline{V}_2) \subseteq V_2$. It follows that $A_2 = \bigcap_k f^{km}(\overline{V}_2)$ is an asymptotically stable set for f^m containing $\omega(f^{m-1}(x), f^m)$, since $f^{hm-1}(x) \in V_2$ and $hm - 1 = (h-1)m + m - 1$. Proceeding in this way, we construct open sets $V_3, \ldots, V_m$ such that $V_j = f^{-1}(V_{j-1}), f^m(\overline{V}_j) \subseteq V_j$ and $A_j = \bigcap_k f^{km}(\overline{V}_j)$ is an asymptotically stable set for f^m containing $\omega(f^{m-j+1}(x), f^m)$ $(j = 3, \ldots, m)$. Evidently $f(V_j) \subseteq V_{j-1}$ $(j = 2, \ldots, m)$ and $f(V_1) \subseteq V_m$.

We began with an asymptotically stable set $\tilde{A}_1$ for f^m containing $\omega(x,f^m)$ but we could equally have begun with an asymptotically stable set $\tilde{A}_i$ for f^m containing $\omega(f^{m-i+1}(x), f^m)$ $(j=1, \ldots, m)$. In this way we would have obtained open sets $V_{i,1}, V_{i,2}, \ldots, V_{i,m}$ such that $f(V_{i,j}) \subseteq V_{i,j-1}$ $(j = 2, \ldots, m)$ and $f(V_{i,1}) \subseteq V_{i,m}$, $f^m(\overline{V}_{i,j}) \subseteq V_{i,j}$, and $A_{i,j} = \bigcap_k f^{km}(\overline{V}_{i,j})$ is an asymptotically stable set for f^m containing $\omega(f^{m-j+1}(x), f^m)$ with $A_{i,i} \subseteq \tilde{A}_i$.

If we now set $W_j = V_{1,j} \cap V_{2,j} \cap \ldots \cap V_{m,j}$ then W_j is an open set, $f^m(\overline{W}_j) \subseteq W_j$, and $f(W_j) \subseteq W_{j-1}$ $(j = 2, \ldots, m)$, $f(W_1) \subseteq W_m$. Put

$$W = W_1 \cup W_2 \cup \ldots \cup W_m.$$

Then W is an open set, $f(W) \subseteq W$ and $f^m(\overline{W}) \subseteq W$. It follows from Proposition 13 that $A = \bigcap f^k(\overline{W})$ is an asymptotically stable set for f. Since also

$$A = \bigcap f^{km}(\overline{W})$$

we have $A \subseteq \tilde{A}_1 \cup \ldots \cup \tilde{A}_m$ and

$$\omega(x,f) = \bigcup_{j=1}^{m} \omega(f^{m-j+1}(x), f^m) \subseteq A.$$

Hence $Q(x,f) \subseteq A$ and actually

$$Q(x,f) \subseteq \bigcup_{i=1}^{m} Q(f^{m-i+1}(x), f^m).$$

Finally, since $Q(x,f^m) \subseteq A_{i,1}$ and $f(A_{i,1}) \subseteq A_{i,m}$ we have

$$f(Q(x,f^m)) \subseteq A_{m,m} \subseteq \tilde{A}_m$$

and hence

$$f(Q(x,f^m)) \subseteq Q(f(x), f^m).$$

The reverse inclusion also holds since, for the same reason,

$$Q(f(x), f^m) = f^m(Q(f(x), f^m)) \subseteq f(Q(f^m(x), f^m)) = f(Q(x,f^m)). \quad \square$$

†LEMMA 33 *If z is a periodic point, then* $W(z,f) \subseteq Q(z,f)$.

Proof Let A be any asymptotically stable set containing z and let U be any open set containing A. Then there exists an open set V containing A such that $\gamma(y) \subseteq U$ for all $y \in V$. If $x \in W(z,f)$ then $x = f^{m_k}(y_k)$, where $y_k \to z$ and $m_k \geq 0$. Since $y_k \in V$ for all large k, it follows that $x \in U$. Since U was arbitrary, this implies $W(z,f) \subseteq A$ and, since A was arbitrary, actually $W(z,f) \subseteq Q(z,f)$. $\square$

For maps of an interval the sets $Q(x,f)$ can be described more explicitly.

LEMMA 34 *For any* $x \in I$, *one of the following statements must hold:*

(i) $Q(x) = \omega(x)$ *is an asymptotically stable periodic orbit,*

(ii) $Q(x)$ *is an asymptotically stable set B which contains* $\omega(x)$ *and is the union of finitely many disjoint closed intervals permuted cyclically by f,*

(iii) $Q(x)$ *is the intersection of a strictly decreasing sequence of asymptotically stable sets* B_n, *where each* B_n *is like B in* (ii).

Proof By Corollary 26 there exists a sequence of asymptotically stable sets A_n with $A_1 \supseteq A_2 \supseteq \ldots$ and $Q(x) = \bigcap A_n$. Moreover, by Proposition 21, we can suppose that each asymptotically stable set A_n is indecomposable. For if A is an asymptotically stable set and V an open set containing A such that $f(\overline{V}) \subseteq V$ and $\bigcap f^n(\overline{V}) \subseteq A$, then $f^n(x) \in V$ for some $n>0$

implies $\omega(x) \subseteq A$. Furthermore, if $\omega(x) \subseteq A$ then also $\omega(x) \subseteq \bigcap f^n(A)$, since $\omega(x)$ is strongly invariant. Hence, by Proposition 21, we can suppose that each set A_n is the union of finitely many disjoint connected closed sets, which are permuted cyclically by f. □

COROLLARY 35 *If A is a closed invariant subset of* $Q(x,f)$, *then every connected component of* $Q(x,f)$ *contains a point of A.* □

In Lemma 34 let d_n denote the number of components of B_n, so that d_n divides d_{n+1} for every n. If the sequence (d_n) is bounded, i.e. if d_n is constant for all large n, then $Q(x)$ is either a periodic orbit or the union of finitely many disjoint closed intervals permuted cyclically by f, although it need not be asymptotically stable. If the sequence (d_n) is unbounded, then we may assume that it is strictly increasing. Since the components of B_n form a cyclic group $C_n = \mathbb{Z}/(d_n)$ under the action of f, the components of $Q(x)$ can be given the structure of the group $C^* = \lim_{\infty \leftarrow n} C_n$ which is their inverse limit.

COROLLARY 36 *For any $x \in I$, the set* $Q(x,f)$ *contains a periodic point if and only if it is either a periodic orbit or the union of finitely many disjoint closed intervals permuted cyclically by f.*
□

The preceding results find an immediate application to the theory of chain recurrence. The original definition of chain recurrence, due to Conley [50], [51] made explicit use of the metric of the underlying space, whereas ours will be purely topological. However, we will establish the equivalence of the two definitions. It should also perhaps be stated here that Conley's definition was for flows, rather than for maps. Thus although he proved Proposition 39 below, with his definition of chain recurrence, the proof made essential use of the fact that orbits can be followed backwards as well as forwards in the case of flows.

We define the set $CR = CR(f)$ of all *chain recurrent* points by $x \notin CR(f)$ if there exists an open set U with $f(\overline{U}) \subseteq U$ such that $x \notin \overline{U}, f(x) \in U$.

In this definition the requirement $x \notin \overline{U}$ can equally well be replaced by $x \notin U$. For if $x \in \overline{U} \setminus U$ we can choose an open set W so that $f(\overline{U}) \subseteq W$ and $\overline{W} \subseteq U$. Then $f(\overline{W}) \subseteq W$ and $x \notin \overline{W}, f(x) \in W$.

†PROPOSITION 37 *The chain recurrent set* $CR(f)$ *is a closed invariant set which contains the non-wandering set* $\Omega(f)$.

Proof It follows at once from the definition that $X\setminus \mathrm{CR}$ is open, and hence CR is closed.

Assume $f(x) \notin \mathrm{CR}$. Thus there exists an open set U with $f(\overline{U}) \subseteq U$ such that $f(x) \notin \overline{U}$, $f^2(x) \in U$. Let V be an open set such that $f(x) \in V$, $V \cap \overline{U} = \varnothing$ and $f(\overline{V}) \subseteq U$. If we put $W = U \cup V$ then W is open, $f(x) \in W$ and $f(\overline{W}) \subseteq U \subseteq W$. If $x \in \overline{W}$ then $f(x) \in U$, which is a contradiction. Hence $x \notin \overline{W}$ and thus $x \notin \mathrm{CR}$. This proves that CR is invariant. Finally it is clear from the definition that if $x \notin \mathrm{CR}$ then $x \notin \Omega$. □

We show next that the chain recurrent set of a map is an upper semi-continuous function of the map, and thus cannot be drastically enlarged by a small change in the map. We will see later that this property is not shared in general by the non-wandering set.

†PROPOSITION 38 *Given any $f \in C(X,X)$ and any open set $U \supseteq \mathrm{CR}(f)$, there exists an open neighbourhood* $\mathbf{V}$ *of f in $C(X,X)$ such that*

$$\mathrm{CR}(g) \subseteq U \text{ for every } g \in \mathbf{V}.$$

Proof The set $Y = X\setminus U$ is closed. If $x \in Y$ then $x \notin \mathrm{CR}(f)$ and thus there exists an open set U_x with $f(\overline{U}_x) \subseteq U_x$ such that $x \notin \overline{U}_x, f(x) \in U_x$. Let W_x be an open neighbourhood of x such that $W_x \cap \overline{U}_x = \varnothing$ and $f(\overline{W}_x) \subseteq U_x$. For each x there exists an open neighbourhood $\mathbf{V}_x$ of f in $C(X,X)$ such that if $g \in \mathbf{V}_x$ then $g(\overline{U}_x) \subseteq U_x$ and $g(\overline{W}_x) \subseteq U_x$. Hence no point of W_x is in $\mathrm{CR}(g)$. But Y is covered by all the open sets W_x. Therefore, since it is compact, it is covered by finitely many of them, say by $W_{x_1}, \ldots, W_{x_m}$. If we take $\mathbf{V} = \mathbf{V}_{x_1} \cap \ldots \cap \mathbf{V}_{x_m}$ then for any $g \in \mathbf{V}$ no point of Y is in $\mathrm{CR}(g)$. □

The following result establishes the connection between chain recurrence and asymptotically stable sets.

†PROPOSITION 39 *A point x is chain recurrent if and only if $x \in Q(x)$.*

Proof Suppose $x \notin \mathrm{CR}$. Then there exists an open set V with $f(\overline{V}) \subseteq V$ such that $f(x) \in V$, $x \notin \overline{V}$. By Corollary 14, $A = \bigcap f^n(\overline{V})$ is an asymptotically stable set. Evidently $\omega(x) \subseteq A$ but $x \notin A$. Hence $x \notin Q(x)$.

Conversely, suppose there exists an asymptotically stable set A which contains $\omega(x)$ but not x. Then, by Proposition 15, there exists an open set V containing A with $f(\overline{V}) \subseteq V$, $\bigcap f^n(\overline{V}) \subseteq A$ and $x \notin \overline{V}$. Since $\omega(x) \subseteq A$ there is a least positive integer m such that

$f^m(x) \in V$. Since $f^{m-1}(x) \notin V$ we have $f^{m-1}(x) \notin$ CR. Since CR is invariant, it follows that also $x \notin$ CR. □

†COROLLARY 40 *The chain recurrent set* CR(f) *is strongly invariant.*

Proof If $x \in Q(x)$ then, by Lemma 28, there exists $y \in Q(x)$ such that $f(y) = x$, and $Q(y) = Q(f(y)) = Q(x)$. □

†COROLLARY 41 *A point $x \in X$ is chain recurrent if there exists a periodic point z such that $x \in W(z,f)$ and $z \in \omega(x,f)$.*

Proof Obviously $z \in Q(x)$ and hence $Q(z) \subseteq Q(x)$. On the other hand, by Lemma 33 we have $x \in Q(z)$ and hence $x \in Q(x)$. □

Thus homoclinic points are chain recurrent. This provides a simple way of constructing maps f for which CR(f) $\neq \Omega(f)$. Indeed in Example IV.21 we have $(1/2, 3/4) \subseteq$ CR(f) $\setminus \Omega(f)$. However, a point in CR(f) $\setminus \Omega(f)$ need not be homoclinic. The following example is due to Block and Franke [34].

EXAMPLE 42 Let $I = [0,4]$ and let $f: I \to I$ be the piecewise linear map defined by

$$f(0) = 0, f(1) = 3/2, f(2) = 2, f(3) = 4, f(4) = 0 .$$

Clearly $(0,2) \cap \Omega(f) = \varnothing$. But $[2,3] \subseteq Q(2)$, by Lemma 33, and hence $Q(2) = I$, by invariance. Since $\omega(x) = \{2\}$ if $x \in (0,2)$, it follows that $(0,2) \subseteq$ CR(f).

Here are some further applications of Proposition 39.

†PROPOSITION 43 *If A is an asymptotically stable set then, for any $x \in X$, either $\omega(x) \subseteq A$ or $\omega(x) \cap A = \varnothing$. Moreover, if $x \in$ CR(f) then either $\overline{\gamma(x)} \subseteq A$ or $\overline{\gamma(x)} \cap A = \varnothing$.*

Proof Let V be an open set containing A such that $f(\overline{V}) \subseteq V$ and $\bigcap f^n(\overline{V}) \subseteq A$. If $\omega(x) \cap A \neq \varnothing$ then $f^n(x) \in V$ for some $n > 0$ and hence $\omega(x) \subseteq A$.

Suppose $x \in$ CR(f) and $y \in \overline{\gamma(x)} \cap A$. Then $\omega(x) \cap A \neq \varnothing$ and hence $\omega(x) \subseteq A$, by what we have already proved. Since $x \in$ CR(f), this implies $x \in A$ and hence $\overline{\gamma(x)} \subseteq A$. □

†PROPOSITION 44 *For any positive integer m,*

$$CR(f^m) = CR(f).$$

Proof It follows at once from Lemma 32 that if $x \in Q(x,f^m)$ then $x \in Q(x,f)$. Thus $CR(f^m) \subseteq CR(f)$. Conversely, if $x \in Q(x,f)$ then $x \in Q(f^j(x), f^m)$ for some j such that $0 \le j < m$. It follows that

$$Q(x,f^m) \subseteq Q(f^j(x),f^m)$$
$$Q(f^j(x),f^m) \subseteq Q(f^{2j}(x),f^m)$$
$$\cdots$$
$$Q(f^{j(m-1)}(x),f^m) \subseteq Q(f^{jm}(x),f^m) = Q(x,f^m).$$

Hence we must actually have $Q(f^j(x),f^m) = Q(x,f^m)$ and $x \in Q(x,f^m)$. Thus $CR(f) \subseteq CR(f^m)$. □

We show next that the definition of chain recurrence adopted here is equivalent to that of Conley. At the same time this will make clear the reason for the name.

Let $\varepsilon > 0$ be given and let x, y be any points of X. An ε – *chain*, or *pseudo–orbit*, from x to y is a finite sequence $\{x_0, x_1, \ldots, x_n\}$ of points of X with $x = x_0$, $y = x_n$ and $d(f(x_{k-1}), x_k) < \varepsilon$ for $k = 1, \ldots, n$.

The notion of pseudo-orbit is quite a natural one, since on account of rounding errors a computer will actually calculate a pseudo-orbit, rather than an orbit. The notion appeared explicitly in the shadowing lemma of Bowen [43], which in turn was distilled from work of Anosov [7] on structural stability.

Let $R_\varepsilon(x)$ denote the set of $y \in X$ such that there is an ε – chain from x to y. It follows at once from the definition that $R_\varepsilon(x)$ is open and that there is no ε–chain from a point in $R_\varepsilon(x)$ to a point outside $R_\varepsilon(x)$. Moreover, as we now show, $f(\bar{R}_\varepsilon(x)) \subseteq R_\varepsilon(x)$.

Let $y \in \bar{R}_\varepsilon(x)$. We can choose $\delta > 0$ so that $d(y,z) < \delta$ implies $d(f(y), f(z)) < \varepsilon$. Since $y \in \bar{R}_\varepsilon(x)$ there is a point $z \in R_\varepsilon(x)$ with $d(y,z) < \delta$. If $\{x_0, \ldots, x_n\}$ is an ε–chain from x to z, then $\{x_0, \ldots, x_n, f(y)\}$ is an ε–chain from x to $f(y)$. Thus $f(y) \in R_\varepsilon(x)$.

Our next result establishes the connection between ε–chains and the sets $Q(x)$.

†PROPOSITION 45 *If $y \in Q(x)$, then $y \in R_\varepsilon(x)$ for every $\varepsilon > 0$.*

Conversely, if $y \in R_\varepsilon(x)$ for every $\varepsilon > 0$, then either $y \in Q(x)$ or $y = f^k(x)$ for some positive integer k.

Proof Suppose $y \notin R_\varepsilon(x)$ for some $\varepsilon > 0$. If we put $W = R_\varepsilon(x)$, then W is open and $f(\overline{W}) \subseteq W$. By Corollary 14, the set $A = \bigcap_{n\geq 0} f^n(\overline{W})$ is asymptotically stable. Furthermore $\omega(x) \subseteq A$, since $f(x) \in W$. It follows that $Q(x) \subseteq A$, and hence $y \notin Q(x)$. This proves the first statement.

Conversely, suppose $y \notin Q(x)$ and $y \neq f^k(x)$ for all $k > 0$. By Lemma 25, there is an asymptotically stable set A which contains $\omega(x)$ but not y. By Proposition 15, there exists an open set $W \supseteq A$ such that $f(\overline{W}) \subseteq W$, $\bigcap_{n\geq 0} f^n(\overline{W}) \subseteq A$, and $y \notin \overline{W}$.

Then $f^N(x) \in W$ for some positive integer N, since $\omega(x) \subseteq A$. Since the points $f^k(x)$ $(1 \leq k \leq N)$ are distinct from y, there is an $\varepsilon_1 > 0$ such that if $1 \leq n \leq N$ and if $\{x_0, \ldots, x_n\}$ is an ε_1-chain with $x_0 = x$, then $x_n \neq y$. Similarly, there is an $\varepsilon_2 > 0$ such that if $n = N$ and if $\{x_0, \ldots, x_n\}$ is an ε_2-chain with $x_0 = x$, then $x_n \in W$. Finally, we set $\varepsilon_3 = d(f(\overline{W}), X \setminus W)$ and denote by ε the minimum of ε_1, ε_2 and ε_3. Then $\varepsilon > 0$.

Suppose $\{x_0, x_1, \ldots, x_n\}$ is an ε-chain with $x_0 = x$. We wish to show that $x_n \neq y$. If $n \leq N$ this follows from the choice of ε_1, so suppose that $n > N$. Then by the choice of ε_2, $x_N \in W$ and hence $f(x_N) \in f(\overline{W})$. By the choice of ε_3 it follows that $x_{N+1} \in W$, and hence $f(x_{N+1}) \in f(\overline{W})$. It follows by induction that $x_n \in W$, and thus $x_n \neq y$. This proves the second statement. □

By combining Propositions 39 and 45 we obtain

†COROLLARY 46 *A point $x \in X$ is chain recurrent if and only if, for every $\varepsilon > 0$, there is an ε-chain from x to itself.* □

In view of possible generalizations, it will now be shown that Conley's definition can itself be reformulated in a purely topological way. If X is a compact metric space with metric d, then the product $X \times X$ is also a compact metric space with metric d* defined by

$$d^*[(x_1,y_1), (x_2,y_2)] = d(x_1,x_2) + d(y_1,y_2).$$

We define the *diagonal* D of $X \times X$ to be the subset consisting of all points (x,y) with $y = x$.

†PROPOSITION 47 *A point $x \in X$ is chain recurrent if and only if, for every open set G in the product space $X \times X$ which contains the diagonal D, there exists a finite sequence of points $\{x_0, x_1, \ldots, x_n\}$ with $x_0 = x_n = x$ and $(f(x_{j-1}), x_j) \in G$ for $j = 1, \ldots, n$.*

Proof Suppose first that $x \in \mathrm{CR}(f)$ and let G be an open set in $X \times X$ with $D \subseteq G$. If K is the complement of G in $X \times X$ then, since D and K are compact, there exists an $\varepsilon > 0$ such that $d^*(z,w) \geq \varepsilon$ for all $z \in D$ and all $w \in K$. There exists a finite sequence $\{x_0, x_1, \ldots, x_n\}$ with $x_0 = x_n = x$ and $d(f(x_{j-1}), x_j) < \varepsilon$ for $j = 1, \ldots, n$. Since

$$d^*[(f(x_{j-1}), x_j), (x_j, x_j)] = d[f(x_{j-1}), x_j] < \varepsilon,$$

it follows that $(f(x_{j-1}), x_j) \in G$ for $j = 1, \ldots, n$.

Conversely, for any given $\varepsilon > 0$ let G denote the set of all points $(x,y) \in X \times X$ with $d(x,y) < \varepsilon$. Then G is open in $X \times X$ and $D \subseteq G$. If $\{x_0, x_1, \ldots, x_n\}$ is a finite sequence of points of X with $x_0 = x_n = x$ and $(f(x_{j-1}), x_j) \in G$ for $j = 1,\ldots,n$ then evidently $d(f(x_{j-1}), x_j) < \varepsilon$ for $j = 1, \ldots, n$. □

The following result illustrates the application of Corollary 46.

†PROPOSITION 48 *Every chain recurrent point remains chain recurrent for the restriction of f to* $\mathrm{CR}(f)$, *i.e.*

$$\mathrm{CR}(f) = \mathrm{CR}(f \mid \mathrm{CR}(f)).$$

Proof Let $x \in \mathrm{CR}(f)$. We show first that for any open set $U \supseteq \mathrm{CR}(f)$ and any $\varepsilon > 0$, there is an ε–chain from x to itself which lies entirely within U.

Assume on the contrary that, for some open set $U \supseteq \mathrm{CR}(f)$ and some $\varepsilon > 0$, every ε–chain from x to itself contains a point in the complement of U. Then there exists an ε_n–chain from x to itself which contains a point z_n in the complement of U, where $\varepsilon_n \to 0$ as $n \to \infty$. Moreover, we may suppose that $z_n \to z$. Evidently $z \notin U$. On the other hand, by replacing z_n by z in the ε_n-chain from x to itself we see that for any given $\varepsilon > 0$ there is an ε-chain from x to itself which contains z. It follows that $z \in \mathrm{CR}(f)$. Since $\mathrm{CR}(f) \subseteq U$, this is a contradiction.

It remains to show that for any $\varepsilon > 0$ there is an ε–chain from x to itself within $\mathrm{CR}(f)$. We can choose δ, with $0 < \delta < \varepsilon/3$, so that $d(x,y) < \delta$ implies $d(f(x), f(y)) < \varepsilon/3$. Let U be the set of all points y such that $d(y, \mathrm{CR}(f)) < \delta$ and let $\{x_0, x_1, \ldots, x_n\}$ be an $\varepsilon/3$–chain from x to itself in U. There exist points $y_0, y_1, \ldots, y_n$ in $\mathrm{CR}(f)$ with $y_0 = y_n = x$ and $d(x_j, y_j) < \delta$ for $1 \leq j \leq n$. Then for each $j=1, \ldots, n$

$$\begin{aligned} d(f(y_{j-1}), y_j) &\leq d(f(y_{j-1}), f(x_{j-1})) + d(f(x_{j-1}), x_j) + d(x_j, y_j) \\ &< \varepsilon. \end{aligned}$$ □

It will now be shown that, for maps of an interval, the chain recurrent set cannot be replaced by the non-wandering set in the statement of Proposition 38 unless the two are equal. We use a type of *closing lemma.*

LEMMA 49 *If* $x \in \mathrm{CR}(f)$, *then any open neighbourhood* $\mathbf{V}$ *of* f *in* $C(I,I)$ *contains a map* g *such that* $x \in \mathrm{P}(g)$.

Proof For any given $\varepsilon > 0$ let $\{x_0, x_1, \ldots, x_n\}$ be an ε–chain from x to x of least possible length n. Since the length is minimal we must have $x_i \neq x_j$ and $f(x_{i-1}) \neq f(x_{j-1})$ for $i \neq j$ and $i, j=1, \ldots, n$. By joining piecewise linearly the points $(f(x_{i-1}), x_i)$ of $I \times I$ we can obtain a map h in the ε–neighbourhood of the identity map in $C(I,I)$ such that

$$h(f(x_{i-1})) = x_i \quad (i = 1, \ldots, n).$$

Thus if we put $g = h \circ f$ then x is a periodic point of g. Moreover $g \in \mathbf{V}$ if ε is sufficiently small. □

Lemma 49 immediately implies

PROPOSITION 50 *If for any open set* $U \supseteq \Omega(f)$ *there exists an open neighbourhood* V *of* f *in* $C(I,I)$ *such that*

$$\Omega(g) \subseteq U \text{ for every } g \in \mathbf{V},$$

then $\mathrm{CR}(f) = \Omega(f)$. □

It is shown in Block and Franke [35] that Proposition 50 continues to hold if the compact interval I is replaced by a compact topological manifold X. They also show by an example that it need not hold if I is replaced by a compact metric space X.

Additional information about the chain recurrent set for maps of an interval is provided by the next result, due to Block and Franke [36].

PROPOSITION 51 *Any isolated point of the chain recurrent set is eventually periodic.*

Proof Let x be an isolated point of the chain recurrent set CR. Obviously we may assume that x is not itself periodic. Then, being isolated, x is also not recurrent. Since every connected component of $Q(x)$ contains a point of $\omega(x)$, by Corollary 35, it follows that $Q(x)$ contains an open interval J with endpoint x. Moreover we can choose J so that $J \cap \mathrm{CR} = \emptyset$, again since x is isolated.

For any $y \in J$ we have $\overline{\gamma(x)} \cap \overline{\gamma(y)} = \emptyset$. Indeed, if $z \in \overline{\gamma(x)} \cap \overline{\gamma(y)}$ then, by Lemma 24, $Q(x) = Q(z) = Q(y)$. Since $y \in Q(x)$ but $y \notin Q(y)$, this is a contradiction.

Thus none of the intervals $f^k(J)$ $(k \geq 0)$ contains a point in the orbit of x. If x is not eventually periodic then the points $f^k(x)$ $(k \geq 0)$ are all distinct. Let $z \in \omega(x)$, let G be any open interval containing z and let $f^i(x) < f^j(x) < f^k(x)$ be three points of $\gamma(x)$ in G. Then the interval $f^j(J)$, with endpoint $f^j(x)$, is contained in G. It follows that $z \in \omega(y)$ for every $y \in J$. But this contradicts what we proved in the preceding paragraph. □

The following example, adapted from Young [136], shows that an isolated point of the chain recurrent set need not itself be periodic.

EXAMPLE 52 Let $I = [0,8]$ and let $f: I \to I$ be the piecewise linear map defined by

$$f(0) = 7, f(1) = 5, f(2) = 7, f(3) = 1, f(5) = 5, f(6) = 7, f(8) = 8\ .$$

Evidently $1 \notin \Omega(f)$ and $[1,8] = f(I)$ is an asymptotically stable set. From Lemma 33 we obtain $Q(5) = [1,8]$. Since $Q(1) = Q(5)$, it follows that $1 \in \mathrm{CR}(f)$. Moreover 1 is an isolated point of $\mathrm{CR}(f)$, since if $x \in (1-\varepsilon, 1+\varepsilon)$ and $x \neq 1$ then $Q(x) = \omega(x) = \{8\}$.

If we restrict f to the interval $[1,8]$ then the endpoint 1 is in $\mathrm{CR}(f) \setminus \Omega(f)$. This shows that in Corollary IV.17 we cannot replace $\Omega(f)$ by $\mathrm{CR}(f)$. On the other hand, if we redefine $f(0) = 3$ in Example 52 then 1 is again an isolated point of $\mathrm{CR}(f)$, but $1 \in \Omega(f)$ and hence $1 \notin \overline{P}(f)$.

It is shown in Block and Coven [28] that if $\mathrm{CR}(f) = I$, for a continuous self-map f of a compact interval I, then either f^2 is the identify map or f^2 is turbulent. If $\mathrm{CR}(f) \neq I$ then, by Lemma 34, there exists either a point or a proper closed subinterval which is an asymptotically stable set for some iterate of f.

Besides the original contributions of Conley, already cited, there are other approaches to the concept of chain recurrence in Šarkovskii [110] , Bronštein and Burdaev [45], Vereikina and Šarkovskii [122], [123], and Bronštein and Kopanskii [46].

VI
Chaotic and Non-chaotic Maps

1 CHARACTERIZATIONS OF CHAOTIC MAPS

In this chapter we examine the many differences between non-chaotic and chaotic maps. We suppose throughout, unless otherwise specified, that I is a compact interval and $f: I \to I$ an arbitrary continuous map of this interval into itself.

We begin by studying a class of maps which are as far as possible from being chaotic – those for which every trajectory is convergent. Clearly for this to be possible it is necessary that the map have no periodic point of period 2. It will now be shown that this necessary condition is also sufficient.

PROPOSITION 1 *If f has no periodic point of period 2 then, for every $c \in I$, the trajectory $\{f^k(c)\}$ is bimonotonic and converges to a fixed point of f.*

Proof To show that every trajectory is bimonotonic it is sufficient to show that, for every $x \in I$ and every $n > 1, f^n(x) > x$ or $< x$ according as $f(x) > x$ or $< x$. By Proposition I.7, the only periodic points of f are fixed points. Thus if the claim is false then for some $c \in I$ and some $n > 1$ we have either $f^n(c) < c < f(c)$ or $f(c) < c < f^n(c)$. Without loss of generality we restrict attention to the first case.

If f has no fixed point less than c then $f^n(x) < x$ for all $x < c$. Hence the sequence $f^{kn}(c)$ decreases to a limit z. Then z is a fixed point of f^n and hence also of f, which is a contradiction. Therefore f has a fixed point less than c; let a be the greatest. Then $f(x) > x$ for $a < x \leq c$. Moreover if $x > a$ is sufficiently close to a then $f^n(x) > x$. Hence $f^n(y) = y$ for some $y \in (a,c)$. Then $f(y) = y$ and we again have a contradiction.

Put $x_k = f^k(c)$ $(k \geq 0)$. If $x_{m+1} = x_m$ for some m then $x_k = x_m$ for all $k > m$ and the trajectory of c certainly converges. We may therefore assume that $x_{k+1} \neq x_k$ for all k. The trajectory also converges if it is ultimately monotonic. We may therefore assume that it contains infinitely many U-points x_p and infinitely many D-points x_q. Since the trajectory is bimonotonic, the subsequence $\{x_p\}$ increases to a limit y and $y = \underline{\lim}\, x_k$. Similarly the subsequence $\{x_q\}$ decreases to a limit z and $z = \overline{\lim}\, x_k$. For infinitely many k, x_k must be a U-point and x_{k+1} a D-point, from which it follows that $z = f(y)$. For infinitely many k also, x_k must be a D-point and x_{k+1} a U-point, from which it follows that $y = f(z)$. Since there are no points of period 2, this implies $z = y$. That is, $\overline{\lim}\, x_k = \underline{\lim}\, x_k$. □

We consider next, under greatly relaxed hypotheses, the behaviour of trajectories which do not converge.

PROPOSITION 2 *Suppose f is not turbulent. Let $c \in I$ and put*

$$\alpha = \underline{\lim}\, f^n(c), \quad \delta = \overline{\lim}\, f^n(c).$$

If $\alpha < \delta$, i.e. if the trajectory $\{f^n(c)\}$ does not converge, then exactly one of the following alternatives holds:

(i) *$\omega(c,f)$ contains a fixed point ζ of f such that $\alpha < \zeta < \delta$ and*

$$f(x) > x \text{ for } \alpha \leq x < \zeta, \quad f(x) < x \text{ for } \zeta < x \leq \delta,$$

(ii) *there exist $\beta, \gamma \in \omega(c,f)$ with $\alpha \leq \beta < \gamma \leq \delta$ such that $\omega(c,f) \subseteq [\alpha,\beta] \cup [\gamma,\delta]$ and*

$$f(x) > x \text{ for } \alpha \leq x \leq \beta, \quad f(x) < x \text{ for } \gamma \leq x \leq \delta.$$

Proof Since $\alpha < \delta$ the trajectory $\{f^n(c)\}$ contains infinitely many U-points and infinitely many D-points. Moreover, by Lemma II.8, there exists a fixed point y of f such that all U-points of the trajectory lie to the left of y and all D-points to the right of y. Evidently $\alpha \leq y \leq \delta$.

Put $x_n = f^n(c)$ and let $\varepsilon_n = (\delta - \alpha)/2^n$. We can choose $n_1 > 0$ so that $x_{n_1+1} \leq \alpha + \varepsilon_1 < x_{n_1}$, then $n_2 > n_1$ so that $x_{n_2+1} \leq \alpha + \varepsilon_2 < x_{n_2}$, and so on. Then $x_{n_k} > y$ and $x_{n_k+1} \to \alpha$. By restriction to a subsequence we may assume that $x_{n_k} \to \mu$. Then $f(\mu) = \alpha$ and $y \leq \mu \leq \delta$. Similarly there exists λ such that $f(\lambda) = \delta$ and $\alpha \leq \lambda \leq y$.

If $f(\alpha) = \alpha$ then $\alpha < \lambda \leq y < \mu$ and

$$[\alpha,\mu] \subseteq f[\alpha,\lambda] \cap f[\lambda,\mu],$$

which is a contradiction. Therefore α is a U-point. Similarly δ is a D-point, and hence $\alpha < y < \delta$. Let y' be the least fixed point of f greater than all U-points of the trajectory and y'' the greatest fixed point less than all D-points of the trajectory. Then $\alpha < y' \leq y'' < \delta$. It follows that $f(x) < x$ for $y'' < x \leq \delta$. For if $x_{n_k} > y''$ and $x_{n_k+1} < y'$ then $f(x) < x$ for $x_{n_k} < x \leq \delta$, by

Lemma II.7, and f has no fixed point between y'' and the infimum of such x_{n_k}. Similarly $f(x) > x$ for $\alpha \leq x < y'$. It follows at once that (ii) holds if the limit set $\omega(c,f)$ does not contain a fixed point of f.

On the other hand, if $\omega(c,f)$ contains a fixed point it can only be y' or y''. If, say, $y' \in \omega(c,f)$ then the fixed point y' is a limit of U-points x_{n_k}. Hence we cannot have $y' < y''$, since then the whole sequence $\{x_n\}$ would converge to y'. Thus (i) holds if $\omega(c,f)$ contains a fixed point of f. Moreover the fixed point is a two-sided limit of points of the trajectory. □

We can obtain a sharper result than Proposition 2 if we require that f^2, rather than f, is non-turbulent.

PROPOSITION 3 *Suppose f^2 is not turbulent. Then for each $c \in I$ exactly one of the following alternatives holds:*

(i) *the trajectory $\{f^k(c)\}$ is bimonotonic and converges to a fixed point of f,*

(ii) *the trajectory $\{f^k(c)\}$ is alternating from some point on and the limit set $\omega(c,f)$ is an f–orbit of period 2,*

(iii) *the trajectory $\{f^k(c)\}$ is alternating from some point on and if we set*

$$\alpha = \min \omega(c,f^2), \quad \beta = \max \omega(c,f^2),$$
$$\gamma = \min \omega(f(c),f^2), \quad \delta = \max \omega(f(c),f^2),$$

then $[\alpha,\beta]$ and $[\gamma,\delta]$ are disjoint non-degenerate intervals which contain no fixed point of f.

Proof Put $x_k = f^k(c)$. Suppose first that the sequence $\{x_k\}$ is bimonotonic. If the sequence $\{x_k\}$ is monotonic from some point on, then it converges and its limit is a fixed point. Otherwise the sequence $\{x_k\}$ contains infinitely many U-points and infinitely many D-points. The U-points increase to a limit λ, the D-points decrease to a limit μ, and $\lambda \leq \mu$. Moreover $\mu = f(\lambda)$, since x_k is a U-point and x_{k+1} a D-point for infinitely many k, and similarly $\lambda = f(\mu)$. If $\lambda = \mu$ then (i) holds and if $\lambda < \mu$ then (ii) holds.

Thus we may suppose that the sequence $\{x_k\}$ is not bimonotonic. Then for some $m \geq 0$ and some $n > 1$ we have either $x_{m+n} \leq x_m < x_{m+1}$ or $x_{m+1} < x_m \leq x_{m+n}$. Hence, by Theorem II.12, the trajectory $\{x_k\}$ is alternating from x_m on. By replacing c by $f^m(c)$ we may suppose that $m = 0$. Moreover, without loss of generality we restrict attention to the case

$x_n \leq x_0 < x_1$. Then there exists a fixed point z of f such that $x_k < z$ or $x_k > z$ according as k is even or odd.

Assume that the limit set $\omega(c,f)$ contains a fixed point y of f. Since $f^{nk}(c) \to y$ implies also $f^{nk+1}(c) \to y$ we must have $y = z$ and

$$\overline{\lim}\, x_{2k} = z = \underline{\lim}\, x_{2k+1}\,.$$

Since $g = f^2$ is not turbulent and the fixed point z is the greatest point of $\omega(c,g)$, it follows from Proposition 2 that also $\underline{\lim}\, x_{2k} = z$. Thus $x_{2k} \to z$ as $k \to \infty$. Hence the g-trajectory of c has U-points in any left neighbourhood of z. Since all U-points of the g-trajectory of c lie to the left of all D-points, it follows that all points x_{2k} are U-points of g. But this contradicts $x_n \leq x_0$.

We conclude that $\omega(c,f)$ contains no fixed point of f. Since also f is not turbulent, it follows that alternative (ii) of Proposition 2 must hold. Moreover all points of $\omega(c,f^2)$ must lie in the interval $[\alpha,\beta]$ and all points of $\omega(f(c),f^2)$ must lie in the interval $[\gamma,\delta]$. Hence $\alpha = \beta$ if and only if $\gamma = \delta$. Thus either (ii) or (iii) holds. □

COROLLARY 4 *If $\omega(c,f)$ is an interval for some $c \in I$, then f^2 is turbulent.* □

COROLLARY 5 *If $\omega(c,f)$ contains an interval for some $c \in I$, then f is chaotic.* □

We can now derive a useful characterization of chaotic maps.

PROPOSITION 6 *A map f is chaotic if and only if, for some $c \in I$, $\omega(c,f)$ properly contains a periodic orbit.*

Proof If $\omega(c,f)$ properly contains an orbit of period n then it is infinite,by Lemma IV.4. Hence $\omega(f^j(c),f^n)$ is also infinite for each j with $0 \leq j < n$. Since at least one of these limit sets contains a fixed point of f^n, it follows from Proposition 3 that f^{2n} is turbulent. Thus f is chaotic. The converse has already been proved (Corollary II.18). □

As an application we prove

PROPOSITION 7 *If f is non-chaotic then, for any $x \in I$, the limit set $\omega(x,f)$ contains a unique minimal set M. Moreover, $M = \omega(y,f)$ for every $y \in \omega(x,f)$.*

Proof It is sufficient to show that if $y_1 \in \omega(x,f)$ and $y_2 \in \omega(x,f)$ then $\omega(y_1,f) = \omega(y_2,f)$. If $\omega(x,f)$ is finite then it is a periodic orbit, by Lemma IV.4, and so the result certainly holds in this case. Suppose now that $\omega(x,f)$ is infinite. Then every point in $\omega(x,f)$ has an infinite orbit, by Proposition 6. Let k be any positive integer and set $g = f^{2^{k-1}}$. Then $y_1 \in \omega(f^i(x), g^2)$ and $y_2 \in \omega(f^j(x), g^2)$ for some i, j with $0 \le i, j < 2^k$. Let $z_k = f^{\ell}(y_2)$, where $\ell = 2^{k+1} - j + i$. Then also $z_k \in \omega(f^i(x), g^2)$. Since g^2 is not turbulent it follows from Proposition 3 that f has no periodic point of period dividing 2^{k-1} between y_1 and z_k.

Let z be a limit point of the sequence $\{z_k\}$. Then $z \in \omega(y_2, f)$ and there is no periodic point of f between y_1 and z. It follows from Lemma IV.20 that $\omega(z,f) = \omega(y_1,f)$. Hence $\omega(y_1,f) \subseteq \omega(y_2,f)$. Since y_1 and y_2 can be interchanged in this argument we must actually have $\omega(y_1,f) = \omega(y_2,f)$. □

COROLLARY 8 *If f is non-chaotic, then* $R(f) = SR(f)$. □

The converses of the last two results are also valid, and actually hold in a stronger form.

PROPOSITION 9 *Suppose f is chaotic. Then*

(i) *for some $x \in I$ the limit set $\omega(x,f)$ contains infinitely many minimal sets,*
(ii) $\overline{R}(f) \setminus R(f)$ *is uncountable,*
(iii) $R(f) \setminus SR(f)$ *is uncountable,*
(iv) $RR(f) \setminus P(f)$ *is uncountable.*

Proof The first statement follows at once from Corollary II.18. To prove the remaining statements we may suppose that f is strictly turbulent, since $P(f) = P(f^n)$, $R(f) = R(f^n)$, $SR(f) = SR(f^n)$ and $RR(f) = RR(f^n)$ for each positive integer n. We now use the set-up of Proposition II.15.

Evidently $x \in P(f)$ implies $h(x) \in P(\sigma)$. In fact $h(P(f)) = P(\sigma)$. For let $y \in P(\sigma)$ and let n be its period. If y is the image under h of a unique point $x \in X$, then x is also periodic and has period n. Suppose, on the other hand, that there exist distinct points $x_1, x_2 \in X$ such that $h(x_1) = h(x_2) = y$. If $f^n(x_1) \neq x_1$ and $f^n(x_2) \neq x_2$, then $f^n(x_1) = x_2, f^n(x_2) = x_1$. Hence at least one of x_1, x_2 is periodic and has period n or $2n$.

Again, since h is continuous, $x \in R(f)$ implies $h(x) \in R(\sigma)$ and $x \in SR(f)$ implies $h(x) \in SR(\sigma)$. If $y \in R(\sigma)$ and there is a unique point $x \in X$ such that $h(x) = y$ then certainly $x \in R(f)$. Suppose on the other hand that there are two points $x_1, x_2 \in X$ with

$h(x_1) = y = h(x_2)$ and $x_1 \notin \omega(x_1)$. Then $x_2 \in \omega(x_1)$ and $x_1 \notin \omega(x_2)$, hence $x_2 \in \omega(x_2)$. This proves that $h(\mathrm{R}(f)) = \mathrm{R}(\sigma)$.

Suppose now that $y \in \mathrm{SR}(\sigma)$. Then $L = \omega(y,\sigma)$ is a minimal set containing y. Since $h^{-1}(L)$ is a closed f-invariant set it contains an f-minimal set M. Since $h(M)$ is closed and invariant we must have $h(M) = L$. Thus for any $y \in \mathrm{SR}(\sigma)$ there exists an $x \in \mathrm{SR}(f)$ such that $h(x) = y$. That is, $h(\mathrm{SR}(f)) = \mathrm{SR}(\sigma)$. Thus to prove the statements (ii) and (iii) we need only prove them with f replaced by σ.

It is obvious that $\overline{\mathrm{P}}(\sigma) = \Sigma$, since any initial block of 0's and 1's can be extended to a periodic sequence. Hence also $\overline{\mathrm{R}}(\sigma) = \Sigma$. On the other hand, let $y = (a_1, a_2, \ldots)$, where $a_1 = a_2 = 0$, $a_3 = 1$ and $a_k = 0$ for some $k > 3$ implies $a_{k+1} = 1$. Then evidently y is not recurrent. Moreover the set of all such y is uncountably infinite, since the blocks of consecutive 1's may be of arbitrary lengths. This proves (ii). Again, in the proof of Proposition II.17 we constructed uncountably many elements $y \in \Sigma$ such that $\omega(y,\sigma) = \Sigma$. Evidently any such $y \in \mathrm{R}(\sigma) \setminus \mathrm{SR}(\sigma)$. This proves (iii).

It remains to prove (iv). Let p be a fixed prime number and let $z = (c_1, c_2, \ldots)$ be any sequence in Σ which contains infinitely many 0's and infinitely many 1's. Set $y = (a_1, a_2, \ldots)$, where $a_n = c_{k+1}$ if p^k is the highest power of p dividing n. Then $a_n = a_{n+jp^{k+1}}$ for all $j > 0$. Hence $a_m = a_{m+jp^{k+1}}$ for $1 \le m \le p^k$ and all $j > 0$. Thus y is regularly recurrent. On the other hand y is not periodic. For if y had period $M = p^i m$, where $i \ge 0$ and p does not divide m, then for any $k > i$ we would have $a_{p^k} = a_{M+p^k}$ and hence $c_{k+1} = c_{i+1}$, which is a contradiction.

Suppose there is a unique point $x \in X$ such that $h(x) = y$, and let U be any open interval containing x. Then there exists an open set V in Σ, containing y, such that if $y' = h(x') \in V$ for some $x' \in X$, then $x' \in U$. Also, there exists a positive integer N such that $\sigma^{kN}(y) \in V$ for all $k > 0$. Since $\sigma^{kN}(y) = h[f^{kN}(x)]$, it follows that $f^{kN}(x) \in U$ for all $k > 0$. Therefore $x \in \mathrm{RR}(f) \setminus \mathrm{P}(f)$.

There exist uncountably many sequences z, and hence also uncountably many sequences y. But the proof of Proposition II.15 shows that, for all but countably many $w \in \Sigma$, there is a unique point $x \in X$ with $h(x) = w$. It follows that there are uncountably many sequences y which are images under h of unique points $x \in X$. This proves (iv). □

A continuous map $f: I \to I$ will be said to be *strongly non-chaotic* if every point $x \in I$ is asymptotically periodic or, equivalently, if $\omega(x,f)$ is finite for every $x \in I$. It follows at once from Proposition 6 that a strongly non-chaotic map is certainly non-chaotic. It will now be shown that strongly non-chaotic maps can be characterized in several different ways.

PROPOSITION 10 *The following statements are equivalent:*

(i) *f is strongly non-chaotic,*
(ii) R(*f*) = P(*f*),
(iii) SR(*f*) = P(*f*),
(iv) Λ(*f*) = P(*f*),
(v) P(*f*) *is a closed set.*

Proof It is clear that (iv) ⇒ (v), since Λ is a closed set, and that (v) ⇒ (ii), since $P \subseteq R \subseteq \bar{R} = \bar{P}$. It is obvious also that (ii) ⇒ (iii), and it follows at once from the definition of a strongly non-chaotic map that (i) ⇒ (ii) and (iv) ⇒ (i). Hence to complete the proof we need only show that (iii) ⇒ (iv).

We note first that (iii) implies *f* is non-chaotic, by Proposition 9(iv). It now follows from Propositions 7 and 6 that (iii) ⇒ (iv). □

On account of Proposition 10 we can require that *f* be strongly non-chaotic, rather than that P(*f*) be compact, in the statements of Theorem II.26 and Proposition II.28. Also, Example I.14 now provides an example of a strongly non-chaotic map with periodic orbits of period 2^d for every $d \geq 0$.

It may have been observed that, although our definitions of such notions as minimal set and chain recurrence were valid in any compact metric space, our definition of a chaotic map was blatantly one-dimensional. At this point it may be worthwhile to place the definition in a more general setting.

Let X be a compact metric space and let $f: X \to X$ be a continuous map of this space into itself. The map *f* will now be defined to be *chaotic* if there exist disjoint closed subsets X_0, X_1 and a positive integer m such that, if $\tilde{X} = X_0 \cup X_1$ and $g = f^m$ then

(i) $g(\tilde{X}) \subseteq \tilde{X}$,
(ii) for every sequence $\alpha = (a_0, a_1, a_2, \ldots)$ of 0's and 1's there exists a point $x = x_\alpha \in \tilde{X}$ such that $g^k(x) \in X_{a_k}$ for all $k \geq 0$.

We can give this definition a more sophisticated form. Let Σ denote the set of all infinite sequences $\alpha = (a_0, a_1, \ldots)$, where $a_k = 0$ or 1. As we saw in Chapter II, by suitably defining the distance between any two elements the set Σ becomes a compact metric space, and the shift map $\sigma: \Sigma \to \Sigma$, defined by $\sigma((a_0, a_1, \ldots)) = (a_1, a_2, \ldots)$, is a continuous 2-1 map of Σ onto itself.

It is easily seen that the preceding definition is equivalent to the following: a continuous map f of a compact metric space X into itself is chaotic if there exists a closed subset $\tilde{X}$, a positive integer m such that $f^m(\tilde{X}) \subseteq \tilde{X}$ and a continuous map h of $\tilde{X}$ onto Σ such that

$$h \circ f^m(x) = \sigma \circ h(x) \text{ for every } x \in \tilde{X}.$$

In other words, the shift σ is a *factor* of $f^m|\tilde{X}$.

A similar argument to that used to prove Proposition II.15 shows that a sufficient condition for f to be chaotic is that there exist disjoint non-empty closed subsets Y_0, Y_1 of X and a positive integer m such that

$$Y_0 \cup Y_1 \subseteq f^m(Y_0) \cap f^m(Y_1).$$

It is clear from Proposition II.15 itself that a continuous map $f: I \to I$ of a compact interval into itself which is chaotic in the sense of Chapter II is also chaotic according to the present definition. However, the converse is also true. For if f is chaotic according to the present definition, choose $\alpha \in \Sigma$ so that $\omega(\alpha,\sigma) = \Sigma$ and $x \in \tilde{X}$ so that $h(x) = \alpha$. If we set $L = \omega(x, f^m)$, then $h(L) = \Sigma$. Thus there exists $x' \in L$ so that $h(x') = \alpha$. But if f were non-chaotic according to the original definition, then $L' = \omega(x', f^m)$ would be a minimal set, by Proposition 7, and hence $L' \subseteq \mathrm{SR}(f)$. Since $h(L') = \Sigma$ and $h(\mathrm{SR}(f)) \subseteq \mathrm{SR}(\sigma)$, this yields a contradiction. In this way chaos for maps of an interval appears as a special case of a much more general concept.

2 REGULAR RECURRENCE FOR NON-CHAOTIC MAPS

The notion of strongly non-chaotic map is in a sense too strong, since it implies that every recurrent point is periodic. We are now going to study in more detail the recurrent points of arbitrary non-chaotic maps. We shall see that the regularly recurrent points have an especially significant role.

PROPOSITION 11 *Suppose f is non-chaotic and let c be a two-sided limit of periodic points which is not itself periodic.*

Then for any open interval J containing c, there exists an open interval K containing c and a positive integer $m = 2^s$ *such that* $f^{km}(K) \subseteq J$ *for all* $k \geq 0$. *In particular, c is regularly recurrent.*

Proof Let a, b be periodic points of J with $a < c < b$. Then a and b are both fixed points of $g = f^m$ for some $m = 2^s$. Since c is not itself periodic, we can suppose a and b chosen so that there are no fixed points of g in (a,b). For definiteness, assume $g(x) > x$ for $a < x < b$. Thus $(a,b) \subseteq W(a,g)$. If $g^k(x) = a$ for some $x \in (a,b)$ and some $k > 0$ then a is a homoclinic point, which contradicts the hypothesis that f is not chaotic. On the other hand, for each $k > 0$ there exists an $x \in (a,b)$ so close to a that $g^k(x) > x$. It follows that $g^k(x) > a$ for all $x \in (a,b)$ and all $k > 0$.

Let d, e be the nearest fixed points of g^2 on either side of c, so that $a \leq d < c < e \leq b$. Suppose first that $g^2(c) < c$ and thus $g^2(x) < x$ for $d < x < e$. Then it follows as above that $g^{2k}(x) < e$ for all $x \in (d,e)$ and all $k > 0$. Consequently we can take $K = (d,e)$.

Suppose next that $g^2(c) > c$ and thus $g^2(x) > x$ for $d < x < e$. If $y \in (d,e)$ is periodic then the g-orbit of y is alternating, by Theorem II.12. Moreover $g(x) > x$ for all x between the least and greatest points of the left half of this orbit, by Lemma II.7. It follows that $g^2(y) < b < g(y)$. Since the g^2-orbit of y is also alternating we obtain in the same way $g^4(y) < e < g^2(y)$. But c lies in Λ and so it has an infinite orbit, by Proposition 6. Since y can be arbitrarily close to c it follows that $g^4(c) < e < g^2(c) < b$. Let d', e' be the nearest fixed points of g^2 on either side of $c' = g^2(c)$, so that $e \leq d' < c' < e' \leq b$. As in the previous part of the proof, we then obtain $g^{2k}(x) < e'$ for all $x \in (d',e')$ and all $k > 0$. Consequently we can take any $K \subseteq (d,e)$ such that $g^2(K) \subseteq (d',e')$. □

COROLLARY 12 *If f is non-chaotic, then* $\overline{R}(f) \setminus RR(f)$ *is countable.* □

PROPOSITION 13 *Suppose f is non-chaotic and let* $d \in \overline{P}(f)$. *If f has no periodic point less than d, or if it has a greatest periodic point less than d, then d is regularly recurrent.*

Similarly if f has no periodic point greater than d, or if it has a least periodic point greater than d, then d is regularly recurrent.

Proof Evidently we may assume that d is not itself periodic. Suppose first that f has no periodic point less than d. Then d is the least point in $\overline{P}(f)$ and hence $f^j(d) > d$ for all $j > 0$. In any right neighbourhood of d there is a point y such that $f^N(y) = y$, where N is a power of 2. If we put $g = f^N$ then there are periodic points of g between d and y and arbitrarily close to d.

For any such point x we have $g^{2k}(x) < y$ for all $k > 0$, since the g-orbit of x is alternating and there are no fixed points of g between the least and greatest points of each half. It follows by continuity that $g^{2k}(d) \leq y$ for all $k > 0$. Therefore d is regularly recurrent.

Suppose next that f has a greatest periodic point c less than d. Then again in any right neighbourhood of d there is a point y such that $f^N(y) = y$, where N is a power of 2. Moreover we may choose N so that also $f^N(c) = c$. Putting $g = f^N$, it follows as in the previous part of the proof that for any periodic point x of g between d and y we have $g^{2k}(x) \in (c,y)$, and hence actually $g^{2k}(x) \in (d,y)$, for all $k > 0$. Therefore $g^{2k}(d) \in (d,y]$ for all $k > 0$, and again d is regularly recurrent.

The remaining statements of the proposition are proved in exactly the same way. □

Nevertheless it is not true in general for non-chaotic maps either that $\overline{\mathrm{R}}(f) = \mathrm{R}(f)$, or that $\mathrm{R}(f) = \mathrm{RR}(f)$. We can make further progress by systematically developing the arguments of Proposition 3.

LEMMA 14 *Suppose f is non-chaotic and $\omega(x,f)$ is an infinite limit set. For any positive integer s and for $i = 0, 1, \ldots, 2^s - 1$ put*

$$J_i^s = [\inf \omega(f^i(x), f^{2^s}), \sup \omega(f^i(x), f^{2^s})].$$

Then J_i^s is a compact interval such that

$$J_i^s \cap J_k^s = \varnothing \quad \text{for } 0 \leq i < k < 2^s$$

and

$$f(J_i^s) \supseteq J_{i+1}^s \text{ for } 0 \leq i < 2^s - 1, \quad f(J_i^s) \supseteq J_0^s \text{ for } i = 2^s - 1.$$

Moreover J_i^s contains a periodic point of period 2^s, but no periodic point of period 2^j with $0 \leq j < s$, and

$$J_i^{s+1} \cup J_{i+2^s}^{s+1} \subset J_i^s \quad \text{for } 0 \leq i < 2^s.$$

Both J_i^{s+1} and $J_{i+2^s}^{s+1}$ have an endpoint in common with J_i^s.

Proof For $s = 1$ this is an immediate consequence of Proposition 3. The general case follows by induction on s. □

It may be noted also that $J_k^{s+1} \cap J_i^s = \varnothing$ if $k \neq i, i+2^s$, since there are 2^{s+1} intervals J_k^{s+1} and 2^s intervals J_i^s. In the following we will refer to the intervals J_i^s as the intervals *associated* with the limit set $\omega(x,f)$.

Lemma 15 *Suppose f is non-chaotic and $\omega(x,f)$ is an infinite limit set. If J_i^s are the intervals associated with $\omega(x,f)$, then for every point $y \in \omega(x,f)$ there is a uniquely determined nested sequence $J_{a_1}^1 \supset J_{a_2}^2 \supset \ldots$ such that $y \in \bigcap_{s=1}^{\infty} J_{a_s}^s$.*

Proof From the definition of the intervals J_i^s we have

$$\omega(x,f) \subset \bigcup_{i=0}^{2S-1} J_i^s .$$

Since $J_i^s \cap J_k^s = \varnothing$ if $i \neq k$, it follows that $y \in J_{a_s}^s$ for a unique a_s with $0 \leq a_s < 2^s$. It follows also that $a_{s+1} = a_s$ or $a_s + 2^s$, and hence $J_{a_{s+1}}^{s+1} \subset J_{a_s}^s$. □

Lemma 16 *Suppose f is non-chaotic and $\omega(x,f)$ is an infinite limit set. Let J_i^s be the intervals associated with $\omega(x,f)$ and for any nested sequence $J_{a_1}^1 \supset J_{a_2}^2 \supset \ldots$ put $K = \bigcap_{s=1}^{\infty} J_{a_s}^s$.*

Then $K \cap \mathrm{P}(f) = \varnothing$. Moreover either $K = \{y\}$ is a singleton and $y \in \omega(x,f)$, or K is a closed interval with both its endpoints but no point of its interior in $\omega(x,f)$.

Proof Since K is a non-empty connected closed set, it is either a point or a closed interval. Moreover in the former case the point is in $\omega(x,f)$, and in the latter case both endpoints are in $\omega(x,f)$. On the other hand K contains no periodic points, since $J_{a_s}^s$ contains no fixed point of $f^{2^{s-1}}$. It follows that the interior of K contains no recurrent point and at most one point of $\omega(x,f)$, by Lemma IV.14. Furthermore, Lemma IV.14 shows that if the interior of K does contain a point $w \in \omega(x,f)$, then both endpoints of K must be in $\overline{\mathrm{P}}(f)$. Since neither endpoint of K has a finite orbit, and $w \in \omega(x,f)$, this contradicts Lemma IV.23. □

The following simple lemma shows that if $K = [y,z]$ is a closed interval, then y and z are not both regularly recurrent.

Lemma 17 *Let y,z be points of $\overline{\mathrm{R}}(f)$, with $y < z$, such that $[y,z] \cap \mathrm{P}(f) = \varnothing$. Then*

(i) *if one of the points y,z is regularly recurrent, then the other is not recurrent,*

(ii) *if (n_i) is any increasing sequence of positive integers, we cannot have both $f^{n_i}(y) \to y$ and $f^{n_i}(z) \to z$.*

Proof No point of (y,z) is recurrent, since $R \subseteq \overline{P}$. Suppose, for example, that y were regularly recurrent. Let V be an open interval containing y but not z. Then there exists a positive integer N such that $f^{kN}(y) \in V$ for all $k > 0$. Then $f^{kN}(y) < y$ for all $k > 0$. Since no point of $[y,z]$ is periodic it follows that $f^{kN}(z) < z$, and actually $f^{kN}(z) \leq y$, for all $k > 0$. Since $R(f^N) = R(f)$, it follows that z is not recurrent.

Suppose now that $f^{n_i}(y) \to y$. Then there exists a positive integer m such that $f^{n_i}(y) < y$ for $i > m$. In the same way as before this implies $f^{n_i}(z) \leq y$ for $i > m$, and hence $f^{n_i}(z) \not\to z$.
□

LEMMA 18 *Suppose f is non-chaotic and $\omega(x,f)$ is an infinite limit set. Let J_i^s be the intervals associated with $\omega(x,f)$, and for any nested sequence $J^1_{a_1} \supset J^2_{a_2} \supset \ldots$ set $K = \bigcap J^s_{a_s}$.*

If $K = \{y\}$, then y is regularly recurrent.

If $K = [y,z]$, then either both endpoints of K are strongly recurrent but not regularly recurrent, or one endpoint is regularly recurrent and the other is not recurrent.

In every case, if $w \in \omega(x,f)$, then w is regularly recurrent if and only if $f^{2^n}(w) \to w$ as $n \to \infty$.

Proof Suppose first that $K = \{y\}$. Then for any open interval V containing y we can choose $s > 0$ so that $J^s_{a_s} \subset V$. Since $y \in J^s_{a_s}$ we must have $y \in \omega(f^{a_s}(x), f^{2^s})$. Hence $f^{k2^s}(y) \in J^s_{a_s}$ for all $k > 0$. Consequently y is regularly recurrent and $f^{2^n}(y) \to y$ as $n \to \infty$.

Suppose next that $K = [y,z]$ and let w be any point of $\omega(x,f)$ in K. We denote the intervals associated with the limit set $\omega(w,f)$ by $J_i^s(w)$ and those associated with the limit set $\omega(x,f)$ by $J_i^s(x)$. Then $J_0^s(w) \subseteq J^s_{a_s}(x)$ and hence $K_0 = \bigcap J^s_0(w) \subseteq K$. By Proposition 7, every point of $\omega(w,f)$ is strongly recurrent. Since no point in the interior of K is recurrent, it follows from Lemma 16 that either $K_0 = \{y\}$, or $K_0 = \{z\}$, or $K_0 = K$.

If $K_0 = \{y\}$ then, by what we have already proved, y is regularly recurrent and $f^{2^n}(y) \to y$ as $n \to \infty$. Moreover by Lemma 17, z is not recurrent.

If $K_0 = \{z\}$ then the same holds with y and z interchanged.

Consider finally the case $K_0 = K$. Then y and z are both strongly recurrent, but not regularly recurrent, by Lemma 17. If $J_0^s(w) = [\alpha_s, \beta_s]$ then, for each s, either $\alpha_{s+1} = \alpha_s$ or $\beta_{s+1} = \beta_s$, but not both. We will show that $\alpha_{s+1} = \alpha_s$ for infinitely many s and $\beta_{s+1} = \beta_s$ for infinitely many s.

Assume, on the contrary, that $\alpha_{s+1} = \alpha_s$ for $s \geq m$. Then $\alpha_m = y$ and $f^{k2^m}(y) > y$ for all $k > 0$. Since the open interval (y,z) contains no point of R(f) we must actually have $f^{k2^m}(y) \geq z$ for all $k > 0$. But, since $R(f) = R(f^{2^m})$, this is a contradiction. Similarly we cannot have $\beta_{s+1} = \beta_s$ for all large s.

If $\alpha_{s+1} = \alpha_s$ then $f^{2^s}(y)$ lies to the right of $J_0^{s+1}(w)$ and hence $f^{2^s}(y) > z$. It follows that $f^{2^n}(y) \not\to y$, and similarly $f^{2^n}(z) \not\to z$. □

PROPOSITION 19 *A map f is non-chaotic if and only if*

$$RR(f) = \{x \in I : f^{2^n}(x) \to x \text{ as } n \to \infty\}.$$

Proof Suppose first that f is non-chaotic. If $x \in P(f)$ then x is regularly recurrent and $f^{2^n}(x) = x$ for all large n. If $x \in R(f) \setminus P(f)$, then $x \in RR(f)$ if and only if $f^{2^n}(x) \to x$ as $n \to \infty$, by Lemma 18 with $w = x$.

Suppose next that f is chaotic. Then f has a periodic point x with period $n = 2^d q$, for some $d \geq 0$ and some odd $q > 1$. Then $x \in RR(f)$, but $f^{2^n}(x)$ does not converge to x as $n \to \infty$. □

We define a *2-adic integer* to be an infinite sequence $\alpha = (a_0, a_1, a_2, \ldots)$, where $a_i = 0$ or 1 for all i. If $\beta = (b_0, b_1, b_2, \ldots)$ is another such sequence the sum

$$\alpha + \beta = (c_0, c_1, c_2, \ldots)$$

is defined in the following way. If $a_0 + b_0 < 2$ then $c_0 = a_0 + b_0$, but if $a_0 + b_0 \geq 2$ then $c_0 = a_0 + b_0 - 2$ and we carry 1 to the next position. The terms $c_1, c_2, \ldots$ are successively determined in the same fashion. With this definition of addition the set J of all 2-adic integers is an abelian group (sometimes called the *adding machine*).

We can also define a metric on J by setting $d(\alpha,\alpha) = 0$ and $d(\alpha, \beta) = 2^{-k}$ if $\alpha \neq \beta$ and k is the least integer such that $a_k \neq b_k$. This metric is invariant and non-archimedean, *i.e.* for all $\alpha, \beta, \gamma \in J$

$$d(\alpha+\gamma, \beta+\gamma) = d(\alpha,\beta),$$

$$d(\alpha+\beta,0) \leq \max\,[d(\alpha,0), d(\beta,0)].$$

Moreover J is now a compact topological group.

If $\mathbf{1} = (1, 0, 0, \dots)$, then the multiples $n\mathbf{1}$ $(n = 0, 1, 2, \dots)$ consist precisely of all $\alpha = (a_0, a_1, a_2, \dots)$ with $a_i = 0$ for all large i. Hence the semigroup J_0 formed by these multiples is dense in J.

The transformation τ of J into itself defined by $\tau(\xi) = \xi + \mathbf{1}$ is evidently a homeomorphism. Actually J is a minimal set for τ, since $\tau^n\xi = \xi + n\mathbf{1}$ and the subset J_0 is dense.

Let J_1 and J_2 be the set of all $\alpha = (a_0, a_1, \dots)$ with $a_0 = 1$ and $a_0 = 0$ respectively. Then $J = J_1 \cup J_2$ and $\tau(J_1) = J_2$, $\tau(J_2) = J_1$. Moreover, if g is the map of J onto J_2 defined by $g((a_0, a_1, \dots)) = (0, a_0, a_1, \dots)$, then $\tau^2 \circ g = g \circ \tau$. Consequently, τ^2 restricted to the subset J_2 (or J_1) is topologically conjugate to τ on the whole set J.

These remarks illuminate the following theorem, in the statement of which we retain the same interpretations of J and τ.

THEOREM 20 *Suppose f is non-chaotic and $Y = \omega(x,f)$ is an infinite limit set. Then there exists a continuous map φ of Y onto the set J of 2–adic integers such that each point of J is the image of at most two points of Y and*

$$\varphi \circ f(y) = \tau \circ \varphi(y) \text{ for every } y \in Y,$$

i.e. the accompanying diagram commutes:

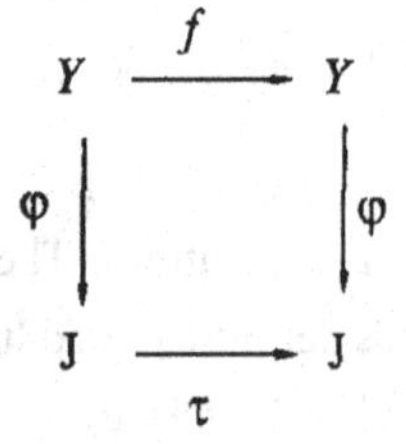

Moreover φ maps Y homeomorphically onto J if and only if every point $y \in Y$ is regularly recurrent.

Proof For any $y \in Y$ there is a uniquely determined nested sequence of intervals $J^1_{a_1} \supset J^2_{a_2} \supset \dots$ such that $y \in K = \bigcap_{s=1}^{\infty} J^s_{a_s}$. We define a map $\varphi: Y \to J$ by setting $\varphi(y) = \beta = (b_0, b_1, b_2, \dots)$, where $b_0 = a_1$ and $b_i = 0$ or 1 according as $a_{i+1} = a_i$ or $a_i + 2^i$ $(i = 1, 2, \dots)$.

We show first that φ maps Y *onto* J. For an arbitrary $\beta = (b_0, b_1, b_2, \ldots) \in$ J set $a_1 = b_0$, $a_{i+1} = a_i + b_i 2^i$ ($i = 1, 2, \ldots$). By induction we see at once that $0 \le a_s < 2^s$ ($s = 1, 2, \ldots$). Moreover $J^{s+1}_{a_{s+1}} \subset J^s_{a_s}$. By Lemma 16 a point $y \in Y$ is contained in $\bigcap J^s_{a_s}$. Then $\varphi(y) = \beta$.

We show next that any point of J is the image of at most two points of Y. Suppose y, $y' \in Y$ and $y \neq y'$. If y and y' belong to different sets $K = \bigcap J^s_{a_s}$ and $K' = \bigcap J^s_{a'_s}$ then there is a least positive integer s for which $a_s \neq a'_s$ and hence $\varphi(y) \neq \varphi(y')$, by the definition of the map φ. Consequently $\varphi(y) = \varphi(y')$ implies that y and y' belong to the same set K and hence, by Lemma 16, that K is a closed interval with endpoints y and y'.

It will now be shown that the map φ is continuous. Suppose $y \in Y$ and let $J^1_{a_1} \supset J^2_{a_2} \supset \ldots$ be the nested sequence such that $y \in \bigcap J^s_{a_s}$. For each positive integer s we can choose $\delta_s > 0$ so that if $y' \in Y$ and $d(y,y') < \delta_s$ then also $y' \in J^s_{a_s}$. It follows that if $\varphi(y) = (b_0, b_1, b_2, \ldots)$ and $\varphi(y') = (b'_0, b'_1, b'_2, \ldots)$ then $b'_i = b_i$ for all $i < s$. Hence $d[\varphi(y), \varphi(y')] \le 2^{-s}$.

We show finally that

$$\varphi \circ f(y) = \tau \circ \varphi(y) \text{ for every } y \in Y.$$

Let $\beta = (b_0, b_1, b_2, \ldots)$ be any element of J and let $\tau(\beta) = (d_0, d_1, d_2, \ldots)$. Then from the definition of the transformation τ:

if $b_0 = 0$ then $d_0 = 1$, $d_i = b_i$ for all $i \ge 1$;
if $b_i = 1$ for all $i \ge 0$ then $d_i = 0$ for all $i \ge 0$;
if $b_0 = \ldots = b_{k-1} = 1$, $b_k = 0$ then $d_0 = \ldots = d_{k-1} = 0$, $d_k = 1$, $d_i = b_i$ for all $i > k$.

If $y \in \bigcap J^s_{a_s}$ then $f(y) \in J^s_{a'_s}$, where $a'_s \equiv a_s + 1 \pmod{2^s}$ and $0 \le a'_s < 2^s$. Set

$$\varphi(y) = \beta = (b_0, b_1, b_2, \ldots), \quad \varphi[f(y)] = \gamma = (c_0, c_1, c_2, \ldots).$$

Then, for all $i \ge 0$,

$$a_{i+1} = b_0 + 2b_1 + \ldots + 2^i b_i, \quad a'_{i+1} = c_0 + 2c_1 + \ldots + 2^i c_i.$$

If $b_0 = 0$ then $c_0 = 1$, $c_i = b_i$ for all $i \ge 1$. If $b_i = 1$ for all $i \ge 0$ then $a_{i+1} = 2^{i+1} - 1$ and $a'_{i+1} = 0$ for all $i \ge 0$. Hence $c_i = 0$ for all $i \ge 0$. If $b_i = 1$ for $i < k$, $b_k = 0$ then $a_{i+1} = 2^{i+1} - 1$ and $a'_{i+1} = 0$ for $i < k$. Hence $c_i = 0$ for $i < k$, $c_k = 1$ and $c_i = b_i$ for $i > k$. In every case $c_i = d_i$ for all $i \ge 0$.

The last statement of the theorem is an immediate consequence of Lemma 18, by what we have already proved. □

In the case that φ is a homeomorphism, the maps $f|\omega(x,f)$ and τ are topologically conjugate. An example of this is given later in the chapter (Example 30).

3 UNIFORMLY NON-CHAOTIC MAPS

We will say that a point $x \in I$ is *approximately periodic* if for every $\varepsilon > 0$ there exists a periodic point y and a positive integer N such that

$$d[f^n(x), f^n(y)] < \varepsilon \text{ for all } n > N.$$

Thus any asymptotically periodic point is also approximately periodic.

We will say that a continuous map $f: I \to I$ is *uniformly non-chaotic* if every point $x \in I$ is approximately periodic. Thus any strongly non-chaotic map is also uniformly non-chaotic. It is easily seen that, for any positive integer m, f is uniformly non-chaotic if and only if f^m is uniformly non-chaotic.

LEMMA 21 *If f is uniformly non-chaotic, then it is also non-chaotic.*

Proof Assume on the contrary that f is chaotic. Then there exists a positive integer m and disjoint compact intervals I_0, I_1 such that

$$I_0 \cup I_1 \subseteq f^m(I_0) \cap f^m(I_1) .$$

We use the same notations as in the proofs of Propositions II.15 and II.17. Let $\alpha = (a_1, a_2, \ldots)$ be a sequence of 0's and 1's which contains every finite sequence of 0's and 1's once, and hence infinitely often. Since there exist uncountably many such α we can choose one for which $I_\alpha = \{x\}$ is a point.

Choose $\varepsilon > 0$ so that $d(I_0, I_1) > 2\varepsilon$. Let J_0 be the compact interval consisting of all points z with $d(z, I_0) \leq \varepsilon$ and let J_1 be the compact interval consisting of all points z with $d(z, I_1) \leq \varepsilon$. Then J_0 and J_1 are disjoint.

Since f is uniformly non-chaotic there exists a periodic point y such that $d[f^n(x), f^n(y)] < \varepsilon$ for all $n > N$. Since $f^{km}(x) \in I_0$ or I_1 according as $a_{k+1} = 0$ or 1, it follows that, for all large k, $f^{km}(y) \in J_0$ or J_1 according as $a_{k+1} = 0$ or 1. But if y has period p this implies $a_{kp+1} = a_{(k+1)p+1}$ for all large k, which contradicts the definition of α. □

Uniformly non-chaotic maps, like strongly non-chaotic maps, can be characterized in several different ways. We first establish two auxiliary results.

LEMMA 22 *Suppose f is non-chaotic and $\omega(x,f)$ is an infinite limit set. If every element of $\omega(x,f)$ is regularly recurrent then, given any $\varepsilon > 0$, there exists a finite set of pairwise disjoint closed intervals $J_1, \ldots, J_m$, all of length at most ε, such that $f(J_i) \supseteq J_{i+1}$ for $i = 1,\ldots,m-1$ and $f(J_m) \supseteq J_1$. Moreover $\omega(x,f) \subset J_1 \cup \ldots \cup J_m$ and f cyclically permutes the sets $\omega(x,f) \cap J_i$ $(i = 1,\ldots,m)$.*

Proof Let J_i^s be the closed intervals associated with $\omega(x,f)$. It is sufficient to show that, given any $\varepsilon > 0$, we can choose s so that J_i^s has length at most ε for $0 \le i < 2^s$. If this is not the case then for each $s > 0$ there exists an i_s with $0 \le i_s < 2^s$ such that $J_{i_s}^s$ has length $> \varepsilon$. It follows that we can choose $a_1 \in \{0,1\}$ so that for each $s > 1$ there exists an i_s with $0 \le i_s < 2^s$ such that $J_{i_s}^s$ has length $> \varepsilon$ and $J_{i_s}^s \subset J_{a_1}^1$. We can now choose $a_2 \in \{a_1, a_1+2\}$ so that for each $s > 2$ there exists an i_s with $0 \le i_s < 2^s$ such that $J_{i_s}^s$ has length $> \varepsilon$ and $J_{i_s}^s \subset J_{a_2}^2$. Proceeding in this way we obtain a nested sequence $J_{a_1}^1 \supset J_{a_2}^2 \supset \ldots$ such that $J_{a_s}^s$ has length $> \varepsilon$ for every s. Hence $K = \bigcap J_{a_s}^s$ is an interval of length at least ε. The endpoints of K are in $\omega(x,f)$ and therefore regularly recurrent. But this contradicts Lemma 17. □

LEMMA 23 *If f is chaotic, then* $SR(f) \ne RR(f)$.

Proof We use the same method of proof as for Proposition 9. Thus it is sufficient to show that $SR(\sigma) \ne RR(\sigma)$. We use for this purpose a sequence $\alpha = (a_0, a_1, a_2, \ldots)$ of 0's and 1's which was studied by Thue [121] and Morse [89], and which is generally known as the *Morse sequence*. It is defined in the following way: if in the binary scale $n = \sum n_i 2^i$, where each $n_i = 0$ or 1, then $a_n \equiv \sum n_i \pmod 2$. Thus

$$\alpha = (0, 1, 1, 0, 1, 0, 0, 1, 1, 0, 0, 1, 0, 1, 1, 0, \ldots).$$

If A_k is the block consisting of the first 2^k terms a_n in α then

$$A_0 = 0, \quad A_{k+1} = A_k \bar{A}_k \quad (k \ge 0),$$

where $\bar{A}_k$ is obtained from A_k by replacing 0's by 1's and 1's by 0's. Hence for any given k we can write

$$\alpha = (B_0, B_1, B_2, \ldots),$$

where $B_0 = A_k$ and $B_n = A_k$ or $\overline{A}_k$ for every $n > 0$. (In fact $B_n = A_k$ or $\overline{A}_k$ according as $a_n = 0$ or 1.)

Let ℓ be any positive integer and let A be any block of ℓ consecutive terms in α. If we define the positive integer k by $2^{k-1} \leq \ell < 2^k$ then there exist consecutive blocks B_n, B_{n+1} of length 2^k such that A is contained in $B_n B_{n+1}$. Thus A is contained in at least one of $A_k A_k$, $A_k \overline{A}_k$, $\overline{A}_k A_k$, $\overline{A}_k \overline{A}_k$. But each of these is contained in

$$A_{k+3} = A_k \overline{A}_k \overline{A}_k A_k \overline{A}_k A_k A_k \overline{A}_k$$

and in

$$\overline{A}_{k+3} = \overline{A}_k A_k A_k \overline{A}_k A_k \overline{A}_k \overline{A}_k A_k .$$

Moreover any block of length 2^{k+4} contains either a block A_{k+3} or a block $\overline{A}_{k+3}$. It follows that each block of $2^5 \ell$ terms in α contains every block of ℓ terms in α. Hence α is strongly recurrent.

To prove that α is not regularly recurrent it is sufficient to show that if $a_N = 0$ for some positive integer N then $a_{kN} = 1$ for some integer $k > 1$. Evidently N cannot be a power of 2. Thus we can write $N = 2^s q$, where $s \geq 0$ and $q \geq 3$ is odd. Then

$$q = 1 + 2^{i_1} + \ldots + 2^{i_r},$$

where $0 < i_1 < \ldots < i_r$. If we take $k = 2^q - 1$ then

$$\begin{aligned} kN &= 2^s(2^{i_1+q} + \ldots + 2^{i_r+q} + 2^q - 1 - 2^{i_1} - \ldots - 2^{i_r}) \\ &= 2^s(2^{i_1+q} + \ldots + 2^{i_r+q} + 1 + 2 + \ldots + 2^{q-1} - 2^{i_1} - \ldots - 2^{i_r}), \end{aligned}$$

which is a sum of $r + (q - r) = q$ distinct powers of 2. Since q is odd, it follows that $a_{kN} = 1$. □

We can now prove the following counterpart to Proposition 10.

THEOREM 24 *The following statements are equivalent:*

(i) *f is uniformly non-chaotic,*
(ii) $\Lambda(f) = \mathrm{RR}(f)$,
(iii) $\Lambda(f) = \{x \in I : f^{2^n}(x) \to x \text{ as } n \to \infty\}$.

Proof Suppose first that (i) holds and let $x \in \Lambda(f)$. That is, $x \in \omega(y,f)$ for some $y \in I$. Given any $\varepsilon > 0$, there exists a periodic point z and a positive integer N such that

$$d[f^n(y), f^n(z)] < \varepsilon/2 \quad \text{for all } n \geq N.$$

Let p be the period of z and put $g = f^p$. Then $x \in \omega(y', g)$, where $y' = f^i(y)$ for some i with $0 \leq i < p$. If we put $z' = f^i(z)$, then z' is a fixed point of g and

$$d[g^n(y'), z'] < \varepsilon/2 \quad \text{for all } n \geq N.$$

It follows that

$$d[g^k(x), z'] \leq \varepsilon/2 \quad \text{for all } k \geq 0$$

and hence

$$d[g^k(x), x] \leq \varepsilon \quad \text{for all } k > 0.$$

Thus $x \in RR(f)$. Hence (ii) holds.

Conversely, suppose that (ii) holds. Then the map f is certainly non-chaotic. We wish to show that (i) holds, i.e. that every point $x \in I$ is approximately periodic. Since x is asymptotically periodic if $\omega(x,f)$ is finite, we may assume that $\omega(x,f)$ is infinite.

Given any $\varepsilon > 0$, we can find disjoint closed intervals $J_1, \ldots, J_m$ satisfying the conditions of Lemma 22 for this ε. Let y be a point of $\omega(x,f)$ in J_1. Thus if $n \equiv i \pmod{m}$, where $0 \leq i < m$, then $f^n(y) \in J_{i+1}$. There exists a periodic point z of period m such that $f^i(z) \in J_{i+1}$ for $0 \leq i < m$. Hence

$$d[f^n(y), f^n(z)] < \varepsilon \quad \text{for all } n \geq 0.$$

Thus y is approximately periodic. It remains to show that also x is approximately periodic.

Let $\delta > 0$ denote the minimum distance between J_i and J_k for $1 \leq i < k \leq m$ and put $\delta' = \min(\varepsilon, \delta/2)$. We can choose $\varepsilon' > 0$, with $\varepsilon' < \delta'$, so that if $x', x'' \in I$ and $d(x', x'') < \varepsilon'$ then

$$d[f^i(x'), f^i(x'')] < \delta' \quad \text{for } 1 \leq i \leq m.$$

There exists a positive integer N such that

$$d[f^n(x), \omega(x,f)] < \varepsilon' \quad \text{for } n \geq N.$$

For some $n' > N$ we have $d[f^{n'}(x), y] < \varepsilon'$. Then

$$d[f^{n'+i}(x), f^i(y)] < \delta' \quad \text{for } 1 \leq i \leq m.$$

There exists a point $y' \in \omega(x,f)$ such that

$$d[f^{n'+m}(x), y')] < \varepsilon'.$$

Evidently we must have $y' \in J_1$. It follows that

$$d[f^{n'+m+i}(x), f^i(y')] < \delta' \quad \text{for } 1 \le i \le m .$$

Continuing in this way, we obtain

$$d[f^{n'+km+i}(x), J_{i+1}] < \delta' \text{ for } 1 \le i < m \text{ and all } k \ge 0.$$

Hence if we choose z' in the orbit of z so that $f^{n'}(z') = z$, we will have

$$d[f^n(x), f^n(z')] < 2\varepsilon \quad \text{for all } n \ge n'.$$

Thus (i) holds.

Finally, if either (ii) or (iii) holds then f is non-chaotic, by Proposition 9. It now follows from Proposition 19 that (ii) and (iii) are equivalent. □

The preceding theorem shows that uniformly non-chaotic maps are rather nice. It will now be shown that non-chaotic maps which are not uniformly non-chaotic are surprisingly nasty.

PROPOSITION 25 *Suppose either that $\Lambda(f) \neq \overline{P}(f)$ or that f is non-chaotic, but not uniformly non-chaotic. Then there exist points y,z with $y < z$ and an increasing sequence (p_i) of positive integers such that, for* every *infinite sequence $\alpha = (a_1, a_2, \dots)$ of* 0's *and* 1's *there is a point $w = w_\alpha \in I$ for which $f^{p_i}(w) < y$ if $a_i = 0$, $f^{p_i}(w) > z$ if $a_i = 1$, and the sequence $\{f^{p_i}(w)\}$ has no limit point different from y,z.*

Proof Suppose first that there exists a point $y \in \Lambda(f) \setminus \overline{P}(f)$, so that $y \in \omega(x,f)$ for some $x \in I$, and let J be the component of $I \setminus \overline{P}(f)$ which contains y. Without loss of generality, assume that J is of increasing type and let z be the right endpoint of J. Then $z \in \overline{P}(f) \setminus P(f)$, by Lemma IV.23. Let V_1 be any open interval with left endpoint z, let z_1 be a periodic point of f in V_1, and let n_1 be its period. If we put $f_1 = f^{n_1}$ then $y \in \Lambda(f_1) \setminus \overline{P}(f_1)$, the interval J is a component of $I \setminus \overline{P}(f_1)$ and is of increasing type also for f_1, and $z \in \overline{P}(f_1) \setminus P(f_1)$. Thus, by replacing f by f_1, we may assume that z_1 is a fixed point of f.

By Proposition V.9, $z \in \omega(y,f)$. Hence $f^k(y) \in (z, z_1)$ for some $k > 0$, again by Lemma IV.23. Let $U_1 \subset J$ be an open interval with right endpoint y such that $f^k(U_1) \subseteq (z, z_1)$. Then $f^j(x) \in U_1$ for some $j \ge 0$ and $f^{j+k}(x) \in (z, z_1)$. Moreover $y_1 = f^{i+j+k}(x) \in (f^j(x), y)$ for some $i > 0$. Hence $f^i[z, z_1] \supseteq [y_1, z_1]$.

If we put $g = f^i$ it follows that

$$g^n [z, z_1] \supseteq [y_1, z_1] \quad \text{for all } n > 0.$$

As we have already seen for f_1, the same hypotheses hold for g as for f. Thus $y \in \omega(x_1, g)$, where $x_1 = f^n(y_1)$ for some $n > 0$ and $y_1 < x_1 < y$. Then $x_1 < g^m(x_1) < y$ for some $m > 0$. It follows that $g^m(y) > y$ and actually $g^m(y) > z$. Hence $g^m[y_1, y] \supseteq [y_2, z_2]$, where $y_2 = g^m(x_1)$ satisfies $y_1 < y_2 < y$ and z_2 is a periodic point of g such that $z < z_2 < z_1$.

Thus if we put $J_i = [y_i, y]$ and $K_i = [z, z_i]$ $(i = 1,2)$, there exists an integer $m_1 > 0$ such that, for $i = 1$,

$$(*) \qquad J_{i+1} \cup K_{i+1} \subseteq f^{m_i}(J_i) \cap f^{m_i}(K_i).$$

However, this construction can be continued inductively. We obtain in this way an increasing sequence (y_i) and a decreasing sequence (z_i) of points, and an increasing sequence (m_i) of positive integers, such that (*) holds for all $i \geq 1$ and $y_i \to y$, $z_i \to z$ as $i \to \infty$.

For each $i \geq 1$ let I_i denote either J_i or K_i. If we put $p_1 = 0$, $p_i = m_1 + \ldots + m_{i-1}$ for $i > 1$, then for each $n > 0$ there exists a point $w_n \in I$ such that

$$f^{p_i}(w_n) \in I_i \quad \text{for } i = 1, \ldots, n.$$

If w is a limit point of the sequence (w_n), then $f^{p_i}(w) \in I_i$ for every $i \geq 1$. Moreover y and z are the only possible limit points of the sequence $\{f^{p_i}(w)\}$. Thus the proposition is proved in the case $\Lambda(f) \neq \overline{P}(f)$.

We now consider the case where f is non-chaotic, but not uniformly non-chaotic. By what we have already proved we may suppose that $\Lambda(f) = \overline{P}(f)$. By Theorem 24 there exists a point $w \in \Lambda(f) \setminus RR(f)$. Thus $w \in \omega(x,f)$ for some $x \in I$ and $\omega(x,f)$ is infinite. If J_i^s are the closed intervals associated to $\omega(x,f)$, there exists a nested sequence $J_{a_1}^1 \supset J_{a_2}^2 \supset \ldots$ such that $w \in K = \bigcap_{s=1}^{\infty} J_{a_s}^s$. Moreover $K \neq \{w\}$, by Lemma 18. Hence $K = [y, z]$ is a closed interval with endpoints $y, z \in \omega(x,f)$. If $J_{a_s}^s = [\alpha_s, \beta_s]$ then, for each s, either $\alpha_{s+1} = \alpha_s$ or $\beta_{s+1} = \beta_s$, but not both. Without loss of generality, assume that $\alpha_{s+1} = \alpha_s$ for infinitely many s. By Lemma 14, for such an s we have $f^{2^s}(y) \geq \beta_{s+1} \geq z$.

Consider first the case where also $\beta_{s+1} = \beta_s$ for infinitely many s. Then there exist infinitely many s such that $\alpha_{s-1} = \alpha_s < \alpha_{s+1}$. For such an s we have $f^{2^s}(y) < y$, since $\beta_{s+1} = \beta_s$. By Lemma 14 there exists a periodic point $y_s \in (\alpha_s, \beta_s)$ of period 2^s whose orbit is contained in the union of the intervals J_i^s $(i = 0,1, \ldots, 2^s-1)$. Thus if we put $z_s = f^{2^{s-1}}(y_s)$, then $z_s \in (\beta_s, \beta_{s-1})$ and, in particular, $z_s > z$. Moreover $y_s \in (\alpha_s, \alpha_{s+1})$, since the interval $[\alpha_{s+1}, \beta_{s+1}]$ contains no point of period 2^s.

Hence there exists an increasing sequence (y_i) of points and an increasing sequence (n_i) of positive integers such that y_i is periodic with period $2^{n_i} = 2m_{i-1}$. In addition, if $z_i = f^{m_{i-1}}(y_i)$, then the sequence (z_i) is decreasing and $y_i \to y$, $z_i \to z$ as $i \to \infty$. It follows that (*) holds for every $i \geq 1$, with $J_i = [y_i, y_{i+1}]$, $K_i = [z_{i+1}, z_i]$. The argument can now be completed in the same way as before.

It remains to consider the case where $\alpha_{s+1} = \alpha_s$ for all $s \geq m$, say, and hence $y = \alpha_m$. Since $y \in \Lambda(f) = \overline{P}(f)$, there exists an increasing sequence (y_i) of points and an increasing sequence (n_i) of positive integers such that y_i is periodic with period $2^{n_i} = 2m_{i-1}$, $y_i \to y$ as $i \to \infty$ and there is no fixed point of $f^{2m_{i-1}}$ between y_i and y. Evidently we may assume that $n_i \geq 2m$ for all i, so that $f^{m_{i-1}}(y_i) > y$. Put $z_i = f^{m_{i-1}}(y_i)$. Since there is no fixed point of $f^{2m_{i-1}}$ in (y_i, y), we must have $z_i > y$ and actually $z_i > z$, since $K \cap P(f) = \emptyset$. Since $f^{m_{i-1}}(z_i) = y_i$, there is a fixed point of $f^{m_{i-1}}$ in (y_i, z_i) and actually in (z, z_i). Let w_i denote the fixed point of $f^{m_{i-1}}$ in (z, z_i) which is nearest to z. Then w_i is also a fixed point of $f^{m_i/2}$, since m_{i-1} divides $m_i/2$. Since $f^{m_i/2}$ has no fixed point between y_{i+1} and z_{i+1}, by Lemma II.7, it follows that $z_{i+1} < w_i$. Hence the sequence (z_i) is decreasing. On the other hand, if $s = n_i - 1$ the interval $[\alpha_s, \beta_s]$ contains a point of period m_{i-1}, and this point is actually in (z, β_s). Thus $w_i < \beta_s$, and hence $z_i \to z$ as $i \to \infty$. We now have the same situation as in the preceding case, and can draw the same conclusions. □

COROLLARY 26 *If f is non-chaotic, but not uniformly non-chaotic, then there exists an uncountable set $S \subset I$ and a positive number δ such that,*

(i) *for every $s_1, s_2 \in S$ with $s_1 \neq s_2$,*

$$\overline{\lim}_{n\to\infty} d[f^n(s_1), f^n(s_2)] \geq \delta,$$
$$\underline{\lim}_{n\to\infty} d[f^n(s_1), f^n(s_2)] = 0,$$

(ii) *for any $s \in S$ and any periodic point z,*

$$\overline{\lim}_{n\to\infty} d[f^n(s), f^n(z)] \geq \delta.$$

Proof We use the notation and proof of the preceding proposition, and in addition we denote by A the set of all sequences $\alpha = (a_1, a_2, \ldots)$ of 0's and 1's. We can make A into an abelian group by defining the sum of two sequences $\alpha' = (a'_1, a'_2, \ldots)$ and $\alpha'' = (a''_1, a''_2, \ldots)$ to be the sequence $\alpha = (a_1, a_2, \ldots)$, where $a_i \equiv a'_i + a''_i \pmod 2$. If A_0 is the subset of A consisting of all sequences α which have only finitely many 0's or only finitely many 1's then A_0 is a countable subgroup of A. Let $\{\alpha_\mu\}$ be a complete set of representatives of the cosets of A_0 in A. Thus $\alpha_\mu \in A$ and $\alpha_\mu - \alpha_\nu \notin A_0$ if $\mu \neq \nu$. In other words, if $\alpha_\mu = (a_1, a_2, \ldots)$, $\alpha_\nu = (b_1, b_2, \ldots)$ and $\mu \neq \nu$ then $a_i \neq b_i$ for infinitely many i and $a_i = b_i$ for infinitely many i. It follows that if w_μ, w_ν are the points of I corresponding to the sequences α_μ, α_ν then

$$\overline{\lim}_{i\to\infty} d[f^{p_i}(w_\mu), f^{p_i}(w_\nu)] = z - y,$$
$$\underline{\lim}_{i\to\infty} d[f^{p_i}(w_\mu), f^{p_i}(w_\nu)] = 0.$$

Both y and z are in $\omega(w_\mu, f)$, except for the single μ for which $\alpha_\mu \in A_0$. If v is a periodic point, with period p say, then no point of its orbit lies in the interval $[y,z]$. Since f is non-chaotic, p divides m_i for all large i. It follows that

$$\overline{\lim}_{n\to\infty} d[f^n(w_\mu), f^n(v)] \geq z - y.$$

Finally, the set $S = \{w_\mu\}$ is uncountable, since there are uncountably many sequences α_μ. □

It will now be shown that there exists a set S with the same properties for any chaotic map f.

PROPOSITION 27 *If f is chaotic then there exists an uncountable set $S \subset I$ and a positive number δ such that,*

(i) *for every $s_1, s_2 \in S$ with $s_1 \neq s_2$,*

$$\overline{\lim}_{n\to\infty} d[f^n(s_1), f^n(s_2)] \geq \delta,$$
$$\underline{\lim}_{n\to\infty} d[f^n(s_1), f^n(s_2)] = 0,$$

(ii) *for any $s \in S$ and any periodic point z,*

$$\overline{\lim}_{n\to\infty} d[f^n(s), f^n(z)] \geq \delta.$$

Proof Since f is chaotic, f^m is strictly turbulent for some positive integer m. We use again the set up of Propositions II.15 and II.17, with f replaced by f^m. Set $\delta = d(I_0, I_1)/2 > 0$ and let $\alpha = (a_1, a_2, \ldots)$ be an element of Σ which contains every finite sequence of 0's and 1's. Since there exist uncountably many such α we can choose one for which $I_\alpha = \{x\}$ is a point. For any $\beta = (b_1, b_2, \ldots) \in \Sigma$ define

$$\begin{aligned}\gamma_\beta &= (a_1, b_1, a_1, a_2, b_1, b_2, a_1, a_2, a_3, b_1, b_2, b_3, \ldots)\\ &= (c_1, c_2, \ldots)\end{aligned}$$

and choose $x_\beta \in I$ so that $h(x_\beta) = \gamma_\beta$. We take $S = \{x_\beta : \beta \in \Sigma\}$.

If $\beta' = (b'_1, b'_2, \ldots)$ and $\beta' \neq \beta$ then $b'_k \neq b_k$ for some k. Hence $c'_{n^2+k} \neq c_{n^2+k}$ for $n \geq k$ and

$$d[f^{m(n^2+k-1)}(x_\beta), f^{m(n^2+k-1)}(x_{\beta'})] \geq 2\delta \text{ for } n \geq k.$$

On the other hand, since I_α is a point and

$$\sigma^{n(n+1)}(\gamma_\beta) = (a_1, a_2, \dots, a_{n+1}, b_1, \dots),$$

it follows that

$$d[f^{mn(n+1)}(x_\beta), f^{mn(n+1)}(x_{\beta'})] \to 0 \text{ as } n \to \infty.$$

For any positive integer p there exist infinitely many k such that $a_{kp+1} = 0$ and infinitely many k such that $a_{kp+1} = 1$. It follows that if z is a periodic point, with period p, then

$$d[f^n(x_\beta), f^n(z)] \geq \delta \quad \text{for infinitely many } n. \quad \square$$

A set S which possesses the properties (i) and (ii) in the statements of Corollary 26 and Proposition 27 is sometimes called a *δ–scrambled set*. It is simply called a *scrambled set* if it possesses the properties

(i)' *for every* $s_1, s_2 \in S$ *with* $s_1 \neq s_2$,

$$\overline{\lim}_{n\to\infty} d[f^n(s_1), f^n(s_2)] > 0,$$
$$\underline{\lim}_{n\to\infty} d[f^n(s_1), f^n(s_2)] = 0,$$

(ii)' *for any* $s \in S$ *and any periodic point* z,

$$\overline{\lim}_{n\to\infty} d[f^n(s), f^n(z)] > 0.$$

It is clear from the definitions that no point of a δ-scrambled set is approximately periodic. The following lemma shows that a scrambled set contains at most one approximately periodic point. It further shows that if a set S has the property (i)', then a scrambled set can be obtained from it by omitting at most one point.

LEMMA 28 *If x and y are approximately periodic, then either*

$$\lim_{n\to\infty} d[f^n(x), f^n(y)] = 0$$

or

$$\underline{\lim}_{n\to\infty} d[f^n(x), f^n(y)] > 0.$$

Proof Assume on the contrary that

$$\underline{\lim}_{n\to\infty} d[f^n(x), f^n(y)] = 0,$$
$$\overline{\lim}_{n\to\infty} d[f^n(x), f^n(y)] = \rho > 0.$$

Choose ε so that $0 < \varepsilon < \rho/5$. There exist periodic points z and w and a positive integer N such that, for all $n \geq N$,

$$d[f^n(x), f^n(z)] < \varepsilon, \quad d[f^n(y), f^n(w)] < \varepsilon.$$

Let m be the least common multiple of the periods of z and w, and choose $\delta > 0$ so small that $d[x_1, x_2] < \delta$ for any x_1, x_2 implies

$$d[f^k(x_1), f^k(x_2)] < \varepsilon \quad \text{for } k = 1,\ldots,m.$$

For some $p \geq N$ we have $d[f^p(x), f^p(y)] < \delta$ and hence

$$d[f^{p+k}(x), f^{p+k}(y)] < \varepsilon \quad \text{for } k = 1,\ldots,m.$$

It follows that

$$d[f^{p+k}(z), f^{p+k}(w)] < 3\varepsilon \quad \text{for } k = 1,\ldots,m.$$

Thus

$$d[f^n(z), f^n(w)] < 3\varepsilon \quad \text{for all } n \geq 0.$$

Hence, for all $n \geq N$,

$$d[f^n(x), f^n(y)] \; < \; 5\varepsilon \; < \; \rho,$$

which is a contradiction. □

In the paper in which the word *chaos* first appeared in connection with maps of an interval, Li and Yorke [77] showed that any map with a point of period 3 has an uncountable scrambled set. Consequently several authors have taken the existence of an uncountable scrambled set to be the definition of a chaotic map. Although this may not be inconsistent with the everyday use of the word, we believe that the definition of a chaotic map which we have adopted is preferable. The notion of a uniformly non-chaotic map also appears to have more theoretical significance than the non-existence of an uncountable scrambled set, even though the two are equivalent. Indeed, a map which is uniformly non-chaotic cannot have a scrambled set containing two or more points, by Lemma 28, and a map which is not uniformly non-chaotic has an uncountable δ-scrambled set for some $\delta > 0$, by Corollary 26 and Proposition 27.

It remains to be shown that the distinction between non-chaotic and uniformly non-chaotic maps is not vacuous. Misiurewicz and Smital [87] give an example of a C^∞– map which is non-chaotic, but not uniformly non-chaotic. They also give the following simple example of a continuous map with the same property.

EXAMPLE 29 Let g be the piecewise linear 'tent' map defined by

$$g(0) = 0, \ g(1/2) = 1, \ g(1) = 0.$$

Since g is turbulent there exists, for each $n > 0$, a least positive number λ_n such that the interval $[0, \lambda_n]$ contains a periodic orbit of g of period 2^n. Moreover $0 < \lambda_1 < \lambda_2 < \ldots$ and $\lambda_n \to \mu < 1$ as $n \to \infty$. We will show that the piecewise linear 'truncated tent' map f defined by

$$f(0) = 0, \ f(\mu/2) = \mu = f(1 - \mu/2), \ f(1) = 0$$

is non-chaotic, but not uniformly non-chaotic.

Evidently λ_n belongs to an orbit of g of period 2^n which is entirely contained in the interval $[0, \lambda_n]$, and therefore contains no point of the interval $(\lambda_n/2, 1 - \lambda_n/2)$. It follows that this orbit is also a periodic orbit of f, and thus $\mu \in \overline{P}(f)$.

No point in the g-orbit of μ can lie in the open interval $J = (\mu/2, 1-\mu/2)$, since $g^k(\mu) \in J$ would imply $g^k(\lambda_n) \in J$ for large n. It follows that no point of J is periodic for f. Hence $\mu \in P(f)$ if and only if $\mu \in P(g)$, and any periodic orbit of f is also a periodic orbit of g.

If a periodic orbit of f is contained in the interval $(0, \lambda_n)$ then, by Šarkovskii's theorem, its period must have one of the values $1,2,\ldots,2^{n-1}$. It follows that f is non-chaotic, since even if μ is periodic its period must be a power of 2.

In fact μ cannot be periodic. For the proof of Proposition II.28 shows that, for a piecewise monotone non-chaotic map, if x_n is the greatest point in a periodic orbit of period m_n, and if $x_n \to x$ and $m_n \to \infty$ as $n \to \infty$, then x is not periodic.

It follows that also $\mu/2$ and $1 - \mu/2$ are not periodic. However, at least one of them is in $\overline{P}(f)$, since $f(x) = \lambda_n$ implies $x = \lambda_n/2$ or $1 - \lambda_n/2$. It will now be shown that they are both in $\overline{P}(f)$. Indeed this is an immediate consequence of the following general statement:

(#) Let $N = 2^n$, where n is a positive integer. Then there is a positive integer $m = m_n$ such that, if $P = \{x_1 < \ldots < x_{2N}\}$ is any periodic orbit of f of period $2N = 2^{n+1}$, then $f(x_m) = x_{2N}$. Furthermore, if n is even then $m_n = (2N + 1)/3$ and x_m lies to the left of J, whereas if n is odd then $m_n = (2N + 2)/3$ and x_m lies to the right of J.

In order to prove (#) we introduce some notation. Let $B_1 = \{x_1, x_2\}, \ldots, B_N = \{x_{2N-1}, x_{2N}\}$ be pairs of adjacent points of P, and let $L_1 = [x_1, x_2], \ldots, L_N = [x_{2N-1}, x_{2N}]$ be the closed intervals with these points as endpoints. Since f is non-chaotic, it permutes the sets B_i. For, by Theorem II.12, f interchanges the left and right halves of P, f^2 interchanges the left and right halves of these halves, and so on.

We will say that f preserves order on B_i if f maps the smaller point of B_i to the smaller point of $f(B_i)$ and that f reverses order on B_i otherwise. Since the period of P is twice the number of pairs, f reverses order on an odd number of pairs B_i.

If $\bar{J} \cap L_i = \emptyset$ for each i, then f maps each L_i linearly onto L_j for some $j = j_i$ and the length of L_j is twice the length of L_i. Since f permutes the pairs B_i, this is impossible. We conclude that $J \subset B_l$ for a unique l, and $f(B_l) = B_N$. Moreover f maps the right endpoint of B_l to x_{2N} if f preserves order on B_l, and the left endpoint of B_l to x_{2N} otherwise. Clearly, if (#) holds then m_n has the opposite parity to n.

Suppose first that $n = 1$, so that $2^{n+1} = 4$. Evidently $J \subset B_1$ and f reverses order on B_2. Hence f preserves order on B_1, $f(x_2) = x_4$, and x_2 lies to the right of J.

Suppose next that $n > 1$ and that (#) holds with $n-1$ in place of n. Evidently f has a periodic orbit Q of period N contained in $L_1 \cup \ldots \cup L_N$ such that, for each $z \in$ Q, if $z \in L_i$ and $f(B_i) = B_j$ then $f(z) \in L_j$. Assume first that n is even. It follows from the induction hypothesis that if $m = m_{n-1} = (N + 2)/3$, then $f(B_m) = B_N$ and m is even. Since there are an even number of pairs to the right of B_m and since f reverses order on these pairs but preserves order on pairs to the left of B_m it follows that f reverses order on B_m. Hence $m_n = 2m_{n-1} - 1 = (2N + 1)/3$. This establishes (#) in this case. If n is odd then $m = m_{n-1} = (N + 1)/3$ and, by similar reasoning, f preserves order on B_m. Thus $m_n = 2m_{n-1} = (2N + 2)/3$, and again (#) holds.

Thus $\mu/2$ and $1 - \mu/2$ are both in $\overline{P}(f)$. Since $\bar{J} \cap P(f) = \emptyset$, it follows from Lemma 17 that they are not both regularly recurrent. Hence, by Theorem 24, f is not uniformly non-chaotic. (However, by Proposition 53 below, $\overline{P}(f) = \overline{R}(f) = R(f)$.) □

Chu and Xiong [48] have given a much more complicated example of a non-chaotic map f for which $\overline{R}(f) \neq R(f)$, thus disproving a claim of Blokh [40]. It may be noted that if $\overline{R}(f) \neq R(f)$ then $\overline{R}(f) \setminus R(f)$ is infinite, since $\overline{R}(f)$ is strongly invariant.

Block and Coven [27] have given an example of a non-chaotic map f for which $\Lambda(f) \neq \overline{P}(f)$. Moreover $f|\overline{P}$ is not one-to-one, thus disproving an assertion of Šarkovskii [111]. By suitably modifying this example, one could obtain a map f with $\Lambda(f) = RR(f)$ but $\Omega(f) \neq \Lambda(f)$. Both these examples depend on elaborations of the construction used in the following example, due to Delahaye [60], of a map which is uniformly non-chaotic, but not strongly non-chaotic.

EXAMPLE 30 Let $I = [0,1]$ and let $f: I \to I$ be the continuous map defined by

$$f(0) = 2/3, \; f(1) = 0,$$
$$f(1 - 2/3^k) = 1/3^{k-1}, \; f(1 - 1/3^k) = 2/3^{k+1} \quad (k \geq 1),$$

and by linearity at intermediate points.

It is readily verified that

$$f(x) = x + 2/3 \quad \text{for } 0 \le x \le 1/3,$$
$$f(x) = 16/9 - 7x/3 \quad \text{for } 1/3 \le x \le 2/3,$$
$$f(x) = f(3x-2)/3 \quad \text{for } 2/3 \le x \le 1.$$

It follows that $f^2(x/3) = f(x)/3$ for $0 \le x \le 1$ and hence, for any $n \ge 1$,

(*) $$f^{2n}(x/3) = f^n(x)/3 \quad \text{for } 0 \le x \le 1.$$

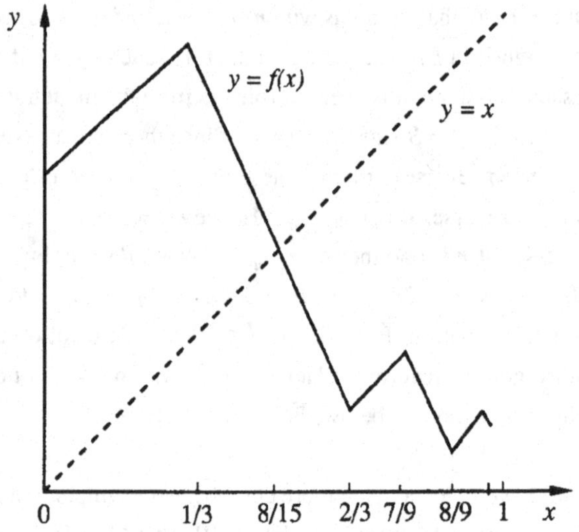

Fig. 1

From its graph (see Figure 1), the map f has the unique fixed point $\bar{x} = 8/15$. Also, it is easily seen that if $x \ne \bar{x}$ then $f^m(x) \in (0, 1/3)$ for some $m = m(x) \ge 0$. Hence any periodic orbit with period > 1 has a point in (0, 1/3). Since f maps the interval [0, 1/3] onto the interval [2/3, 1] and vice versa, it follows that the orbit must have even period. Again, by (*), $x/3$ is periodic if and only if x is periodic. Moreover, if x has period n then $x/3$ has period $2n$.

Suppose there exists a point y of period $n = 2^d q$, where $q > 1$ is odd and $d \ge 1$. We may assume that $y \in (0, 1/3)$. Then $x = 3y$ has period $n/2$. By repeating this argument we ultimately obtain a point of odd period q, which is a contradiction. Hence f is non-chaotic.

Similar reasoning shows that if there exists a unique orbit of period n then there exists a unique orbit of period $2n$. Hence f has a unique orbit of period 2^d for every $d \geq 0$.

Set $J_0^1 = [0, 1/3]$, $J_1^1 = [2/3, 1]$ and suppose we have defined 2^s pairwise disjoint closed intervals $J_i^s = [\alpha_i^s, \alpha_i^s + 1/3^s]$ $(0 \leq i < 2^s)$ of length $1/3^s$. Then we define 2^{s+1} pairwise disjoint closed intervals $J_i^{s+1} (0 \leq i < 2^{s+1})$ of length $1/3^{s+1}$ by setting

$$J_i^{s+1} = [\alpha_i^s, \alpha_i^s + 1/3^{s+1}], \quad J_{i+2s}^{s+1} = [\alpha_i^s + 2/3^{s+1}, \alpha_i^s + 1/3^s]$$

for $0 \leq i < 2^s$. In this way the intervals J_i^s $(0 \leq i < 2^s)$ are defined inductively for every positive integer s. The left endpoints α_i^s of the intervals J_i^s are actually the rational numbers $b_1/3 + b_2/3^2 + \ldots + b_s/3^s$ $(b_j = 0$ or $2)$. It follows from the definition of f that the 2^s intervals J_i^s are permuted cyclically by f. Hence the orbit of period $n = 2^s$ $(s > 0)$ is contained in $\bigcup_i J_i^s$.

For any $x_0 \in I$ with $x_0 \neq \bar{x}$ we can choose $m_0 \geq 0$ so that $x_1 = f^{m_0}(x_0) \in J_0^1$. If x_1 is not periodic, then neither is $3x_1$ and we can choose $m_1 \geq 0$ so that $f^{m_1}(3x_1) \in J_0^{\,1}$. Then $x_2 = f^{m_0+2m_1}(x_0) = f^{2m_1}(x_1) = f^{m_1}(3x_1)/3 \in J_0^{\,2}$. In this way we see that, for any $x \in I$, either x is eventually periodic or, for every positive integer $s, f^n(x) \in \bigcup_i J_i^s$ for all large n. It follows that x is approximately periodic, and hence that the map f is uniformly non-chaotic.

The first 2^s points in the trajectory of $x = 0$ are the numbers α_i^s $(0 \leq i < 2^s)$. Hence $0 \in R(f) \setminus P(f)$ and the intervals J_i^s defined above are precisely the intervals associated with the limit set $\omega(0,f)$. Since the point 0 is not asymptotically periodic, the map f is not strongly non-chaotic.

The classical Cantor set C consists of all real numbers in I whose ternary expansions contain only 0's and 2's (including $1 = 0.222\ldots$). It is readily seen that if $J_{a_1}^1 \supset J_{a_2}^2 \supset \ldots$ is a nested sequence of intervals J_i^s then $\bigcap J_{a_s}^s = \{y\}$, where y is a point of C. Moreover every point of C is obtained in this way. Hence $\omega(0,f) = C$, and in fact $\omega(x,f) = C$ for every x which is not eventually periodic. Thus C is a minimal set and

$$R(f) = \Lambda(f) = P(f) \cup C.$$

Since each point of C is regularly recurrent, by Lemma 18, it follows from Theorem 20 that $f|C$ is topologically conjugate to the adding machine transformation τ. Explicitly, if $x \in C$ has the ternary expansion $x = \sum_{i=0}^{\infty} 2b_i/3^{i+1}$ then $f(x)$ has the ternary expansion $f(x) = \sum_{i=0}^{\infty} 2c_i/3^{i+1}$, determined by the relation $\gamma = \beta+1$ between the 2-adic integers $\beta = (b_0, b_1, \ldots)$ and $\gamma = (c_0, c_1, \ldots)$.

4 CHAIN RECURRENCE FOR NON-CHAOTIC MAPS

We propose to study next the chain recurrent points of non-chaotic maps. In particular we will show that if f is non-chaotic and $x \in \mathrm{CR}(f) \setminus \mathrm{P}(f)$, then $\omega(x,f)$ is an infinite minimal set.

†LEMMA 31 *Let A and B be closed subsets of the compact metric space X such that $A \cup B = X$. If $A \cap B \cap \mathrm{CR}(f) \neq \emptyset$, then $f(A) \cap B \neq \emptyset$ and likewise $f(B) \cap A \neq \emptyset$.*

Proof It is sufficient to prove that $f(A) \cap B \neq \emptyset$. Let $x \in A \cap B \cap \mathrm{CR}(f)$. Then for any $\varepsilon > 0$ there exist points $x_0, x_1, \ldots, x_n$ with $x_0 = x_n = x$ and $d(f(x_{k-1}), x_k) < \varepsilon$ for $k = 1, \ldots, n$. Since $x_0 \in A$, $x_n \in B$ and $A \cup B = X$, we have $x_{k-1} \in A$, $x_k \in B$ for some k with $1 \leq k \leq n$. Hence

$$d[f(A), B] := \inf_{y \in A,\, z \in B} d[f(y), z] < \varepsilon .$$

Since ε was arbitrary, it follows that $d[f(A), B] = 0$. Since $f(A)$ and B are compact sets, this implies $f(A) \cap B \neq \emptyset$. □

LEMMA 32 *Suppose $x \in \mathrm{CR}(f)$ and $f^2(x) \neq x$.*
If $f(x) > x$ there exists a point $y \in I$ such that $x = f^2(y) < y < f(y)$.
If $f(x) < x$ there exists a point $y \in I$ such that $x = f^2(y) > y > f(y)$.

Proof We restrict attention to the case $f(x) > x$ and $I = [0,1]$. Since $[0,x] \cup [x,1] = I$ and $x \in [0,x] \cap [x,1] \cap \mathrm{CR}(f)$, we have $[0,x] \cap f[x,1] \neq \emptyset$, by Lemma 31. Since $f(x) > x$ it follows that $f(z) = x$ for some $z \in (x,1]$. If z_0 is the infimum of all $z \in (x,1]$ such that $f(z) = x$ then $f(z_0) = x$ and $z_0 \in (x,1]$. Put

$$A = [x,z_0], \quad B = [0,x] \cup [z_0, 1] .$$

Then $A \cup B = I$ and $x \in A \cap B \cap \mathrm{CR}(f)$. Since $\mathrm{CR}(f) = \mathrm{CR}(f^2)$ we have $f^2(A) \cap B \neq \emptyset$, by Lemma 31 again.

We will show that $z_0 \in f(A)$. If this is not the case then, since $f(A)$ is connected and $f(z_0) = x \in [0,z_0)$, we have $f(A) \subseteq [0,z_0)$. But $x \notin f[x,z_0)$, by the definition of z_0. Since $f[x,z_0)$ is connected and $f(x) > x$, we have $f[x,z_0) \subseteq (x,1]$. Thus

$$f[x,z_0) \subseteq [0,z_0) \cap (x,1] = (x,z_0)$$

and $f(A) \subseteq [x,z_0)$. Hence

$$f^2(A) \subseteq f[x,z_0) \subseteq (x, z_0),$$

which contradicts $f^2(A) \cap B \neq \varnothing$.

Thus $f(y) = z_0$ for some $y \in A$. Evidently $y \neq z_0$. Since x is not of period 2, also $y \neq x$. Thus

$$x = f^2(y) < y < f(y) = z_0 . \quad \square$$

PROPOSITION 33 *Suppose f^2 is not turbulent and $c \in$ CR(f).*
If $f(c) > c$, then $f(x) > x$ for $\inf f^{2k}(c) \le x \le \sup f^{2k}(c)$.
If $f(c) < c$, then $f(x) < x$ for $\inf f^{2k}(c) \le x \le \sup f^{2k}(c)$.

Proof It will be sufficient to consider the case $f(c) > c$. Moreover we may assume $f^2(c) \neq c$, since otherwise there is nothing to prove. Then, by Lemma 32, there exists a point d such that

$$c = f^2(d) < d < f(d).$$

By Theorem II.12 the orbit of d, and hence also the orbit of c, is alternating. Thus

$$f^{2j}(c) < f^{2k+1}(d) \quad \text{for all } j, k \ge 0.$$

On the other hand, the orbit of d is not bimonotonic. Hence the limit set $\omega(c,f)$ cannot contain a fixed point of f, by Proposition 3.

We claim that if $0 \le j < k$ then $f(x) > x$ for all $x \in \langle f^{2j}(c), f^{2k}(c)\rangle$. Suppose first that $f^{2k}(c) < f^{2j}(c)$. Then $f(x) > x$ for $f^{2k}(c) \le x \le f^{2j}(c)$ by Lemma II.7, with c replaced by $f^{2j}(c)$. Suppose next that $f^{2j}(c) < f^{2k}(c)$. By Lemma 32 there exists a point y such that

$$f^{2j}(c) = f^2(y) < y < f(y).$$

Hence $f(x) > x$ for $f^{2j}(c) \le x \le y$, by Lemma II.7. Thus it only remains to consider the case where $f^{2k}(c) > y$. If we put $g = f^2$ and $m = k - j$ then

$$f^{2j}(c) = g(y) < y < g^{m+1}(y) = f^{2k}(c).$$

Hence $g(x) < x$ for $y \le x \le f^{2k}(c)$ by Lemma II.7, with f replaced by g. Consequently f has no fixed points in $[y, f^{2k}(c)]$. Since $f(y) > y$ it follows that $f(x) > x$ for $y \le x \le f^{2k}(c)$. Thus our claim is now completely established.

Put

$$\alpha_0 = \inf f^{2k}(c), \quad \beta_0 = \sup f^{2k}(c) .$$

If $\alpha_0 \notin \gamma(c)$ then $\alpha_0 \in \omega(c)$. Thus α_0 is not a fixed point of f and hence $f(\alpha_0) > \alpha_0$. Similarly if $\beta_0 \notin \gamma(c)$ then $\beta_0 \in \omega(c)$ and $f(\beta_0) > \beta_0$. The result follows. □

THEOREM 34 *Suppose f is non-chaotic. If $x \in \mathrm{CR}(f) \setminus \mathrm{P}(f)$, then $\omega(x,f)$ is an infinite minimal set.*

Proof Let m be any positive integer and put $g = f^m$. Then $x \in \mathrm{CR}(g) \setminus \mathrm{P}(g)$, since $\mathrm{CR}(f) = \mathrm{CR}(g)$ and $\mathrm{P}(f) = \mathrm{P}(g)$. Without loss of generality assume $g(x) > x$. Then, by Lemma 32, there exists a point y such that

$$x = g^2(y) < y < g(y) .$$

Then the g-orbit of x is alternating, by Theorem II.12. Since the g-orbit of y is not bimonotonic it cannot converge to a fixed point of g, by Proposition 3. This already proves that x is not asymptotically periodic and hence that $\omega(x,f)$ is infinite. Thus

$$J_0 = [\inf g^{2k}(x), \sup g^{2k}(x)], \ J_1 = [\inf g^{2k+1}(x), \sup g^{2k+1}(x)]$$

are non-degenerate intervals and, by Proposition 33, contain no fixed point of g.

To show that $\omega(x,f)$ is a minimal set it is sufficient to prove that every point $y \in \omega(x,f)$ is recurrent, by Proposition 7.

Let $y \in \omega(x,f)$. If $y = f^m(x)$ for some $m \geq 0$ then $\omega(y,f) = \omega(x,f)$ and hence $y \in \mathrm{R}(f)$. Consequently we may suppose $y \notin \gamma(x)$. Without loss of generality assume that every open interval G with right endpoint y contains a point of $\gamma(x)$. Then there exist points $y_i = f^{n_i}(x) \in G$ $(i = 1, 2, 3)$, where

$$y_1 < y_2 < y_3 \quad \text{and} \quad n_1 < n_2 < n_3 .$$

Hence $f^{m_1}(y_1) = y_2$ and $f^{m_2}(y_2) = y_3$, where $m_1 = n_2 - n_1 > 0$ and $m_2 = n_3 - n_2 > 0$. Since y_1 and y_2 are points of $\mathrm{CR}(f) \setminus \mathrm{P}(f)$, it follows as in the beginning of the proof that the orbits of these points under any iterate of f are alternating. Hence the interval (y_1, y_2) contains a point z_1 such that $f^{m_1}(z_1) = z_1$ and the interval (y_2, y_3) contains a point z_2 such that $f^{m_2}(z_2) = z_2$. If we put $m = m_1 m_2$ and $g = f^m$ then, by Proposition 33, the g^2–orbit of y_2 is contained in a closed interval which contains no fixed point of g. Consequently the g^2–orbit of y_2 is entirely contained in (z_1, z_2). Since $\omega(y_2, f) = \omega(x,f)$ we have $y \in \omega(f^j(y_2), g^2)$ for some j with $0 \leq j < 2m$. Then

$$f^{2m-j}(y) \in \omega(f^{2m}(y_2), g^2) = \omega(y_2, g^2).$$

Hence the interval G contains a point in the f-orbit of y. Since G was arbitrary, it follows that $y \in R(f)$. □

It was pointed out in Chapter III that the definition of a homoclinic point adopted there was more restrictive than the original definition of Poincaré. We will say that a point x is *homoclinic in the sense of Poincaré* if there exists a periodic point z, not containing x in its orbit, such that $x \in W(z,f)$ and $z \in \omega(x,f)$.

Clearly any homoclinic point, as previously defined, is also homoclinic in the sense of Poincaré. Therefore, by Proposition III.21, any chaotic map has a point which is homoclinic in the sense of Poincaré. As an application of Theorem 34 we now show that the converse is also true. At the same time we establish some other equivalents of chaos.

PROPOSITION 35 *The following statements are equivalent:*

(i) *f is chaotic,*
(ii) *f has a nonwandering point which is also a homoclinic point,*
(iii) *f has a nonwandering point which has a finite orbit but is not periodic,*
(iv) *there exists a point which is homoclinic in the sense of Poincaré.*

Proof It follows immediately from the definitions that (ii) $\Rightarrow$ (iii). We now show that (iii) $\Rightarrow$ (iv). Let x be a non-periodic nonwandering point with a finite orbit. If z is a periodic point in the orbit of x, then clearly $z \in \omega(x,f)$. Since x is not periodic, it follows from Proposition III.15 that $x \notin \overline{W(z,f)} \setminus W(z,f)$. We claim that $x \in W(z,f)$. Suppose the claim is false. Then, by Proposition III.14, there is an open interval G containing z such that, if $H = \bigcup_{n\geq 0} f^n(G)$, then $x \notin \overline{H}$. Hence there is an open interval V containing x such that $\overline{V} \cap \overline{H} = \emptyset$. Finally, there is an open interval V_1 containing x with $V_1 \subseteq V$ such that, if m is the least positive integer with $f^m(x) = z$, then $f^m(V_1) \subseteq G$ and $f^k(V_1) \cap V_1 = \emptyset$ for $k = 1, \ldots, m-1$. It follows that $f^k(V_1) \cap V_1 = \emptyset$ for every positive integer k. This contradicts $x \in \Omega(f)$. Thus our claim that $x \in W(z,f)$ is established. By definition, x is homoclinic in the sense of Poincaré.

We show next that (iv) $\Rightarrow$ (i). By hypothesis, there exists a periodic point z and a point x not in the orbit of z such that $x \in W(z,f)$ and $z \in \omega(x,f)$. Then $x \in CR(f)$, by Corollary V.41. On the other hand $x \notin P(f)$, since x is not in the orbit of z but $z \in \omega(x,f)$. It now follows from Theorem 34 that f is chaotic.

Finally we show that (i) $\Rightarrow$ (ii). Suppose that f is chaotic. Then f^n is turbulent for some positive integer n. By Lemma II.2, we may assume that there exist points a,b,c such that $f^n(b) = f^n(a) = a$, $f^n(c) = b$ and

$$a < c < b,$$
$$f^n(x) > a \text{ for } a < x < b,$$
$$x < f^n(x) < b \text{ for } a < x < c.$$

Then $c \in W(a, f^n, R)$, by Proposition III.4. Since $f^{2n}(c) = a$, it follows that c is a homoclinic point. On the other hand, if V is any open interval containing c then $f^n(V)$ contains an interval $[b-\delta, b]$ for some $\delta>0$, and $f^{2n}(V)$ contains an interval $[a, a+\varepsilon]$ for some $\varepsilon > 0$. Since $c \in W(a, f^n, R)$, it follows that $c \in \Omega(f^n) \subseteq \Omega(f)$. □

As another application of Theorem 34 we can obtain a new characterization of strongly non-chaotic maps.

PROPOSITION 36 *The following statements are equivalent:*

(i) *f is strongly non-chaotic,*
(ii) CR(f) = P(f),
(iii) RR(f) = P(f) .

Proof It follows at once from Proposition 10 that (ii) $\Rightarrow$ (i) and (i) $\Rightarrow$ (iii).

Suppose f is strongly non-chaotic. Then $\omega(x,f)$ is finite for every $x \in I$ and hence, by Theorem 34, CR(f) = P(f).

Suppose finally that RR(f) = P(f). Then f is non-chaotic, by Proposition 9. Assume that there exists a point $x \in$ SR(f)\ P(f). Then $M = \omega(x,f)$ is an infinite minimal set. Thus M is closed and dense in itself, by Lemma V.4, and hence uncountably infinite (cf. Rudin [101], Theorem 2.43). But if $y \in M$ then $y \in$ SR(f)\ P(f). Since this contradicts Corollary 12, we conclude that SR(f) = P(f). Hence, by Proposition 10, f is strongly non-chaotic. □

5 TRANSITIVITY AND TOPOLOGICAL MIXING

In this section we consider two important concepts which are related to chaos. Although we will mainly be concerned with maps of an interval, the definitions and some basic equivalences will be given in a more general setting.

Let X be a compact metric space and $f: X \to X$ a continuous map of this space into itself. We say that f is *(topologically) transitive* if for every pair of non-empty open sets U and V in X, there is a positive integer k such that $f^k(U) \cap V \neq \emptyset$. The next two results follow easily from this definition.

†LEMMA 37 *If $f: X \to X$ is a continuous map, then the following statements are equivalent:*

(i) *f is transitive,*
(ii) *for every non-empty open set W in X, $\bigcup_{n=1}^{\infty} f^n(W)$ is dense in X,*
(iii) *for every pair of non-empty open sets U and V in X, there is a positive integer k such that $f^{-k}(U) \cap V \neq \emptyset$,*
(iv) *for every non-empty open set W in X, $\bigcup_{n=1}^{\infty} f^{-n}(W)$ is dense in X,*
(v) *every proper closed invariant subset of X has empty interior.*

†LEMMA 38 *If $f: X \to X$ is transitive, then $f(X) = X$ and $\Omega(f) = X$.*

It may be noted that the definition of transitivity, and Lemmas 37 and 38, remain valid if X is any topological space. The following useful characterization of transitivity lies somewhat deeper.

†PROPOSITION 39 *A continuous map $f: X \to X$ of a compact metric space into itself is transitive if and only if there exists a point $x \in X$ such that $\omega(x,f) = X$.*

Proof Suppose first that there is a point $x \in X$ such that $\omega(x,f) = X$. Let U and V be arbitrary non-empty open sets in X. Then $f^m(x) \in U$ for some positive integer m and $f^n(x) \in V$ for some integer $n > m$. Hence $f^{n-m}(U) \cap V \neq \emptyset$. Thus f is transitive.

Suppose next that f is transitive. For each positive integer n the space X is covered by finitely many open balls of radius $1/n$. The collection of such balls, for all n, can be enumerated as a sequence: $U_1, U_2, U_3, \ldots$. For each positive integer k, the set $G_k = \bigcup_{n=1}^{\infty} f^{-n}(U_k)$ is open and dense in X. Since X is certainly complete, it follows from the Baire Category Theorem, see e.g. Rudin [100], that there exists a point $x \in X$ which is contained in G_k for all k. Since the orbit of x intersects each U_k, this orbit is dense in X. By Lemma 38, there exists a point $y \in X$ such that $f(y) = x$. If $y \in \gamma(x,f)$ then x is periodic and $\omega(x,f) = X$. Otherwise $y \in \omega(x,f)$; then $x \in \omega(x,f)$ and again $\omega(x,f) = X$. □

COROLLARY 40 *The shift map* $\sigma: \Sigma \to \Sigma$ *is transitive.* □

It may be noted that, with the same hypotheses, Proposition 39 admits the alternative formulation: f is transitive if and only if f is 'onto' and there exists a point $x \in X$ whose orbit is dense in X.

A continuous map $f: X \to X$ is said to be *topologically mixing* if, for every pair of non-empty open sets U and V, there is a positive integer N such that $f^k(U) \cap V \neq \emptyset$ for all $k > N$.

Clearly if f is topologically mixing then it is also transitive. In the special case in which X is an interval there is an intimate connection between the two concepts, as we shall see. Thus we return to the case of a continuous map $f: I \to I$, where I is a compact interval.

LEMMA 41 *If* $f: I \to I$ *is transitive, then* $\overline{P}(f) = I$.

Proof In fact $\Omega(f) = I$, by Lemma 38, and hence $\overline{P}(f) = \Omega(f|\Omega(f)) = I$. □

PROPOSITION 42 *Let* $f: I \to I$ *be transitive and let* $x \in I$ *be such that* $\omega(x,f) = I$. *Then exactly one of the following alternatives holds*:

(i) $\omega(x,f^s) = I$ *for every positive integer* s,

(ii) *there exist non-degenerate closed intervals* J, K *with* $J \cup K = I$ *and* $J \cap K = \{y\}$, *where* y *is a fixed point of* f, *such that* $f(J) = K$ *and* $f(K) = J$.

Proof Fix an arbitrary positive integer s and set $B_i = \omega(f^i(x), f^s)$ for $0 \leq i < s$. Since $B_0 \cup ... \cup B_{s-1} = I$, at least one B_i has non-empty interior. Moreover, since the orbit of x cannot contain a periodic point, f cannot collapse an interval to a point. Since $f(B_i) = B_{i+1}$ for $0 \leq i \leq s-2$ and $f(B_{s-1}) = B_0$, it follows that each B_i has non-empty interior.

We claim that if the interiors of B_i and B_j intersect, then $B_i = B_j$. To see this, suppose that $\text{int}\, B_i \cap \text{int}\, B_j \neq \emptyset$. Then for some positive integer n, $f^{sn+i}(x) \in \text{int}\, B_i \cap \text{int}\, B_j$. It follows that $B_i \subseteq B_j$, since B_j is f^s-invariant and $B_i = \omega(f^{sn+i}(x), f^s)$. Since i and j can be interchanged in this argument we must actually have $B_i = B_j$.

Let **A** denote the collection of all subsets of I which are components of int B_i for some $i \in \{0, ..., s-1\}$. Thus **A** is a collection of disjoint open intervals whose union is dense in I. If $A_1 \in$ **A** then $f(\overline{A}_1) \subseteq \overline{A}_2$ for some $A_2 \in$ **A**, again because f cannot collapse an interval to a point. Moreover, since the orbit of x is dense in I, for every $A_2 \in$ **A** there is a positive integer k

such that $f^k(\bar{A}_1) \subseteq \bar{A}_2$. It follows that A is finite, say $A = \{A_1, \ldots, A_h\}$, and that the sets $C_i = \bar{A}_i$ $(i = 1, \ldots, h)$ are cyclically permuted by f.

If $h = 1$ then $B_i = I$ for $i = 0, \ldots, s-1$ and, in particular, $\omega(x,f^s) = I$. Suppose $h > 1$. We claim that $h = 2$. To see this, let y be a fixed point of f. If $y \in A_i$ for some i then $f(C_i) = C_i$, which is impossible. Similarly y cannot be an endpoint of I. The only other possibility is that y is a common endpoint of C_i and C_j, where $i \neq j$. Then $f(C_i) = C_j$ and $f(C_j) = C_i$, which is possible only if $h = 2$. Thus we have the situation in (ii), and hence f^2 is not transitive.

Since s was arbitrary, this proves that (i) holds if f^2 is transitive and (ii) holds if it is not.

□

COROLLARY 43 *Let $f: I \to I$ be a continuous map such that f^2 is transitive. Then f^s is transitive for every positive integer s.* □

PROPOSITION 44 *Let $f: I \to I$ be a continuous map. Then f^2 is transitive if and only if, for every open subinterval J and every closed subinterval H which does not contain an endpoint of I, there is a positive integer N such that $H \subseteq f^n(J)$ for every $n > N$.*

Proof Suppose first that f^2 is transitive. Then f is transitive and $\bar{P}(f) = I$. Thus the open interval J contains a periodic point x of f. Let y and z be any periodic points in the interior of I with $y < z$. Let m be a common multiple of the periods of x, y, and z and set $g = f^m$. Then $K = \bigcup_{n=1}^{\infty} g^n(J)$ is an interval, since x is a fixed point of g. On the other hand, g is transitive, by Corollary 43, and hence K is dense in I. It follows that K contains the interior of I. Thus $y \in g^i(J)$, $z \in g^j(J)$ for some positive integers i,j. Since y and z are fixed points of g, it follows that y and z both belong to $g^k(J)$, where $k = \max(i, j)$. Consequently the interval $f^{km}(J) = g^k(J)$ contains the interval $[y,z]$.

In particular, if y and z are the least and greatest points of the same periodic orbit π, then $f^{km}(J)$ contains all points of π and hence $f^n(J)$ contains the interval $[y,z]$ for all $n > km$. Now let π_1, π_2 be periodic orbits contained in the interior of I such that the least point y_1 of π_1 lies to the left of H and the greatest point z_2 of π_2 lies to the right of H. Then there exists a positive integer N such that y_1 and z_2 lie in $f^n(J)$ for all $n > N$. Hence $H \subset f^n(J)$ for all $n > N$.

Conversely, suppose that for each closed subinterval H in the interior of I and each open subinterval J there is a positive integer N such that $H \subset f^n(J)$ for all $n > N$. Since we can take n even, it follows at once from the definition that f^2 is transitive. □

For piecewise monotone maps, Proposition 44 can be strengthened in the following way.

PROPOSITION 45 *Let $f: I \to I$ be piecewise monotone. If f^2 is transitive, then for every interval $J \subseteq I$ there is a positive integer n such that $f^n(J) = I$.*

Proof Let $I = [a,b]$. Suppose first that there is a closed interval $K \subset (a,b)$ such that $f^2(K) = I$. If $J \subseteq I$ is any interval then, by Proposition 44, $K \subset f^n(J)$ for some positive integer n. It follows that $f^{n+2}(J) = I$, so that the proposition holds in this case. Thus we suppose from now on that $f^2(K) \neq I$ for every closed interval $K \subset (a,b)$.

Since f is onto, it follows that also $f(K) \neq I$ for every closed interval $K \subset (a,b)$. Hence either $f^{-1}(a) \subseteq \{a,b\}$ or $f^{-1}(b) \subseteq \{a,b\}$. Without loss of generality, assume that $f^{-1}(a) \subseteq \{a,b\}$.

We claim that $f^{-2}(a) = \{a\}$. Since this is certainly true if $f^{-1}(a) = \{a\}$, we may assume that $f^{-1}(a) \neq \{a\}$. Then $f(b) = a$. Moreover, since $f^{-1}(b) = a$ implies $f^{-2}(a) = \{a\}$, we may assume that $f(c) = b$ for some $c \in (a,b)$. Then $f(d) = c$ for some $d \in (c,b)$. If we put $K = [c,d]$, it follows that $f^2(K) = I$, which is a contradiction. This establishes our claim.

Let T denote the set of points in (a,b) which are endpoints of maximal intervals on which f^2 is monotonic. Then T is non-empty and finite. Let t denote the smallest element of T. If $f^2(p) = p$ for some $p \in (a,t)$ then, since f^2 is monotonic on $[a,p]$, we would have $f^2(a,p] = (a,p]$, contrary to the hypothesis that f^2 is transitive. Hence f^2 has no fixed points in (a,t). Similarily we obtain a contradiction if $f^2(x) < x$ for every $x \in (a,t)$. It follows that $f^2(x) > x$ for every $x \in (a,t)$.

Let s denote the minimum value of $f^2(x)$ for $x \in [t,b]$. Since $f^2(t,b]$ is not a subset of $[t,b]$ and $f^{-2}(a) = \{a\}$, we must have $a < s < t$. Since $f^2(x) > x$ for $x \in (a,t)$, it follows that $f^2(s,b] \subseteq [s,b]$. This again contradicts the hypothesis that f^2 is transitive. □

We can now easily derive the following theorem, which summarizes the connection between transitivity and topological mixing for maps of an interval.

THEOREM 46 *For any continuous map $f: I \to I$ the following statements are equivalent:*

(i) *f is transitive and has a periodic point of odd period greater than one,*
(ii) *f^2 is transitive,*
(iii) *f is topologically mixing.*

Proof It follows immediately from the definitions that (iii) $\Rightarrow$ (ii), and it follows from Proposition 44 that (ii) $\Rightarrow$ (iii).

If f is transitive, but f^2 is not transitive, then Proposition 42 implies that f has no periodic point of odd period greater that one. Thus (i) $\Rightarrow$ (ii).

Finally, suppose that (ii) holds. Let J and K be disjoint closed subintervals of the interior of I. By Proposition 44, there is a positive integer N such that $J \cup K \subset f^n(J) \cap f^n(K)$ for all $n > N$. Thus f^n is strictly turbulent. If we choose n odd then f has a periodic point of period n, by Theorem II.14. Thus (ii) $\Rightarrow$ (i), and the periodic points of odd period are actually *dense* in I. □

It follows directly from Theorem 46 that in case (i) of Proposition 42 *f is topologically mixing*, whereas in case (ii) *$f^2|J$ and $f^2|K$ are topologically mixing*. In connection with Theorem 46, it is also worth noting that, by Corollary 4, f^2 is turbulent for *any* transitive map $f: I \to I$.

Examples of topologically mixing maps are easily constructed. Let $I = [0,1]$ and let f be the piecewise linear map defined by

$$f(0) = 0, \; f(1/2) = 1, \; f(1) = 0.$$

Then f is topologically mixing, since any open interval J contains a subinterval $[k/2^m, (k+1)/2^m]$ and hence $f^m(J) = I$.

Similarly it may be seen that the map f of Example I.12 is topologically mixing. This shows that in Theorem 46 (i) the *least* odd period greater than one can be arbitrary.

Our next objective is to characterize continuous maps of an interval for which every point is non-wandering. It is convenient to introduce the following definition.

A continuous map $f: I \to I$ will be said to *separate* I if there are closed connected subsets L, C and R, with pairwise disjoint interiors, such that L and R are intervals, $I = L \cup C \cup R$, and $f(L) \subseteq R, f(C) = C, f(R) \subseteq L$. It is easy to see that if f separates $I = L \cup C \cup R$, then C lies between L and R.

LEMMA 47 *If f maps I onto I, then f^2 does not separate I.*

Proof Suppose that f^2 separates I and let $I = L \cup C \cup R$, where L, C and R are as in the definition. If $I = [a, b]$, we may assume that $a \in L$ and $b \in R$. Let $z \in C$ be a fixed point of f. Since f is onto, there are points a_1 and b_1 in I with $f(a_1) = a$ and $f(b_1) = b$. Then $a_1 > z$, since otherwise $a_1 \in f\,[a_1, z]$ and hence

$$a \in f^2[a_1, z] \subseteq f^2(L \cup C) = C \cup R.$$

Since L is nondegenerate, this is a contradiction. Similarly, $b_1 < z$. It follows that

$$b_1 \in [a, z] \subseteq f[z, a_1],$$

and hence

$$b \in f^2[z, a_1] \subseteq f^2(C \cup R) = L \cup C.$$

Since R is nondegenerate, this is again a contradiction. □

THEOREM 48 *Let $f: I \to I$ be a continuous map such that $\Omega(f) = I$. If f does not separate I, then I admits a decomposition*

$$I = K \cup J_1 \cup J_2 \cup \dots,$$

where the terms of the decomposition have pairwise disjoint interiors, K is a closed set of fixed points of f and each J_n is a nondegenerate closed invariant subinterval of I such that $f|J_n$ is topologically mixing. The collection $\{J_1, J_2, \dots\}$ is countable, finite or empty.

Even if f separates I, f^2 does not, and the preceding conclusions hold with f replaced by f^2.

Proof Since $\Omega(f) = I$ implies $\overline{P}(f) = \Omega(f|\Omega(f)) = I$, it is clear that f maps I onto I. Hence, by Lemma 47, f^2 does not separate I. Moreover $\Omega(f^2) = I$, since $\overline{P}(f) = \overline{P}(f^2)$. Consequently the second assertion of the theorem follows at once from the first.

Suppose now that f does not separate I. Let $I = [a,b]$ and, for any nondegenerate subinterval J of I, let J^* denote the closure of the set $\bigcup_{k=0}^{\infty} f^k(J)$. We claim that J^* is a (closed) subinterval. To show this we first note that J^* has finitely many components, since $\Omega(f) = I$. Each component is a nondegenerate closed interval, and the components are cyclically permuted by f.

Assume that J^* has at least three components. Then there are two adjacent components whose images under f are not adjacent. Let G be the smallest closed interval containing the two adjacent components, and let H be the smallest closed interval containing their images. Then $f(G) \supseteq H$ and H contains a component E of J^* which is disjoint from the images of the adjacent components in G. Then $f(D) = E$ for some nondegenerate closed interval $D \subseteq G$. Since $D \cap J^* = \varnothing$, but $f^n(D) \subseteq J^*$ for all positive integers n, this contradicts $\Omega(f) = I$.

Hence, if J^* is not an interval, it is the union of two disjoint intervals $J_1 = [a_1, b_1]$ and $J_2 = [a_2, b_2]$. We may assume that $b_1 < a_2$. Since $\Omega(f) = I$, a point in the open interval (b_1, a_2) cannot be mapped by f into a point in (a_1, b_1) or (a_2, b_2). Hence $f[b_1, a_2] = [b_1, a_2]$. Similarly a point in (a, a_1) or (b_2, b) cannot be mapped into a point in (b_1, a_2). It follows that f separates $I = [a, b_1] \cup [b_1, a_2] \cup [a_2, b]$. This contradiction establishes the claim that J^* is a closed interval.

Next we make the following claim:

(#) If J is a closed invariant subinterval of I, then any endpoint c of J which is interior to I is a fixed point of f.

To prove this we suppose (without loss of generality) that c is the left endpoint of J, and we assume that $f(c) > c$. Then $[a,c]^*$ is an interval containing the points c and $f(c)$, and so $[a,c]^*$ contains a point of int J. Hence, for some $x < c$ and some positive integer $k, f^k(x) \in \text{int } J$. Since $x \in \Omega(f)$, this is a contradiction. This proves (#).

If K is the set of fixed points of f, then $I \setminus K$ is the union of at most countably many pairwise disjoint open intervals $G_1, G_2, \ldots$. We claim that $f|G_n^*$ is transitive for every n. To prove this, assume that $f|G_n^*$ is not transitive for some n. Then G_n^* contains a proper closed invariant set with nonempty interior. Let J be a nondegenerate component of that set, and let $H = J^*$. Then H is a proper closed invariant subinterval of G_n^*. It follows that at least one endpoint of H is interior to G_n^*, and hence to I. By (#), that endpoint is a fixed point of f. Furthermore, the other endpoint of H is either a fixed point of f or an endpoint of I. Since $G_n \subseteq I \setminus K$, it follows that either $G_n \subset H$ or $G_n \cap H = \emptyset$. In the first case $G_n^* \subseteq H$, a contradiction. In the second case there is a point $x \in G_n$ with $f^k(x) \in \text{int } H$ for some $k > 0$, which contradicts $x \in \Omega(f)$. This establishes our claim that each $f|G_n^*$ is transitive.

Let $J_1, J_2, \ldots$ be the distinct sets in the collection $\{G_n^*\}$. Since $f|J_n$ is transitive for each positive integer n, it follows that the intervals $J_1, J_2, \ldots$ have pairwise disjoint interiors. If we redefine K to be the closure of the complement of $\bigcup J_n$ in I, then every point of K is a fixed point of f. By Theorem 46, to complete the proof we need to show that $f^2|J_n$ is transitive for each positive integer n.

Fix a positive integer n, and let $J = J_n$. Assume $f^2|J$ is not transitive. Since $f|J$ is transitive, it follows from Proposition 42 that $J = L \cup R$, where L and R are nondegenerate closed intervals with exactly one common point such that $f(L) = R$ and $f(R) = L$. In particular, neither endpoint of J is a fixed point of f. If $J \neq I$ then at least one endpoint of J is interior to I, which contradicts (#). Hence $J = I$. It follows that f separates $I = L \cup C \cup R$, where $C = L \cap R$. Thus we again have a contradiction. □

It will be shown next that Theorem 48 is sharp in some sense.

LEMMA 49 *For any interval $I = [a,b]$, there is a continuous map $f: I \to I$ with $f(a) = a$ and $f(b) = b$ such that f is topologically mixing.*

Proof Choose a_0, a_1, a_2, a_3 so that $a_0 = a$, $a_3 = b$, $a_0 < a_1 < a_2 < a_3$, and the intervals $[a_0, a_1]$, $[a_1, a_2]$, $[a_2, a_3]$ have equal length. Let f be the piecewise linear function defined by

$$f(a_0) = a_0, \; f(a_1) = a_3, \; f(a_2) = a_0, \; f(a_3) = a_3 .$$

For any subinterval K of an interval $[a_i, a_{i+1}]$, the length of $f(K)$ is three times the length of K. It follows that $f^n(K) = I$ for some positive integer n. Hence f is topologically mixing. □

PROPOSITION 50 *Let I be a compact interval and let $\{J_1, J_2, \ldots\}$ be any countable, finite or empty collection of closed nondegenerate subintervals of I with pairwise disjoint interiors. Then there is a continuous map $f: I \to I$ with $\Omega(f) = I$ such that, for each n, $f(J_n) = J_n$ and $f|J_n$ is topologically mixing. Moreover every point of $I \setminus \bigcup J_n$ is a fixed point of f.*

Proof Define f on each J_n as in Lemma 49 and set $f(x) = x$ for all other $x \in I$. □

The scope of Theorem 48 is considerably extended by the following simple remark.

LEMMA 51 *Let $f: I \to I$ be a continuous map and let J be a nondegenerate component of $\Omega(f)$. Then, for some positive integer n, J is invariant under f^n and $J \subseteq \Omega(f^n)$, so that Theorem 48 holds for $f^n|J$.*

Proof Let n be the least positive integer such that $f^n(J) \cap J \neq \emptyset$. Since $f^n(J) \subseteq \Omega(f)$, it follows that $f^n(J) \subseteq J$. Moreover $J \subseteq \Omega(f^n)$, since $f^k(J) \cap J \neq \emptyset$ only if k is a multiple of n. □

We conclude this discussion with a characterization of chaos in terms of topological mixing.

PROPOSITION 52 *A map f is chaotic if and only if there exists a positive integer n and an infinite closed set X, such that X is invariant under f^n and the restriction of f^n to X is topologically mixing.*

Proof Suppose first that f is not chaotic, and assume that for some $n > 0$ there exists an infinite closed set X such that X is invariant under $g = f^n$ and $g|X$ is topologically mixing. Then g is not chaotic and $X = \omega(x,g)$ for some $x \in X$, by Proposition 39. Let $X_0 = X \cap J_0^1$ and

$X_1 = X \cap J_1^1$, where J_0^1 and J_1^1 are intervals associated with $\omega(x,g)$ as in Lemma 14. Then $g(X_0) = X_1$, $g(X_1) = X_0$ and X_0, X_1 are disjoint closed subsets of X whose union is X. It follows that X_0 and X_1 are open as subsets of the metric space X. Since $g^{2k}(X_0) = X_0$ for every positive integer k, this contradicts the assumption that $g|X$ is topologically mixing.

Suppose next that f is chaotic. By replacing f by some iterate f^n we may assume that f is strictly turbulent. We now use the set-up of Proposition II.15.

Let Y denote the set of points $x \in X$ such that, if $x \in I_\alpha$ where $\alpha \in \Sigma$, then $I_\alpha = \{x\}$. Then Y contains all but countably many points of X, and $f(Y) \subseteq Y$. Let $Z = \overline{Y}$. Then Z is a closed, invariant subset of X. We will show that $f|Z$ is topologically mixing. It is sufficient to establish the following two claims:

(i) any non-empty open subset of Z contains a set of the form $Z \cap I_{a_1 \dots a_k}$ for some $k \geq 1$ and some $a_i \in \{0, 1\}$ $(i = 1,\dots,k)$,

(ii) $f^k(Z \cap I_{a_1 \dots a_k}) = Z$ for every $k \geq 1$ and every $a_i \in \{0, 1\}$ $(i = 1,\dots,k)$.

For, by (ii) with $k = 1$, we have $f(Z) = Z$. Let V be an arbitrary open subset of Z. It follows from (i) and (ii) that $f^k(V) = Z$ for some positive integer k. In particular, $f|Z$ is topologically mixing.

We now prove (i). Let V be any non-empty open subset of Z. Then there is an open subset J of I such that $V = Z \cap J$. It follows that there is an element x of Y in J. Let $\alpha = (a_1, a_2, \dots)$ be the sequence of 0's and 1's such that $I_\alpha = \{x\}$. Then, for k sufficiently large, $I_{a_1 \dots a_k} \subseteq J$ and hence $Z \cap I_{a_1 \dots a_k} \subseteq Z \cap J = V$. This establishes (i).

Finally, we prove (ii). Let $K = I_{a_1 \dots a_k}$ and let $y \in Y$. Then $I_\beta = \{y\}$ for some sequence $\beta = (b_1, b_2, \dots)$ of 0's and 1's. If $\alpha = (a_1, \dots, a_k, b_1, b_2, \dots)$, then $f^k(I_\alpha) = I_\beta$. Let W be an open subset of I which contains I_α. Then $I_{a_1 \dots a_k b_1 b_2 \dots b_n} \subseteq W$ for n sufficiently large. But any interval $I_{c_1 \dots c_j}$ contains uncountably many points of X and hence also of Y. Thus W contains a point of Y. Since W was arbitrary, it follows that if I_α is not a point then at least one of its endpoints is in $Z = \overline{Y}$. Thus, whether or not I_α is a single point, there is a point $x \in Z \cap I_\alpha$ such that $f^k(x) = y$. Since $Z \cap I_\alpha \subseteq Z \cap I_{a_1 \dots a_k}$, this shows that $f^k(Z \cap I_{a_1 \dots a_k}) \supseteq Y$. Since $Z = \overline{Y}$ is invariant, we must have $f^k(Z \cap I_{a_1 \dots a_k}) = Z$. □

6 PIECEWISE MONOTONE MAPS

Some important additional results hold for maps which are not only non-chaotic, but also piecewise monotone.

PROPOSITION 53 *If f is piecewise monotone and non-chaotic, then* $\Lambda(f) = R(f)$.

Proof By Proposition IV.22, $\Lambda(f) = \overline{P}(f)$. We will assume that there exists a point $x \in \Lambda(f) \setminus R(f)$ and derive a contradiction. By Proposition 11, x is not a two-sided limit of periodic points. Without loss of generality, assume that x is a left limit of periodic points. Then we can choose $\varepsilon > 0$ so small that $f^n(x) \notin [x-\varepsilon, x+\varepsilon]$ for all $n > 0$ and $[x, x+\varepsilon] \cap P(f) = \emptyset$.

If $z \in (x-\varepsilon, x)$ is periodic with period $N = 2^s$ then, by Lemma 14, $f^{2kN}(x) > x+\varepsilon$ for all $k > 0$. Since f^{2N} is piecewise monotone and non-chaotic, and since $\Lambda(f) = \Lambda(f^{2N})$, $R(f) = R(f^{2N})$, $P(f) = P(f^{2N})$, we may simply replace f by f^{2N}. Then $f^k(x) > x+\varepsilon$ for all $k > 0$.

There exists a point $y \in I$ and an increasing sequence (n_k) of positive integers such that $f^{n_k}(y) \to x$. We may suppose that $f^{n_1}(y) \in (x-\varepsilon, x+\varepsilon)$ and that the sequence $f^{n_k}(y)$ converges to x monotonically. Then

$$x-\varepsilon < f^{n_1}(y) < f^{n_2}(y) < \ldots < x,$$

since $[x, x+\varepsilon] \cap P(f) = \emptyset$.

Denote by $E = E_0$ the set of all points $v \in I$ such that $f^j(v) = y$ for some $j > 0$. We consider first the case where x is a left accumulation point of E. Then we may assume that each interval $(f^{n_k}(y), f^{n_{k+1}}(y))$ contains at least two distinct points of E. Let C_n be the connected component of $I \setminus E$ which contains $f^n(y)$. Then $f^n(C_0) \subseteq C_n$, since $f^n(C_0)$ is a connected subset of $I \setminus E$ containing $f^n(y)$, and hence $f^n(\overline{C}_0) \subseteq \overline{C}_n$. We will show that $\overline{C}_0 \cap \overline{P} \neq \emptyset$.

Obviously we may assume that $\overline{C}_0 \cap P = \emptyset$. Then $\overline{C}_0$ is contained in some connected component D of $I \setminus P$, and it is sufficient to show that $\overline{C}_0$ and D have at least one common endpoint. Let a and b be the left and right endpoints of C_0, so that $a \leq y \leq b$ (and possibly $a = b$), and let a' and b' be the left and right endpoints of D. If $a' < a$ and $b < b'$, then $(a', a]$ and $[b, b')$ contain points of E. Hence the interval (a', b') is neither of increasing type nor of decreasing type. Since $f^k(a', b') \cap (a', b') \neq \emptyset$ for some $k > 0$, this is a contradiction. Similarly a' and a cannot both be the left endpoint of I, and b' and b cannot both be the right endpoint of I.

Hence there exists a point $u \in \overline{C}_0 \cap \overline{P}$. Then $f^{n_k}(u) \in \overline{C}_{n_k}$. Since the intervals $\overline{C}_{n_k}$ are pairwise disjoint, the length ℓ_k of $\overline{C}_{n_k}$ tends to zero as $k \to \infty$. It follows that $f^{n_k}(u) \to x$, and thus $x \in \omega(u,f)$. Since $u \in \Lambda(f)$, this implies $x \in R(f)$, by Proposition 7. Again we have a contradiction.

We obtain a contradiction in the same way if x is a left accumulation point of the set E_k of all points $v \in I$ such that $f^j(v) = f^{n_k}(y)$ for some $j > 0$. It remains to consider the case where, for every $k \geq 0$, x is not a left accumulation point of E_k.

Then, for each $k \geq 0$, we can choose $z_k \in (f^{n_k}(y), x)$ so that $f^j(t) \neq f^{n_k}(y)$ for all $t \in [z_k, x]$ and all $j > 0$. Since $x \in \Omega(f)$ and $[x, x+\varepsilon] \cap P(f) = \emptyset$, there exists a point $w_k \in (z_k, x)$ and an integer $m_k > 0$ such that $f^{m_k}(w_k) = x$. Moreover we may choose w_k so that $f^{m_k}(t) > x$ for all $t \in (w_k, x]$. Since $f^{n_i}(y) \in (w_k, x)$ for all large i, we can choose $v_k \in (w_k, x)$ and $p_k > m_k$ so that $v_k < f^{p_k}(v_k) < x$. Then f^{p_k} takes its minimum value in the interval $[w_k, x]$ at an interior point u_k, since $f^{p_k}(w_k) = f^{p_k - m_k}(x) > x$. If the minimum is assumed at more than one point, we choose u_k to be the nearest such point to x. Then $f^{n_k}(y) < f^{p_k}(u_k) < x$ and $c_k = f^{j_k}(u_k)$ is a turning-point of f for some j_k with $0 \leq j_k < p_k$.

Since f has only finitely many turning-points, we can pick a subsequence so that $c_k = c$ for all k. Evidently $u_k \to x$ and $p_k \to \infty$. Moreover $f^{p_k - j_k}(c) < x$ and $f^{p_k - j_k}(c) \to x$. Hence $p_k - j_k \to \infty$ and $x \in \omega(c,f)$. Since x is now a left accumulation point of the set $E(c)$ of all points $v \in I$ such that $f^j(v) = c$ for some $j > 0$, this yields a contradiction in the same way as before. □

PROPOSITION 54 *If f is piecewise monotone and non-chaotic, then*

$$\Lambda(f) = P(f) \cup \omega(c_1, f) \cup \ldots \cup \omega(c_m, f),$$

where $c_1, \ldots, c_m$ are the turning-points of f.

Proof Assume, on the contrary, that there exists a point $x \in \Lambda(f) \setminus P(f)$ such that $x \notin \omega(c_i, f)$ for $1 \leq i \leq m$. Since $x \in R(f)$, by Proposition 53, this implies that $x \notin \overline{\gamma(c_i, f)}$ for $1 \leq i \leq m$. Let J be an open subinterval of I containing x such that $J \cap \overline{\gamma(c_i, f)} = \emptyset$ for $1 \leq i \leq m$. Then J contains five distinct points $u < r < y < s < v$ in the orbit of x. Since r and s are recurrent, and hence in $\overline{P}(f)$, f has a periodic point $z_1 \in (u, y)$ and a periodic point $z_2 \in (y, v)$.

If we put $g = f^r$, where r is the larger of the periods of z_1 and z_2, then z_1 and z_2 are fixed points of g. Moreover $g^2(y) \in (z_1, z_2)$, since by Proposition 3 we cannot have $g^2(y) < z_1$ or $g^2(y) > z_2$. Without loss of generality, suppose $g^2(y) > y$. Then, by Proposition 3 again, the interval $(y, g^2(y))$ contains a fixed point w of g^2 and w has period 2 for g. Moreover g^2 attains its maximum value in the interval $[z_1, w]$ at an interior point x_1, since $g^2(y) \in (w, z_2)$, and attains its minimum value in the interval $[w, z_2]$ at an interior point x_2, since $g^4(y) \in (z_1, w)$. If the maximum (resp. minimum) is attained at more than one point, we take x_1 (resp. x_2) to be the nearest such point to w. Then x_1, x_2 are turning-points of g^2. Hence $f^j(x_1)$ is a turning-point of f for some j with $0 \leq j < 2r$. Thus $g^2(x_1) = f^p(c_s)$ for some $p > 0$ and some $s \in \{1,\ldots,m\}$. Similarly $g^2(x_2) = f^q(c_t)$ for some $q > 0$ and some $t \in \{1,\ldots,m\}$. It follows from the definition of J that $g^2(x_1) > v$ and $g^2(x_2) < u$. Hence

$g^2(y_1) = z_2$ for some $y_1 \in (z_1, w)$ and $g^2(y_2) = z_1$ for some $y_2 \in (w, z_2)$. Since w is a fixed point of g^2, we have $g^2(y_3) = y_2$ for some $y_3 \in (y_1, w)$. Thus

$$g^6(y_3) = z_1 < y_3 < y_2 = g^2(y_3).$$

Since g^2 is non-chaotic, this contradicts Theorem II.9. □

COROLLARY 55 *If f is piecewise monotone and non-chaotic, then the number of infinite minimal sets of f is at most equal to the number of its turning-points.* □

NOTES

Proposition 1 was first proved by Coppel [52]. Proposition 6 is due to Šarkovskii [107]. For Corollary 8 and Proposition 9 see Xiong [129]. The equivalence of (i) and (v) in Proposition 10 was first proved by Šarkovskii [104]; the equivalence of (iii) and (v) is due to Xiong [129]. For Proposition 11 and Corollary 12 see Xiong [127].

For Lemmas 14–16 and Theorem 20 see Tang [120]. However, the special role of regularly recurrent points does not appear there. The notion of a uniformly non-chaotic map appears implicitly in Theorem 2.4 of Smital [116].

The proof of Proposition 25 uses ideas from Liao [80]. The proofs of Corollary 26 and Proposition 27 are taken from Xiong [130] and Osikawa and Oono [97] respectively. Another proof of the existence of scrambled sets for chaotic maps is given by Jankova and Smital [73].

For Lemmas 31 and 32 see Liao [78]. Theorem 34 is a stronger formulation of a result of Liao [79], which in turn improved an earlier result of Xiong [128]. The equivalence of (i) and (ii) in Proposition 36 is due to Block and Franke [34].

Proposition 42 is due to Barge and Martin [12]. Proposition 44 appeared in Barge and Martin [14], and Proposition 45 in Coven and Mulvey [56]. Theorem 48 is originally due to Barge and Martin [13]; the more direct approach given here follows Coven and Mulvey [57], who also give Lemma 51. Results related to Proposition 52 have been obtained by Shihai Li (unpublished) and Xiong [134]. Propositions 53 and 54 were contributed by Xiong [135].

VII
Types of Periodic Orbits

1 FORCING

We may define a periodic orbit to be an ordered pair (P, Π), where P is a finite subset of the real line and Π is a cyclic permutation of P. We sometimes call P the periodic orbit, with Π understood. Thus it makes sense to talk about periodic orbits without reference to any map *f* of an interval. We say a periodic orbit P is an orbit of such a map *f* if $f(x) = \Pi(x)$ for every $x \in$ P, and hence *f*(P) = P.

Let P = $\{p_1 < p_2 < ... < p_n\}$, i.e. let P = $\{p_1, p_2, ..., p_n\}$ where $p_1 < p_2 < ... < p_n$, and let Q = $\{q_1 < q_2 < ... < q_m\}$, and suppose (P, Π) and (Q, Ψ) are periodic orbits. We say P and Q have the same *type*, in symbols P $\approx$ Q, if $n = m$ and $\Pi(p_i) = p_j$ if and only if $\Psi(q_i) = q_j$. If P and Q are of different types, we say P *forces* Q, in symbols P $\prec$ Q, if any continuous map of a compact interval to itself which has a periodic orbit of the same type as P also has a periodic orbit of the same type as Q.

For example, let Π be the cyclic permutation $\Pi(p_i) = p_{i+1}$ for $1 \le i < n$, $\Pi(p_n) = p_1$ and let Ψ be the cyclic permutation $\Psi(q_i) = q_{i+1}$ for $1 \le i < m$, $\Psi(q_m) = q_1$. Using Lemma I.6, we see at once from the Markov graph of Π that Π forces Ψ if $1 < m < n$.

Forcing is really a relation on equivalence classes of periodic orbits, but it is easier and more natural to talk about one periodic orbit forcing another. It is obvious that this relation is transitive, i.e. if P forces Q and Q forces R, then P forces R. The next lemma, due to Baldwin [11], shows that the relation is anti-symmetric.

LEMMA 1 *If* P *forces* Q, *then* Q *does not force* P.

Proof Let *f*: $[a,b] \to [a,b]$ be a polynomial, of degree at least two, which has an orbit of the same type as P. Such a polynomial may be explicitly constructed by Lagrange's interpolation formula. Then *f* has only finitely many orbits of the same type as P, since f^n has only finitely many fixed points for any n.

Let L denote the set of $x \in [a,b]$ such that x is the smallest element of a periodic orbit of f of the same type as P, and let c be the largest element of L. Let d be the largest element of the orbit of c. We define g: $[a,b] \to [a,b]$ by $g(x) = f(x)$ if $x \in [c,d]$, $g(x) = f(c)$ if $x \in [a,c]$, and $g(x) = f(d)$ if $x \in [d,b]$. Then g has exactly one orbit of the same type as P.

Since P and Q do not have the same type, g has a periodic orbit of the same type as Q in the open interval (c,d). Let v and w denote the smallest and largest elements of this orbit. Define h: $[a,b] \to [a,b]$ by $h(x) = g(x)$ if $x \in [v,w]$, $h(x) = g(v)$ if $x \in [a,v]$, and $h(x) = g(w)$ if $x \in [w,b]$. Then h does not have a periodic orbit of the same type as P. Since h has a periodic orbit of the same type as Q, this proves that Q does not force P. □

It follows from Lemma 1 that we can define a partial order on equivalence classes of periodic orbits by writing $P \leq Q$ if $P \approx Q$ or $P \prec Q$. The reader should be warned that some authors take this to be the definition of forcing.

The following example shows that this ordering of equivalence classes of periodic orbits is not a total order.

EXAMPLE 2 Let $I = [1,8]$ and let f,g be the piecewise linear maps of I onto itself defined respectively by

$$f(1) = 5,\ f(2) = 6,\ f(3) = 7,\ f(4) = 8,$$
$$f(5) = 3,\ f(6) = 4,\ f(7) = 2,\ f(8) = 1,$$

and

$$g(1) = 6,\ g(2) = 5,\ g(3) = 8,\ g(4) = 7,$$
$$g(5) = 3,\ g(6) = 4,\ g(7) = 2,\ g(8) = 1.$$

Evidently $\{1,2,...,8\}$ is a periodic orbit of period 8 for both f and g, but their types Π_f and Π_g are different. Setting $I_k = [k, k+1]$ for $1 \leq k \leq 7$, we readily verify that the Markov graph of f contains no primitive cycle of length 8, and $I_1 \to I_5 \to I_3 \to I_7 \to I_1$ is the only primitive cycle of length 4. Hence, in any periodic orbit $x_1 < ... < x_8$ of f of period 8, x_1 is mapped into x_5 and x_2 into x_6. Similarly the Markov graph of g contains no primitive cycle of length 8, and $I_1 \to I_5 \to I_3 \to I_7 \to I_1$ is the only primitive cycle of length 4. Hence, in any periodic orbit $y_1 < ... < y_8$ of g of period 8, y_1 is mapped into y_6 and y_2 into y_5. Consequently Π_f does not force Π_g, and Π_g does not force Π_f. □

Let $P = \{p_1 < p_2 < ... < p_n\}$ be a periodic orbit. There is a unique map f: $[p_1, p_n] \to [p_1, p_n]$ satisfying

(i) $f(p_i) = \Pi(p_i)$ for $i = 1, ..., n$,
(ii) $f \mid [p_i, p_{i+1}]$ is linear for $i = 1, ..., n-1$.

This map f is called the *linearization* of P. We will subsequently show that P forces Q whenever the linearization of P has an orbit of the same type as Q. This was proved, but not explicitly stated, in Baldwin [11] and also in Jungreis [74]. Our proof will be based on some considerations of independent interest.

A periodic orbit (P, Π) of period n, with P = $\{x_1 < ... < x_n\}$, will be said to be a *k-tuple cover*, where $k > 1$, of a periodic orbit (Q, Ψ) of period m if $n = km$ and if the m blocks $\{x_{(j-1)k+1} < x_{(j-1)k+2} < ... < x_{jk}\}$ ($j = 1, ..., m$) of k consecutive terms in P are permuted by Π, the permutation having the same type as Q. We will say simply that a periodic orbit is a k-tuple cover if it is a k-tuple cover of some other periodic orbit, but we do not need to specify which. [Misiurewicz and Nitecki [86] call P a *k-extension*, rather than a k-tuple cover, of Q.]

We will later have occasion to consider k-tuple covers for arbitrary $k > 1$, but for the present we will be interested in *double covers* ($k = 2$). It may be noted that, in Example 2, Π_f and Π_g are double covers of the same periodic orbit of period 4.

Let P be a periodic orbit and let f be its linearization. By Lemma I.6, if $J_1 \to J_2 \to ... \to J_m \to J_1$ is a primitive cycle of the Markov graph of f, then f has a periodic point y of period m such that $f^{k-1}(y) \in J_k$ for $1 \leq k \leq m$. Since f is strictly monotonic on each interval J_k, the type of the orbit of y is uniquely determined by the cycle. To see this, suppose y and z are periodic points of period m with $f^{k-1}(y), f^{k-1}(z) \in J_k$ for $1 \leq k \leq m$. Then $f^{k-1} < y, z> \subset J_k$ for $1 \leq k \leq m$ and $f^m < y, z > = < y, z >$. There are points $y' \in \gamma(y)$ and $z' \in \gamma(z)$ in $< y, z >$ such that there are no points of $\gamma(y) \cup \gamma(z)$ in the interior of $< y', z' >$. Since $f^m < y', z' > = < y', z' >$, it follows that there are no points of $\gamma(y) \cup \gamma(z)$ in the interior of $f^{k-1} < y', z' >$ for $1 \leq k \leq m$. Thus the orbits of y' and z' have the same type.

Conversely, let y be any periodic point of f of period m, which is not in P, and let J_k be the uniquely determined interval such that $f^{k-1}(y) \in J_k$ ($1 \leq k \leq m$). Then $J_1 \to J_2 \to ... \to J_m \to J_1$ is a cycle of length m in the Markov graph of f. By Lemma I.11, if this cycle is not primitive then $m = 2k$ is even, $k > 1$, $J_{i+k} = J_i$ for $1 \leq i \leq k$ and $J_1 \to J_2 \to ... \to J_k \to J_1$ is a primitive cycle of length k. It follows at once that f has a periodic point z of period k such that the orbit of y is a double cover of the orbit of z.

A similar argument can be applied to the fundamental cycle itself. Hence the fundamental cycle is not primitive if and only if the periodic orbit P is a double cover.

LEMMA 3 *Let P be a periodic orbit and f its linearization. Let Q be a periodic orbit of f, with period m, and suppose there are adjacent points x and y of Q such that none of the intervals $f^i[x, y]$ $(0 \le i < m)$ contains a point of P in its interior.*

Then P is a double cover, and P and Q have the same type.

Proof The hypothesis implies that f^i maps the interval $L = [x,y]$ linearly, for each positive integer i, and that $f^m(L) = L$. Let k be the least positive integer such that either $f^k(x) = y$ or $f^k(y) = x$, so that $k \le m/2$. Without loss of generality we may assume that $x < y$ and $f^k(y) = x$. We will show that also $f^k(x) = y$.

First assume that $x < f^k(x) < y$. Then $f^k(L) = [x, f^k(x)]$ and so the interval $[x, f^k(x)]$ is invariant under f^k. Since y is periodic, this is a contradiction.

Assume next that $f^k(x) > y$. Then we can define by induction points $y_i \in (x,y)$ such that $f^k(y_1) = y, f^k(y_i) = y_{i-1}$ $(1 < k \le m)$. Then f^{km} maps the interval L linearly onto L and collapses the subinterval $[y_m, y]$ to the point $\{y\}$. This is a contradiction.

Assume finally that $f^k(x) < x$, and suppose first that $x \in$ P. If $f^{jk}(y) < x$ for some $j \ge 2$ then $f^{jk}(x) \le x$, since the interval $f^{jk}(L)$ contains no point of P in its interior. Since $f^{jk}(y) \ne y$ implies also $f^{jk}(x) \ne x$, we must actually have $f^{(j+1)k}(y) = f^{jk}(x) < x$. It follows by induction that $f^{jk}(y) < x$ for all $j \ge 2$, a contradiction. Suppose now that $x \notin$ P, so that P and Q are disjoint. Then f^{jk} maps linearly the interval

$$[f^k(x), y] = f^k(L) \cup L.$$

It follows by induction that the sequence $f^{jk}(x)$ $(j = 0,1,2, \ldots)$ is strictly decreasing, a contradiction.

The only remaining possibility is $f^k(x) = y$. It follows from the choice of k that the intervals $L, f(L), \ldots, f^{k-1}(L)$ are disjoint and $f^k(L) = L$. Hence $m = 2k$. The interval $f^{i-1}(L)$ is contained in an interval J_i whose endpoints are adjacent points of P. Then f is linear on J_i and the Markov graph of f contains the cycle $J_1 \to J_2 \to \ldots \to J_m \to J_1$.

The restriction of the piecewise linear map f^{2k} to L is the identity. Let K denote the maximal interval containing L on which f^{2k} is linear. If w is an endpoint of K, then the orbit of w must contain a point in P. Since f^{2k} is the identity on K, we must actually have $w \in$ P. Hence f^{2k} is the identity on J_1, and similarly also on $J_2, \ldots, J_m$. Since f^i is not the identity on J_k for $0 < i < 2k$, it now follows that P also has period $2k$ and that P and Q have the same type. Moreover they are double covers. □

Let P be a periodic orbit and f its linearization. We will denote by S(P) the set of closed intervals whose endpoints are adjacent points of P, and we will say that an interval $J \in$ S(P) is *periodic* if $f^n(J) = J$ for some positive integer n. Our next result, which is essentially a corollary of Lemma 3, shows that a periodic interval exists only if P is a double cover.

LEMMA 4 *Let* P *be a periodic orbit and f its linearization. Suppose $J \in$ S(P) is periodic and let n be the least positive integer such that $f^n(J) = J$. Then*

(i) *$f^k(J) \in$ S(P) and $f^k(J)$ is periodic for each positive integer k,*
(ii) *the intervals $J, f(J), \ldots, f^{n-1}(J)$ are pairwise disjoint,*
(iii) *f^n maps J linearly onto J, and the restriction of f^n to J is strictly decreasing,*
(iv) P *has period $2n$,*
(v) *the points of* P *are precisely the endpoints of the intervals $J, f(J), \ldots, f^{n-1}(J)$,*
(vi) *if $K \in$ S(P) is periodic, then $K = f^k(J)$ for some k with $0 \le k < n$.*

Proof If $f^k(J) \notin$ S(P) for some positive integer k, then $f^k(J)$ contains at least three elements of P. Hence $f^i(J)$ contains at least three elements of P for each integer $i \ge k$. This is a contradiction, since $f^i(J) = J$ whenever i is a multiple of n. This proves (i). The remaining statements now follow immediately from Lemma 3, with Q = P. □

The theorem alluded to earlier will now be established.

THEOREM 5 *Let* P *and* Q *be periodic orbits of different types. Then* P *forces* Q *if and only if the linearization of* P *has an orbit of the same type as* Q.

Proof Let f denote the linearization of P and suppose that Q is a periodic orbit of f. Let $g: I \to I$ be continuous and suppose P is a periodic orbit of g. We must show that g has a periodic orbit of the same type as Q.

For each non-negative integer n, let $A_n = \{x: f^n(x) \in \text{P}\}$. Evidently

$$A_n = A_{n-1} \cup f^{-1}(A_{n-1}) = f^{-1}(A_{n-1})$$

for each positive integer n. It follows from Lemma 3 that if Q has period N, each interval joining adjacent points of A_N contains at most one point of Q.

Let $B_0 = \text{P}$, so that $B_0 = A_0$. We define $B_1, B_2, \ldots, B_N$ inductively as follows. Assuming B_{n-1} has been defined, we take B_n to be a subset of $g^{-1}(B_{n-1})$ such that:

(1) $B_n \supseteq B_{n-1}$,
(2) B_n and A_n have the same cardinality,
(3) if $A_n = \{y_1 < ... < y_m\}$ and $B_n = \{z_1 < ... < z_m\}$, then $z_i \in B_{n-1}$ if and only if $y_i \in A_{n-1}$ and $g(z_i) = z_j$ if and only if $f(y_i) = y_j$.

The finite sets A_N, B_N are invariant under f, g respectively, and by construction the Markov graphs of (A_N, f) and (B_N, g) are the same. By following the orbit of Q, we obtain a cycle in the Markov graph of (A_N, f). This same cycle in the Markov graph of (B_N, g) yields a periodic orbit $\overline{\mathrm{Q}}$ of g. Since each interval joining adjacent points of A_N (respectively B_N) contains at most one point of Q (respectively $\overline{\mathrm{Q}}$), it follows that Q and $\overline{\mathrm{Q}}$ have the same type. □

Using Theorem 5, one can write down an algorithm for determining whether or not P forces Q. Details may be found in Baldwin [11]; see also Jungreis [74].

In Theorem II.19 we showed that for any positive integers m,n

$$\mathbb{P}_n \subset \text{int}\, \mathbb{P}_m \quad \text{if } n \prec m,$$

where $\mathbb{P}_n$ denotes the set of all $f \in C(I,I)$ with a periodic orbit of period n. Our next goal is to derive a corresponding statement for periodic orbits of different types. For any periodic orbit P, let F(P) denote the set of all $f \in C(I,I)$ with a periodic orbit of the same type as P. Thus if P and Q are periodic orbits of different types, then $\mathbf{F}(\mathrm{P}) \subseteq \mathbf{F}(\mathrm{Q})$ if and only if $\mathrm{P} \prec \mathrm{Q}$. We are going to prove that actually

$$\mathbf{F}(\mathrm{P}) \subset \text{int}\, \mathbf{F}(\mathrm{Q}) \quad \text{if } \mathrm{P} \prec \mathrm{Q}.$$

This was first proved by Block and Hart [39]. We give here a different proof, using ideas from Block and Coven [29].

Let P be a periodic orbit and f its linearization. If P is a double cover, of period $2n$ say, then the interval $J \in$ S(P) which is furthest to the left satisfies the hypotheses of Lemma 4 and we set

$$\hat{\mathrm{P}} = J \cup f(J) \cup \ ... \cup f^{n-1}(J) .$$

If P is not a double cover then no interval $J \in$ S(P) is periodic and we set $\hat{\mathrm{P}} = \mathrm{P}$. In either case, $\mathrm{P} \subseteq \hat{\mathrm{P}}$.

The following lemma is a special case of Lemma 2.5 in Block and Coven [29].

LEMMA 6 *If f is the linearization of a periodic orbit* P, *then* $\bigcup_{n=0}^{\infty} f^{-n}(\hat{\mathrm{P}})$ *is dense in the convex hull of* P.

Proof We can write $\mathbf{S}(P) = \mathbf{S}_1 \cup \mathbf{S}_2$, where $\mathbf{S}_1$ is the set of all $J \in \mathbf{S}(P)$ for which also $f(J) \in \mathbf{S}(P)$ and $\mathbf{S}_2$ is the set of all $J \in \mathbf{S}(P)$ for which $f(J)$ is the union of at least two elements of $\mathbf{S}(P)$. For any closed interval K, let $|K|$ denote the length of K and let

$$c = \min |f(J)| / |J'| ,$$

where the minimum is taken over all pairs (J,J') with $J \in \mathbf{S}_2$, $J' \in \mathbf{S}(P)$ and $J' \subset f(J)$. Thus $c > 1$.

Let K be an arbitrary closed interval contained in the interior of an interval from $\mathbf{S}(P)$. It suffices to show that $\bigcup_{n=0}^{\infty} f^{-n}(\hat{P}) \cap K \neq \varnothing$. Hence we may assume that $f^n(K) \cap P = \varnothing$ for every positive integer n. Then, for each positive integer n, there exists $J_n \in \mathbf{S}(P)$ such that $f^n(K) \subset J_n$. If $J_n \in \mathbf{S}_1$ then, since f is linear on J_n,

$$|f^{n+1}(K)| / |J_{n+1}| = |f^n(K)| / |J_n| .$$

If $J_n \in \mathbf{S}_2$, then

$$|f^{n+1}(K)| / |J_{n+1}| \geq c\, |f^n(K)| / |J_n| .$$

Since $c > 1$ and $f^n(K) \cap P = \varnothing$ for every n, it follows that $J_n \in \mathbf{S}_2$ for at most finitely many n. Thus there exists a positive integer N such that $J_n \in \mathbf{S}_1$ for all $n > N$. Since $\mathbf{S}(P)$ is finite, there exist integers $n > N$ and $j > 0$ such that $J_n = J_{n+j}$. Then

$$f^j(J_n) = J_{n+j} = J_n ,$$

and thus J_n is a periodic interval containing $f^n(K)$. Hence $K \subseteq f^{-n}(\hat{P})$. □

The following theorem, whose statement has already been announced, can now be established.

THEOREM 7 *Let* P *and* Q *be periodic orbits such that* P *forces* Q. *Then*

$$\mathbf{F}(P) \subset \operatorname{int} \mathbf{F}(Q) .$$

Proof Let $g_0 \in \mathbf{F}(P)$. We must show that there is a neighbourhood $\mathbf{N}$ of g_0 in $C(I,I)$ such that $\mathbf{N} \subset \mathbf{F}(Q)$.

Let f denote the linearization of P. By hypothesis f has a periodic orbit Q′ of the same type as Q. We consider separately two cases.

Case I: Some point of Q′ lies in a periodic interval of $\mathbf{S}(P)$.

Then every point of Q′ lies in a periodic interval of S(P). Let K be one such interval, and let m be the least positive integer such that $f^m(K) = K$. Then $K, f(K),..., f^{m-1}(K)$ are pairwise disjoint, by Lemma 4. Furthermore, since P and Q are of different types, it follows from Lemma 4 that Q has period m.

But g_0 has a periodic orbit P′ of the same type as P, since $g_0 \in$ **F**(P). Let $J_1, J_2, ..., J_m$ be the intervals in S(P′) corresponding to the intervals $K, f(K), ..., f^{m-1}(K)$ in S(P). Then $J_1,...,J_m$ are pairwise disjoint and g_0 maps the endpoints of J_1 onto the endpoints of J_2, and so on. By Lemma I.3 and its proof, there are closed subintervals $L_1, L_2,..., L_m$ of $J_1, J_2, ..., J_m$ such that $g_0(L_1) = L_2,..., g_0(L_{m-1}) = L_m$, $g_0(L_m) = J_1$, and (setting $L_{m+1} = J_1$) g_0 maps the endpoints of L_i onto the endpoints of L_{i+1} for $i = 1,..., m$, preserving the order of the endpoints of L_i if and only if g_0 preserves the order of the endpoints of J_i. Since the two endpoints of J_1 are interchanged by $g_0{}^m$, the following two properties hold for $g = g_0$:

(i) if $L_1 = [c,d]$, then $g^m(c) > c$ and $g^m(d) < d$;
(ii) $L_1, g(L_1), ..., g^{m-1}(L_1)$ are pairwise disjoint and these intervals lie on the real line in the same order as the intervals $J_1, J_2, ..., J_m$.

Evidently (i) and (ii) continue to hold for all g in a neighbourhood **N** of g_0 in $C(I,I)$. Furthermore, if g satisfies these conditions then g^m has a fixed point $y \in L_1$, by (i), and the orbit of y under g has the same type as Q, by (ii). Thus **N** $\subset$ **F**(Q).

Case II: No point of Q′ lies in a periodic interval of S(P).

For each positive integer n, let $A_n = \{x: f^n(x) \in \text{P}\}$, as in the proof of Theorem 5. Let S_n(P) denote the set of closed intervals whose endpoints are adjacent points of A_n. It follows from Lemma 6 that if N is sufficiently large each of the following holds:

(iii) each interval in S_N(P) contains at most one point of Q′,
(iv) if an interval in S_N(P) contains a point of Q′, then the endpoints of this interval are not in P,
(v) any two intervals in S_N(P) which contain distinct points of Q′ are disjoint.

Let B_N be defined for g_0 in the same manner as in the proof of Theorem 5, and let $S_N(g_0)$ denote the set of closed intervals whose endpoints are adjacent points of B_N.

Let $I_1, I_2 ,..., I_m$ denote the distinct intervals in S_N(P) which contain points of Q′. These intervals are pairwise disjoint and may be numbered so that $f(I_1) \supset I_2,..., f(I_{m-1}) \supset I_m$, and $f(I_m) \supset I_1$.

Let $K_1, K_2, ..., K_m$ denote the intervals in $S_N(g_0)$ which correspond to $I_1, I_2, ..., I_m$. Then $g_0(K_1) \supset K_2 ,..., g_0(K_{m-1}) \supset K_m$, and $g_0(K_m) \supset K_1$. As in Lemma I.3, there are closed subintervals $L_1, L_2, ..., L_m$ of $K_1, K_2, ..., K_m$ respectively such that $g_0(L_1) = L_2, ..., g_0(L_{m-1}) = L_m$, $g_0(L_m) = K_1$, and (setting $L_{m+1} = K_1$) g_0 maps the endpoints of L_i onto the endpoints of L_{i+1} for $i = 1, ..., m$.

Let $L_1 = [c,d]$. By construction, g_0^m maps the endpoints of L_1 onto the endpoints of $K_1 \supseteq L_1$ and neither c nor d is periodic. Then the following two properties hold for $g = g_0$:

(vi) either $g^m(c) > c$ and $g^m(d) < d$, or $g^m(c) < c$ and $g^m(d) > d$;

(vii) the intervals $L_1, g(L_1), ..., g^{m-1}(L_1)$ are pairwise disjoint and these intervals lie on the real line in the same order as $K_1, K_2, ..., K_m$.

As in Case I, we now obtain a neighbourhood N of g_0 with $\mathrm{N} \subset \mathbf{F}(\mathrm{Q})$. □

The following result, due to Bernhardt [17], provides another connection between covers and forcing.

PROPOSITION 8 *Suppose the periodic orbit* P *is a double cover of the periodic orbit* Q. *Then* P *forces* Q. *Moreover, if* P *forces* R *and if* R *does not have the same type as* Q, *then* Q *forces* R.

Proof Suppose the periodic orbit P has period $n = 2m$ and let $I_1, ..., I_{2m-1}$ be the closed intervals whose endpoints are adjacent points of P, numbered from left to right. Also let $I_1', ..., I'_{m-1}$ be the corresponding intervals for Q. If f is the linearization of P, then f has an orbit of type Q, by the definition of a double cover, and thus P forces Q. The Markov graph of P contains a cycle $J_1 \to J_3 \to ... \to J_{2m-1} \to J_1$ of length m, where $J_1, J_3, ..., J_{2m-1}$ is a permutation of $I_1, I_3, ..., I_{2m-1}$ and the arrows shown are the only arrows leaving the I's with odd subscripts. Hence the cycle is primitive, and the points of a periodic orbit of f of type R must all lie in I's with even subscripts. But it follows from the definition of a double cover that in the Markov graph of P there is an arrow from I_{2j} to I_{2k} if and only if in the Markov graph of Q there is an arrow from I'_j to I'_k. Hence the Markov graph of Q contains a primitive cycle for which the corresponding periodic orbit is either of type R or is doubly covered by an orbit of type R. In either case, Q forces R. □

This result shows that if a periodic orbit is a double cover, then it has an immediate successor in the partial ordering of all orbit types. Bernhardt [17] also proved a converse of Proposition 8. However, Jungreis [74], and Misiurewicz and Nitecki [86], have established the following stronger result, which is stated here without proof.

PROPOSITION 9 *If a periodic orbit* P *is not a double cover, then there exists a sequence* $\{Q_n\}$ *of periodic orbits such that*

(i) $P \prec Q_{n+1} \prec Q_n$ *for* $n = 1, 2, \ldots,$
(ii) *if* $M \prec Q_n$ *for all* $n \geq 1$ *and* M *does not have the same type as* P, *then* $M \prec P$,
(iii) *if* $P \prec R$, *then* $Q_n \prec R$ *for some* $n \geq 1$.

It was already shown by Baldwin [11] that any two periodic orbits have a common upper bound in the partial ordering of all periodic orbits. On the other hand, it follows from Proposition 9 that two non-equivalent periodic orbits, neither of which forces the other, cannot have a *least* upper bound. Thus the partial ordering of orbit types is rather complicated.

Although the study of unimodal maps lies outside our frame of reference, it is worth mentioning that the restriction of this partial ordering to the set of periodic orbits with a unique turning point, which is a maximum (or minimum), is a total order. Baldwin [11] sketches a proof, based on the kneading theory of Milnor and Thurston [81]. This theory is also set out in Collet and Eckmann [49].

2 PRIMARY PERIODIC ORBITS

We say a periodic orbit is *primary* if it forces no periodic orbit with the same period. A main goal of this section is to identify the primary orbits.

The following proposition, which is an extension of a result of Coppel [53], is our first step in this direction. In this proposition, and all other results of this section, f denotes an arbitary continuous map of a compact interval into itself.

PROPOSITION 10 *Suppose* f *has a periodic orbit* $P = \{x_1 < \ldots < x_n\}$ *of period* $n = qm$, *where* $q > 3$ *is odd, such that for* $k=1, \ldots, m$,

(i) $\{x_{(k-1)q+1} < x_{(k-1)q+2} < \ldots < x_k\}$ *is an orbit of* f^m,
(ii) *the interval* $[x_{(k-1)q+1}, x_{kq}]$ *contains no periodic orbit of* f^m *of period* $q-2$.

Then each orbit of f^m *in* (i) *is a Stefan orbit and* f *maps each such orbit monotonically onto another such orbit, with one exception.*

Proof Let

$$\begin{array}{llll} J_1 & \to J_2 & \to \dots \to J_m & \\ \to J_{m+1} & \to J_{m+2} & \to \dots \to J_{2m} & \\ \to & & \dots & \\ \to J_{(q-1)m+1} & \to J_{(q-1)m+2} & \to \dots \to J_{qm} & \to J_1 \end{array}$$

be the fundamental cycle for f. It follows from the first hypothesis that if the endpoints of J_i lie in the same orbit of f^m, then the same holds for the endpoints of J_{i+1}. Therefore, since $[x_1, x_2]$ belongs to the fundamental cycle, the endpoints of each J_i lie in the same orbit of f^m. Moreover all intervals J_i in the same column of the above array have their endpoints in the same orbit of f^m. Since there are only $q-1$ distinct intervals corresponding to a given orbit of f^m, it follows that in each column at least one interval occurs twice.

Without loss of generality we may assume that in the first column the interval J_1 occurs twice. If $J_{km+1} = J_1$ for some odd k with $1 \le k < q$ then by terminating the fundamental cycle at J_{km+1} we obtain a primitive cycle of length km. Hence there exists a point x of period km with $x \in J_1, f^m(x) \in J_{m+1}, \dots, f^{(k-1)m}(x) \in J_{(k-1)m+1}$. Since x has period k for f^m, this contradicts the second hypothesis unless $k = 1$. If $J_{km+1} = J_1$ for some even k with $1 < k < q$ then by commencing in the fundamental cycle at J_{km+1} we obtain a primitive cycle of length $(q-k)m$, which similarly gives a contradiction unless $k = q-1$. It follows that if two intervals in a given column are the same then they are adjacent elements of that column, viewed cyclically.

Without loss of generality assume $J_{(q-1)m+1} = J_1$. If for some k $(2 \le k \le m)$ we have $J_{(i-1)m+k} = J_{im+k}$ for some i such that $1 \le i \le q-2$ then by omitting two strings of m consecutive terms from the fundamental cycle we obtain a primitive cycle

$$J_1 \to J_2 \to \dots \to J_{(i-1)m+k} \to J_{im+k+1} \to \dots \to J_{(q-1)m+1}$$

of length $(q-2)m$, which gives a contradiction in the same way as before. If $J_{(q-1)m+k} \ne J_k$ for some k with $2 \le k \le m$, we may choose the notation so that $J_{(q-2)m+2} = J_{(q-1)m+2}$, as well as $J_{(q-1)m+1} = J_1$. A similar argument to that just used then shows that we cannot have $J_{(q-1)m+j} = J_j$ for some j with $2 < j \le m$.

Hence either $J_{(q-1)m+k} = J_k$ for $1 \le k \le m$, or $J_{(q-1)m+1} = J_1$ and $J_{(q-2)m+k} = J_{(q-1)m+k}$ for $2 \le k \le m$. In either case we may now choose the notation so that the fundamental cycle has the form

$$\begin{array}{llll} J_1 & \to J_2 & \to \dots \to J_m & \\ \to J_1 & \to J_2 & \to \dots \to J_m & \\ \to J_{2m+1} & \to J_{2m+2} & \to \dots \to J_{3m} & \\ \to & \quad \dots & & \\ \to J_{(q-1)m+1} & \to J_{(q-1)m+2} & \to \dots \to J_{qm} & \to J_1 . \end{array}$$

By the definition of a fundamental cycle each interval J_k $(1 \le k \le m)$ has the form $<c, f^m(c)>$. But by the hypothesis (ii) each orbit of f^m is a Stefan orbit, and in a Stefan orbit of odd period $q > 3$ the only interval of this form is the one for which c is the midpoint of the orbit. Also, if $J_1 = < a, f^m(a) >$ then a must be the endpoint which generates the fundamental cycle, since $f^{2m}(a) \neq a$. Similarly, if $J_k = < c, f^m(c) >$ then $c = f^{k-1}(a)$ $(1 < k \le m)$. Thus each of the points $f^{k-1}(a)$ $(1 \le k \le m)$ is the midpoint of an orbit of f^m in (i). It now follows from the structure of a Stefan orbit that f maps the orbit of f^m containing $f^{k-1}(a)$ monotonically onto the orbit of f^m containing $f^k(a)$ for $k = 1, \dots, m-1$. □

It may be noted that, in the terminology introduced in Section 1, the hypothesis (i) of Proposition 10 says that the periodic orbit P is a q-tuple cover. Also it is clear that f cannot map *every* orbit of f^m in (i) monotonically onto another such orbit, since the midpoints of these orbits are not fixed points of f^m.

We now define the notion of a *simple* periodic orbit. The definition is by induction on the highest power of 2 which divides the period. Let (P,Π) be a periodic orbit of period n. When n is odd, we say that P is simple if $n = 1$ or if $n > 1$ and P is of Stefan type. When $n = qm$, where q is odd and $m = 2^s > 1$, we say that P is simple if the left and right halves of P each form simple orbits of Π^2 with period $n/2$. It follows from this definition that, if $q > 1$ and $m > 1$, then P $= \{x_1 < \dots < x_n\}$ is a simple orbit if and only if the m blocks $\{x_{q(k-1)+1}, \dots, x_{qk}\}$ $(k = 1, \dots, m)$ of q consecutive points each form a simple orbit of Π^m with period q, and the blocks themselves are permuted by Π like a simple orbit with period m.

Let (P, Π) be a periodic orbit of period $n = qm$, where q is odd and $m = 2^s$. We say P is *strongly simple* if it is simple and either n is a power of 2 $(q = 1)$, or n is odd $(m = 1)$, or $q > 1$, $m > 1$, and Π maps each of the m blocks of q consecutive points monotonically onto another such block, with one exception.

Since P is simple, the last condition is equivalent to requiring that Π maps the midpoint of each of the m blocks of q consecutive points into the midpoint of another such block, with one exception.

To illustrate the definitions, we observe that for both f and g in Example 2 the periodic orbit $\{1,2,...,8\}$ of period 8 is simple, and hence also strongly simple.

We are going to show that the primary orbits are precisely the strongly simple ones. First we give a result of Coppel [53], which relates the notions of simple and strongly simple orbit to the Šarkovskii stratification. Similar results were obtained independently by Ho [67] and Alseda *et al.* [6]. [In the latter paper the authors use 'weakly simple' in place of our 'simple' and 'simple' in place of our 'strongly simple'.]

THEOREM 11 *Let $n = qm$, where q is odd and $m = 2^s$. Suppose $f \in \mathbb{P}_n$, but $f \notin \mathbb{P}_k$ whenever $\mathbb{P}_k \subset \mathbb{P}_n$ and $1 < k < n$. Then every periodic orbit of f of period n is simple. Furthermore, if $q \neq 3$ then every orbit of f of period n is strongly simple.*

Proof The proof of the first assertion is by induction on s. If $s = 0$, then n is odd, and the conclusion follows from Proposition I.8. Now suppose that $s > 0$ and that the conclusion holds when $m = 2^{s-1}$. Let P be an orbit of f of period n. Since f has no periodic point of odd period less than n, it follows from Theorem II.9 that the left and right halves of P are interchanged by f. By the induction hypothesis, each half forms a simple orbit of f^2. Hence P is simple.

Now suppose that $q \neq 3$, and again let P be an orbit of f of period n. Then P is simple by what we have already proved. If $q = 1$ then P is strongly simple by definition, so we may assume $q > 3$. Then it follows from our hypothesis that f^m does not have a periodic point of period $q - 2$. This implies, by Proposition 10, that P is strongly simple. □

We will see later, in Theorem 22, that the exceptional case ($q = 3$) in Theorem 11 is removed by using the turbulence stratification instead of the Šarkovskii stratification.

However, we will first generalize Proposition I.15 by showing (Theorem 18) that if f has an orbit of period n, then f has a strongly simple orbit of period n. This result, proved by Block and Coppel [26], immediately implies that primary orbits are strongly simple. A number of preliminary results will be used in the proof. They are stated in greater generality than is actually necessary for Theorem 18, so that they may also be applied in other situations, for example to lifts of maps of the circle as in Chapter IX. In particular, Theorem 14 and Proposition 15 are extensions of results in Block and Coppel [26], where m was a power of 2.

LEMMA 12 *Suppose f has a periodic orbit* $P = \{x_1 < ... < x_n\}$ *of period* $n = 3m$, *where* $m > 1$. *Furthermore, suppose f cyclically permutes the sets* $\alpha_k = \{x_{3(k-1)+1}, x_{3(k-1)+2}, x_{3k}\}$ $(k = 1, ..., m)$. *If there are at least two values of k for which f does not map* α_k *monotonically onto* $f(\alpha_k)$, *then for some j with* $1 \le j \le m$ *we have*

$$J \cup K \subseteq f^m(J) \cap f^m(K),$$

where

$$J = [x_{3(j-1)+1}, x_{3(j-1)+2}], \; K = [x_{3(j-1)+2}, x_{3j}].$$

Proof Suppose the triple α_i is not mapped monotonically onto the triple $f(\alpha_i)$. Then the midpoint of α_i is not mapped onto the midpoint of $f(\alpha_i)$. It follows that we can write $\alpha_i = \{y_1, y_2, y_3\}$ and $f(\alpha_i) = \{z_1, z_2, z_3\}$, where y_2 is the midpoint of α_i, z_2 is the midpoint of $f(\alpha_i)$, and

$$f(y_2) = z_1, \; f(y_1) = z_2, \; f(y_3) = z_3.$$

Then $f\langle y_2, y_3\rangle \supseteq \langle z_1, z_3\rangle$ and hence $f^m\langle y_2, y_3\rangle \supseteq \langle y_1, y_3\rangle$. If $f^k(y_1)$ and $f^k(y_2)$ are both endpoints of $f^k(\alpha_i)$ for some k with $1 < k \le m$, then also $f^m\langle y_1, y_2\rangle \supseteq \langle y_1, y_3\rangle$.

We may therefore assume that for each k with $0 < k \le m$ either $f^k(y_1)$ or $f^k(y_2)$ is the midpoint of $f^k(\alpha_i)$, and hence $f^k(y_3) = f^{k-1}(z_3)$ is an endpoint of $f^k(\alpha_i)$. In particular this implies that

$$f^m(y_3) = f^{m-1}(z_3) = y_1, \; f^m(y_1) = f^{m-1}(z_2) = y_2,$$
$$f^m(y_2) = f^{m-1}(z_1) = y_3,$$

and hence $f^m\langle z_1, z_2\rangle \supseteq f\langle y_2, y_3\rangle \supseteq \langle z_1, z_3\rangle$. But by hypothesis there is a least k with $1 < k \le m$ such that $f^{k-1}(z_2)$ is not the midpoint of $f^k(\alpha_i)$. Then $f^{k-1}\langle z_2, z_3\rangle \supseteq f^k(\alpha_i)$, and hence $f^m\langle z_2, z_3\rangle \supseteq \langle z_1, z_3\rangle$. □

COROLLARY 13 *If f has an orbit of period 3m which is not strongly simple, where* $m = 2^s > 1$, *then* f^m *is turbulent.*

Proof If some orbit of period $3m$ is not simple, then f has an orbit of period $qm/2$ for some odd $q > 1$, by Theorem 11, and hence f^m is turbulent, by the turbulence stratification. We may therefore assume that all orbits of period $3m$ are simple. The result now follows at once from Lemma 12. □

THEOREM 14 *Suppose f has a periodic orbit* $P = \{x_1 <\ldots< x_n\}$ *of period* $n=3m$, *where* $m>1$. *Furthermore, suppose f cyclically permutes the sets* $\alpha_k = \{x_{3(k-1)+1}, x_{3(k-1)+2}, x_{3k}\}$ $(k = 1, \ldots, m)$.

Then f has a periodic orbit $Q = \{y_1 <\ldots< y_n\}$ *of period n such that f cyclically permutes the sets* $\beta_k = \{y_{3(k-1)+1}, y_{3(k-1)+2}, y_{3k}\}$ $(k = 1, \ldots, m)$ *and in addition:*

(i) $\beta_k \subseteq [x_{3(k-1)+1}, x_{3k}]$ *for* $k = 1, \ldots, m$,

(ii) *f maps* β_k *monotonically onto* $f(\beta_k)$ *for all but one k.*

Proof By Theorem 5 we may assume that f is the linearization of P. We may also assume that there are at least two values of k such that f does not map α_k monotonically onto $f(\alpha_k)$, since otherwise we can take $Q = P$. Thus the hypotheses of Lemma 12 are satisfied. Let J and K be the intervals whose existence is guaranteed by the conclusions of that lemma. If we set $I_k = [x_{3(k-1)+1}, x_{3k}]$, then $J \cup K = I_j$ for some j with $1 \le j \le m$.

It follows from Lemma II.2 that there exist points $a,b,c \in I_j$ such that $f^m(b) = f^m(a) = a$, $f^m(c) = b$ and either

$$(*)\qquad \begin{gathered} a < c < b, \\ f^m(x) > a \text{ for } a < x < b, \\ x < f^m(x) < b \text{ for } a < x < c, \end{gathered}$$

or the same with all inequalities reversed. We may renumber the α_k's so that $j = 1$, $f(\alpha_k) = \alpha_{k+1}$ for $k = 1,\ldots,m-1$, and $f(\alpha_m) = \alpha_1$. Then $f(I_k) = I_{k+1}$ for $k = 1,\ldots, m-1$ and $f(I_m) = I_1$. Set

$$c_k = f^{k-1}(c),\ b_k = f^{k-1}(b),\ a_k = f^{k-1}(a) \quad (k = 1, \ldots, m).$$

Then $a_k, b_k, c_k \in I_k$ for $k = 1, \ldots, m$.

First suppose that $c_k \in \operatorname{int} <a_k, b_k>$ for $k = 1,\ldots,m$. Choose $d_k \in <a_k, c_k>$ so that $f(d_k) = d_{k+1}$ for $k = 1,\ldots,m-1$ and $f(d_m) = c_1$. Then the cycle

$$\begin{gathered} <a_1, d_1> \to <a_2, d_2> \to \ldots \to <a_m, d_m> \\ \to <d_1, c_1> \to <d_2, c_2> \to \ldots \to <d_m, c_m> \\ \to <c_1, b_1> \to <c_2, b_2> \to \ldots \to <c_m, b_m> \to <a_1, d_1> \end{gathered}$$

yields a periodic orbit Q with the desired properties.

Next suppose that $c_j \notin \text{int} <a_j, b_j>$ for some j. We may assume that j is minimal. Then $j \geq 2$ and $c_i \in \text{int} <a_i, b_i>$ for $1 \leq i < j$. We claim that $b_j \notin \text{int} <a_j, c_j>$. In fact if $b_j \in \text{int} <a_j, c_j>$, then $f^{j-1}(x) = b_j$ for some $x \in \text{int} <a_1, c_1>$ and hence $f^m(x) = a_1$, which contradicts (*).

Similarly we can show that $a_j \notin \text{int} <b_j, c_j>$. Indeed if $a_j \in \text{int} <b_j, c_j>$ then $f^{j-1}(x) = a_j$ for some $x \in <b_1, c_1>$, since $f^{j-1}(c_1) = c_j$ and $f^{j-1}(b_1) = b_j$. Thus $f^m(x) = a_1$, which again contradicts (*).

Also $c_j \neq b_j$, since $c_j = b_j$ would imply $f^m(c_1) = a_1$, which contradicts (*) . Similarly $a_j \neq c_j$. The only remaining possibility is $a_j = b_j$. This implies that $f(b_{j-1}) = a_j$.

Since $f(a_m) = a_1$ and $f(c_m) = b_1$, we have $f(d_m) = c_1$ for some $d_m \in <a_m, c_m>$. For $i = j, \ldots, m-1$ choose $d_i \in <a_i, c_i>$ so that $f(d_i) = d_{i+1}$. Also choose $e_{j-1} \in <a_{j-1}, c_{j-1}>$ so that $f(e_{j-1}) = d_j$ and for $i = 1, \ldots, j-2$ choose $e_i \in <a_i, c_i>$ so that $f(e_i) = e_{i+1}$. Finally, choose $e_m \in <a_m, d_m>$ so that $f(e_m) = e_1$, and for $i = j, \ldots, m-1$ choose $e_i \in <a_i, d_i>$ so that $f(e_i) = e_{i+1}$. Then the cycle

$$\begin{aligned} &<a_j, e_j> \to \ldots \to <a_m, e_m> \to <a_1, e_1> \to \ldots \to <a_{j-1}, e_{j-1}> \\ &\to <e_j, d_j> \to \ldots \to <e_m, d_m> \to <e_1, c_1> \to \ldots \to <e_{j-1}, c_{j-1}> \\ &\to <d_j, c_j> \to \ldots \to <d_m, c_m> \to <c_1, b_1> \to \ldots \to <c_{j-1}, b_{j-1}> \\ &\to <a_j, e_j> \end{aligned}$$

yields a periodic orbit Q with the desired properties. □

PROPOSITION 15 *Suppose f has a periodic orbit* $P = \{x_1 < \ldots < x_n\}$ *of period* $n = qm$, *where* $q \geq 3$ *is odd and* $m > 1$. *Furthermore suppose that f cyclically permutes the sets* $\alpha_k = \{x_{q(k-1)+1}, \ldots, x_{qk}\}$ $(k = 1, \ldots, m)$, *that each α_k is a periodic orbit of f^m of Stefan type, and that f maps α_k monotonically onto $f(\alpha_k)$ for all but one k.*

Then f has a periodic orbit $Q = \{y_1 < \ldots < y_{(q+2)m}\}$ *of period $(q+2)m$ such that f cyclically permutes the sets* $\beta_k = \{y_{(q+2)(k-1)+1}, \ldots, y_{(q+2)k}\}$ $(k = 1, \ldots, m)$, *and in addition:*

(i) $\beta_k \subset [x_{q(k-1)+1}, x_{qk}]$ *for* $k = 1, \ldots, m$,

(ii) *each β_k is a periodic orbit of f^m of Stefan type, and f maps β_k monotonically onto $f(\beta_k)$ for all but one k.*

Proof For each $k = 1, \ldots, m$, let $z_1(k)$ denote the midpoint of the q points in α_k and let $z_j(k) = f^{m(j-1)}(z_1(k))$ $(1 < j \leq q)$. Then, for each k, the points of α_k have either the order

$$z_q(k) < z_{q-2}(k) < \ldots < z_3(k) < z_1(k) < z_2(k) < \ldots < z_{q-3}(k) < z_{q-1}(k)$$

or the reverse order. Furthermore, the blocks α_k can be renumbered so that $f(z_1(k)) = z_1(k+1)$ for $1 \le k < m$. Then

$$f(z_j(k)) = z_j(k+1) \text{ for } 1 \le k < m,\ 1 \le j \le q,$$
$$f(z_j(m)) = z_{j+1}(1) \text{ for } 1 \le j < q,\ f(z_q(m)) = z_1(1).$$

Since $f\langle z_1(k), z_2(k)\rangle \supseteq \langle z_1(k+1), z_2(k+1)\rangle$ for $1 \le k < m$ and $f\langle z_1(m), z_2(m)\rangle \supseteq \langle z_3(1), z_2(1)\rangle$, there exists a point w such that $f^m(w) = w$ and $f^{k-1}(w) \in \langle z_1(k), z_2(k)\rangle$ for $1 \le k \le m$. Similarly there exists a point v such that $f^m(v) = z_1(1)$ and $f^{k-1}(v) \in \langle f^{k-1}(w), z_2(k)\rangle$ for $1 \le k \le m$. Finally there exists a point u such that $f^m(u) = v$ and $f^{k-1}(u) \in \langle z_1(k), f^{k-1}(w)\rangle$ for $1 \le k \le m$. For $1 \le k \le m$ put

$$J_1(k) = \langle f^{k-1}(u), f^{k-1}(w)\rangle,\quad J_2(k) = \langle f^{k-1}(w), f^{k-1}(v)\rangle,$$
$$J_3(k) = \langle z_1(k), f^{k-1}(u)\rangle,\quad J_4(k) = \langle f^{k-1}(v), z_2(k)\rangle,$$
$$J_i(k) = \langle z_{i-4}(k), z_{i-2}(k)\rangle \quad (5 \le i \le q+2).$$

Then we have a cycle

$$\begin{aligned} & J_1(1) \to J_1(2) \to \dots \to J_1(m) \\ \to\ & J_2(1) \to J_2(2) \to \dots \to J_2(m) \\ & \cdots \\ \to\ & J_{q+2}(1) \to J_{q+2}(2) \to \dots \to J_{q+2}(m) \to J_1(1). \end{aligned}$$

This cycle yields a fixed point of $f^{(q+2)m}$, by Lemma I.19. In fact it has period $(q+2)m$, since its orbit Q cannot contain an endpoint of $J_1(1)$. It now follows from the ordering of the intervals $J_i(k)$ on the real line that Q satisfies the requirements of the proposition. □

An immediate consequence of Proposition 15 is the following

COROLLARY 16 *If f has a strongly simple orbit of period $n = qm$, where $q \ge 3$ is odd and $m = 2^s$, then f has a strongly simple orbit of period $(q+2)m$.* □

For the proof of Theorem 18 one more preliminary result is needed.

PROPOSITION 17 *If f has a strongly simple orbit of period $n = qm$, where $q > 1$ is odd and $m = 2^s$, then f also has a strongly simple orbit of period $6m$.*

Proof By Corollary 16 we may assume $q \geq 9$. In the given strongly simple orbit of period n there are m Stefan blocks, each containing q elements. We order the blocks so that the first block is mapped monotonically by f onto the second block, the second is mapped monotonically onto the third, ..., and the m-th is mapped non-monotonically onto the first. The intervals between consecutive points in the k-th block will be denoted by $J_1(k),\ldots, J_{q-1}(k)$, with the same system of numbering as in the statement of Proposition I.8. Then in the Markov graph of the strongly simple orbit of period n we have a cycle

$$\begin{aligned} &J_1(1) \to J_1(2) \to \ldots \to J_1(m)\\ \to &J_2(1) \to J_2(2) \to \ldots \to J_2(m)\\ &\quad\cdot \quad \cdot \quad \cdot\\ \to &J_{q-1}(1) \to J_{q-1}(2) \to \ldots \to J_{q-1}(m) \to J_1(1). \end{aligned}$$

Since the image of $J_{q-1}(m)$ under f contains $J_{q-6}(1)$, the graph also has a primitive cycle

$$\begin{aligned} &J_{q-6}(1) \to J_{q-6}(2) \to \ldots \to J_{q-6}(m)\\ \to &J_{q-5}(1) \to J_{q-5}(2) \to \ldots \to J_{q-5}(m)\\ &\quad\cdot \quad \cdot \quad \cdot\\ \to &J_{q-1}(1) \to J_{q-1}(2) \to \ldots \to J_{q-1}(m) \to J_{q-6}(1) \end{aligned}$$

of length $6m$. Consider the corresponding periodic orbit of period $6m$. Under f^m, distinct orbits of period 6 lie in the convex hulls of distinct Stefan blocks of the original strongly simple orbit. Moreover each f^m-orbit of period 6 is strongly simple, since the triple in the intervals $J_{q-6}(k), J_{q-4}(k), J_{q-2}(k)$ on one side of the midpoint of the Stefan block is mapped monotonically to the triple in the intervals $J_{q-5}(k), J_{q-3}(k), J_{q-1}(k)$ on the other side. Since the orbits of f^m are permuted by f like a simple orbit of period m, and the two triples in each such orbit are interchanged by f^m, the $2m$ triples are permuted like a simple orbit of period $2m$. Thus the orbit of period $6m$ is simple. Furthermore all sub-orbits of f^m are mapped monotonically by f onto their images, except the one contained in the intervals $J_i(m)$. Since in addition the triple in the intervals $J_{q-6}(m), J_{q-4}(m), J_{q-2}(m)$, is mapped monotonically to the triple in the intervals $J_{q-5}(1), J_{q-3}(1), J_{q-1}(1)$, it follows that the orbit of period $6m$ is strongly simple. □

THEOREM 18 *If f has an orbit of period n, then f has a strongly simple orbit of period n.*

Proof Let $n = qm$, where q is odd and $m = 2^s$. First assume that $q > 1$. Let j be the smallest non-negative integer such that f has a periodic point of period $2^j k$, where k is odd and $k \geq 3$. We may assume that $k \geq 3$ (odd) is also minimal. By Theorem 11 f has a simple periodic orbit of period $2^j k$. By this same theorem if $k > 3$, or by Theorem 14 if $k = 3$, we see that f has a strongly simple orbit of period $2^j k$. Now, by repeatedly applying Proposition 17 and then Corollary 16, we see that f has a strongly simple orbit of period n.

Finally, suppose that $q = 1$, i.e. $n = 2^s$. If f does not have any orbit of period k, where $\mathbb{P}_k \subset \mathbb{P}_n$ and $1 < k < n$, then by Theorem 11 every orbit of f of period n is strongly simple. Hence, we may assume that for some k, with $\mathbb{P}_k \subset \mathbb{P}_n$ and $1 < k < n$, f has an orbit of period k. Then $k = 2^j t$, where $t \geq 3$ is odd. Since $j < s$, it follows from Šarkovskii's theorem that f has an orbit of period tn; hence by the previous case f has a strongly simple orbit of period tn. Since the Stefan blocks of this orbit are permuted like a simple orbit of period n, it follows that f has a (strongly) simple orbit of period n. □

The primary periodic orbits can now be characterized.

THEOREM 19 *Let* (P,Π) *be a periodic orbit of period n. Then the following statements are equivalent:*

(i) P *is primary,*

(ii) P *is strongly simple,*

(iii) *there is a map f of a compact interval to itself whose only periodic orbit of period n is* P.

Proof (i) ⇒ (ii) follows from Theorem 18 and (iii) ⇒ (i) is immediate. We show that (ii) ⇒ (iii).

Let $P = \{p_1 < p_2 < \ldots < p_n\}$. If $n = 1$, let f be the constant map $f(x) = p_1$ defined on $[p_1-1, p_1+1]$. If $n \geq 2$ is a power of two, let f: $[p_1, p_n] \to [p_1, p_n]$ agree with Π on P, be monotone quadratic on $[p_1, p_2]$ and linear on $[p_i, p_{i+1}]$ for $i \neq 1$. If n is not a power of two, let f: $[p_1, p_n]$ agree with Π on P and be linear on each interval $[p_i, p_{i+1}]$. In each case it is easy to check from the Markov graph that P is the only periodic orbit of f of period n. □

The following generalization of Theorem 14 is useful in extending Theorem 19 to maps of the circle.

PROPOSITION 20 *Suppose f has a periodic orbit* $P = \{x_1 < \ldots < x_n\}$ *of period* $n = qm$, *where* $q \geq 3$ *is odd and* $m > 1$. *Furthermore, suppose f cyclically permutes the sets* $\alpha_k = \{x_{q(k-1)+1}, \ldots, x_{qk}\}$ $(k = 1, \ldots, m)$.

Then f has a periodic orbit $Q = \{y_1 < \ldots < y_n\}$ *of period n, such that f cyclically permutes the sets* $\beta_k = \{y_{q(k-1)+1}, \ldots, y_{qk}\}$ $(k = 1, \ldots, m)$, *and in addition:*

(i) $\beta_k \subseteq [x_{q(k-1)+1}, x_{qk}]$ *for* $k = 1, \ldots, m$,
(ii) *each* β_k *is a periodic orbit of* f^m *of Stefan type, and f maps* β_k *monotonically onto* $f(\beta_k)$ *for all but one k.*

Proof Let j denote the least odd integer greater than 1 such that the restriction of f^m to the convex hull of α_1 has a periodic point of period j. Then $3 \leq j \leq q$. If $j = 3$, the conclusion follows from Theorem 14 and Proposition 15, the latter applied inductively. If $j > 3$, the conclusion follows from Proposition 10 and Proposition 15, the latter again applied inductively. □

Our next goal is to connect the notion of strongly simple orbit with the turbulence stratification.

PROPOSITION 21 *If, for some* $s > 1$, *f has an orbit of period* 2^s *which is not simple, then f also has an orbit of period* 3.2^{s-2}.

Proof If $s = 2$, then f has an orbit of period 3, by Theorem II.9. Proceeding by induction, we assume that $s > 2$ and that the statement holds for $s - 1$. Let P be a periodic orbit of f, of period 2^s, which is not simple. Let L denote the set of 2^{s-1} points forming the left half of P, and let R denote the set of 2^{s-1} points forming the right half of P. First suppose that L and R are not interchanged by f. Then, by Theorem II.9 again, f has a point of odd period > 1. Hence by Šarkovskii's theorem, f has a point of period 3.2^{s-2}.

Finally, suppose that L and R are interchanged by f. Then L and R are orbits of f^2 of period 2^{s-1}, and at least one of them is not simple. By the induction hypothesis, f^2 has a periodic orbit of period 3.2^{s-3}. Hence f has an orbit of period 3.2^{s-2}. □

THEOREM 22 *If* $f \in \mathbb{T}_n$, *but* $f \notin \mathbb{T}_k$ *whenever* $\mathbb{T}_k \subset \mathbb{T}_n$ *and* $1 < k < n$, *then every orbit of f of period n is strongly simple.*

Proof Let $n = 2^s q$, where q is odd. If $q > 1$ then $\mathbb{T}_n = \mathbb{P}_n$ by Theorem II.14. Thus the conclusion follows from Corollary 13 if $q = 3$ and from Theorem 11 if $q > 3$. If $q = 1$ then the result is trivial for $s = 0,1$ and it follows from Proposition 21 if $s > 1$. □

Theorem 22 was obtained in Block and Coppel [26]. It shows that the exceptional case in Theorem 11 is removed by considering the turbulence stratification, instead of the Šarkovskii stratification.

A somewhat different formulation of Theorem 11 is given by Alseda *et al.* [6] and Ho [67]. A periodic orbit P, of period n, may be said to be *minimal* if there is a map f of an interval into itself which has an orbit equivalent to P, but which has no orbit of period k with $k \prec n$. It follows at once from Theorem 11 that if $n = qm$, where q is odd and m is a power of 2, then a minimal orbit of period n is strongly simple if $q \neq 3$ and simple if $q = 3$. Conversely, however, an orbit P of period $n = qm$ is minimal if $q \neq 3$ and P is strongly simple, or if $q = 3$ and P is simple. This is readily seen by taking f to be the linearization of P and considering the Markov graph of P.

We conclude this chapter with a result of Block [22], which gives a necessary and sufficient condition for a map to be chaotic in terms of simple orbits.

PROPOSITION 23 *If $f \in T_{2^d}$ for some $d \geq 0$ then, for every $n > 3$, f has a periodic orbit of period $2^d n$ which is not simple.*

Proof Suppose first that f is turbulent ($d = 0$). Let J, K be closed subintervals of I with at most one common point, and no common periodic point, such that

$$J \cup K \subseteq f(J) \cap f(K).$$

Then there exists a point x of period n such that $x \in J, f(x) \in J$, and $f^k(x) \in K$ for $1 < k < n$. Since $n \geq 4$, the orbit of x is not simple.

Suppose next that a map g has a Stefan orbit P of odd period $n > 3$. We will show that g has an orbit of period $2n$ which is not simple. Let $J_1, \ldots, J_{n-1}$ be the intervals whose endpoints are adjacent points of P, with the same system of numbering as in the statement of Proposition I.8. For definiteness we will assume that J_1 lies to the right of the adjacent interval J_2. Corresponding to the cycle

$$\begin{gathered} J_1 \to J_2 \to \ldots \to J_{n-1} \\ \to J_{n-2} \to J_{n-1} \to \ldots \to J_{n-1} \to J_{n-2} \to J_{n-1} \to J_1 \end{gathered}$$

of length $2n$ in the Markov graph of P, there is an orbit Q of g of period $2n$. If y is the unique point of Q in J_1, then y is the left endpoint of its orbit Q′ under g^2, and Q′ is the right half of Q. Moreover Q′ is not of Stefan type, since g^2 maps y onto the adjacent point of Q′. It follows that Q is not simple.

Suppose now that $f \in \mathrm{T}_{2^d}$ for a minimal value $d > 0$. Then a point x of period n for f^{2^d}, constructed as in the first part of the proof, has period $2^d n$ for f and an orbit which is not simple, except possibly if n is odd when it may have period $2^{d-1}n$. But then x has period n for $g = f^{2^{d-1}}$. Consequently, by the second part of the proof, g has a periodic point y of period $2n$ whose orbit is not simple. Since y has period $2^d n$ for f, the result follows. □

By combining Propositions 21 and 23 we immediately obtain the result of Block:

THEOREM 24 *A map f is chaotic if and only if it has a periodic orbit, of period a power of two, which is not simple.* □

VIII
Topological Entropy

The notion of topological entropy, introduced by Adler *et al.* [1], provides a numerical measure for the complexity of an endomorphism of a compact topological space. We intend to consider here some results which hold for the special case of a compact interval, in particular a theorem of Misiurewicz [83], which implies that a continuous map is chaotic if and only if its topological entropy is positive. However, we first present the definition and basic properties of topological entropy in the general case, since these are still not widely known.

1 DEFINITION AND GENERAL PROPERTIES

Let X be a compact topological space. An *open cover* of X is a collection of open sets whose union is X. An open cover β is said to be a *refinement* of an open cover α, in symbols $\alpha < \beta$, if every open set of β is contained in some open set of α. We say that β is a *subcover* of α if every open set of β actually is an open set of α.

If α and β are two open covers, their *join* $\alpha \vee \beta$ is the open cover consisting of all sets $A \cap B$ with $A \in \alpha$, $B \in \beta$. Thus $\alpha \vee \beta$ is a refinement of both α and β.

Since X is compact, every open cover has a finite subcover. The *entropy* of an open cover α is defined to be

$$H(\alpha) = \log N(\alpha),$$

where $N(\alpha)$ is the minimum number of open sets in any finite subcover. Evidently $H(\alpha) \geq 0$, with equality if and only if $X \in \alpha$. Moreover it is easily seen that

(i) if $\alpha < \beta$, then $H(\alpha) \leq H(\beta)$ and $H(\alpha \vee \beta) = H(\beta)$,

(ii) $H(\alpha \vee \beta) \leq H(\alpha) + H(\beta)$.

Let $f: X \to X$ be a continuous map. For any open cover α we denote by $f^{-1}\alpha$ the open cover consisting of all sets $f^{-1}(A)$ with $A \in \alpha$. Evidently

(iii) if $\alpha < \beta$, then $f^{-1}\alpha < f^{-1}\beta$,
(iv) $f^{-1}(\alpha \vee \beta) = f^{-1}\alpha \vee f^{-1}\beta$.

Moreover it is easily seen that

(v) $H(f^{-1}\alpha) \leq H(\alpha)$, with equality if f is surjective.

From (iv), (ii), (v) we obtain for any positive integers m, n

$$\begin{aligned} H(\alpha \vee \ldots \vee f^{-m-n+1}\alpha) &= H(\alpha \vee \ldots \vee f^{-m+1}\alpha \vee f^{-m}(\alpha \vee \ldots \vee f^{-n+1}\alpha)) \\ &\leq H(\alpha \vee \ldots \vee f^{-m+1}\alpha) + H(f^{-m}(\alpha \vee \ldots \vee f^{-n+1}\alpha)) \\ &\leq H(\alpha \vee \ldots \vee f^{-m+1}\alpha) + H(\alpha \vee \ldots \vee f^{-n+1}\alpha). \end{aligned}$$

Consequently,

$$h(f, \alpha) = \lim_{n\to\infty} H(\alpha \vee f^{-1}\alpha \vee \ldots \vee f^{-n+1}\alpha)/n$$

exists, by virtue of the following simple result from real analysis.

LEMMA 1 *Let* $\{a_n\}$ *be a sequence of real numbers which is subadditive, i.e.*

$$a_{m+n} \leq a_m + a_n \text{ for all } m, n.$$

Then $\lim_{n\to\infty} a_n/n$ *exists and has the value* $c = \inf a_n/n$. (Note that the limit may be $-\infty$.)

Proof For any fixed m set $n = qm + r$, where q, r are non-negative integers and $r < m$. It follows from the subadditivity that $a_n \leq qa_m + a_r$. If $n \to \infty$ for a fixed m then $q/n \to 1/m$ and a_r takes only finitely many values. Hence

$$\overline{\lim}_{n\to\infty} a_n/n \leq a_m/m.$$

Since this holds for arbitrary m we have

$$\overline{\lim}_{n\to\infty} a_n/n \leq c.$$

But, since $a_n/n \geq c$ for every n, we also have

$$c \leq \underline{\lim}_{n\to\infty} a_n/n.$$

The result follows. □

The limit $h(f, \alpha)$ is called the *entropy of f relative to the cover* α. It is evident that $0 \leq h(f, \alpha) \leq H(\alpha)$. From (i) and (iii) we obtain

(vi) if $\alpha < \beta$, then $h(f, \alpha) \leq h(f, \beta)$.

Furthermore

(vii) if f is a homeomorphism, then $h(f^{-1}, \alpha) = h(f, \alpha)$,

since

$$\begin{aligned} H(\alpha \vee \ldots \vee f^{-n+1}\alpha) &= H(f^{n-1}(\alpha \vee \ldots \vee f^{-n+1}\alpha)) \\ &= H(\alpha \vee f\alpha \vee \ldots \vee f^{n-1}\alpha) \\ &= H(\alpha \vee (f^{-1})^{-1}\alpha \vee \ldots \vee (f^{-1})^{-n+1}\alpha). \end{aligned}$$

The *topological entropy* of a continuous map $f: X \to X$ is defined to be

$$h(f) = \sup_\alpha h(f, \alpha),$$

where the supremum is taken over all open covers α. Actually, by (vi), one need only take the supremum over all finite open covers. Evidently $0 \leq h(f) \leq +\infty$. The main properties of topological entropy will now be derived.

PROPOSITION 2 *If $f: X \to X$ is a continuous map then, for any positive integer k,*

$$h(f^k) = k\, h(f).$$

Proof For any open cover α we have

$$\begin{aligned} h(f^k) &\geq h(f^k, \alpha \vee f^{-1}\alpha \vee \ldots \vee f^{-k+1}\alpha) \\ &= \lim_{n\to\infty} k\, H(\alpha \vee f^{-1}\alpha \vee \ldots \vee f^{-k+1}\alpha \vee f^{-k}\alpha \vee \ldots \vee f^{-nk+1}\alpha)/nk \\ &= k\, h(f, \alpha). \end{aligned}$$

Hence $h(f^k) \geq k\, h(f)$. On the other hand, since

$$\alpha \vee (f^k)^{-1}\alpha \vee \ldots \vee (f^k)^{-n+1}\alpha < \alpha \vee f^{-1}\alpha \vee \ldots \vee f^{-nk+1}\alpha,$$

we have

$$\begin{aligned} h(f, \alpha) &= \lim_{n\to\infty} H(\alpha \vee f^{-1}\alpha \vee \ldots \vee f^{-nk+1}\alpha)/nk \\ &\geq \lim_{n\to\infty} H(\alpha \vee (f^k)^{-1}\alpha \vee \ldots \vee (f^k)^{-n+1}\alpha)/nk \\ &= h(f^k, \alpha)/k. \end{aligned}$$

Hence $h(f^k) \leq k\, h(f)$. □

PROPOSITION 3 *If $f: X \to X$ is a homeomorphism, then $h(f^{-1}) = h(f)$.*

Proof This follows at once from (vii) . □

PROPOSITION 4 *Let X, Y be compact topological spaces and let $f: X \to X$, $g: Y \to Y$ be continuous maps. If there exists a continuous map $\varphi: X \to Y$ such that $\varphi(X) = Y$ and the diagram*

$$
\begin{array}{ccc}
X & \xrightarrow{\ f\ } & X \\
\varphi \downarrow & & \downarrow \varphi \\
Y & \xrightarrow[\ g\ ]{} & Y
\end{array}
$$

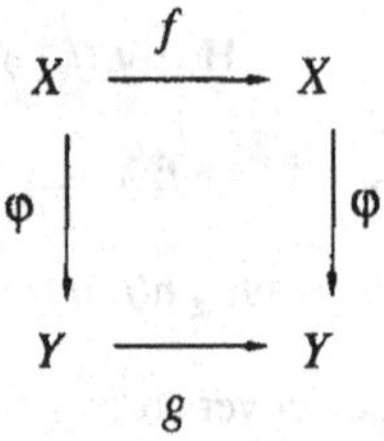

commutes, then $h(g) \leq h(f)$. Moreover, if φ is a homeomorphism, then $h(g) = h(f)$.

Proof If α is any open cover of Y then, since φ is surjective, $\varphi^{-1}\alpha$ is an open cover of X . Moreover, since $\varphi \circ f^k = g^k \circ \varphi$,

$$
\begin{aligned}
h(g, \alpha) &= \lim\ H(\alpha \vee g^{-1}\alpha \vee ... \vee g^{-n+1}\alpha)/n \\
&= \lim\ H(\varphi^{-1}(\alpha \vee g^{-1}\alpha \vee ... \vee g^{-n+1}\alpha))/n \\
&= \lim\ H(\varphi^{-1}\alpha \vee \varphi^{-1}g^{-1}\alpha \vee ... \vee \varphi^{-1}g^{-n+1}\alpha)/n \\
&= \lim\ H(\varphi^{-1}\alpha \vee f^{-1}\varphi^{-1}\alpha \vee ... \vee f^{-n+1}\varphi^{-1}\alpha)/n \\
&= h(f, \varphi^{-1}\alpha).
\end{aligned}
$$

Hence $h(g) \leq h(f)$.

If φ is a homeomorphism then $\varphi^{-1} \circ g = f \circ \varphi^{-1}$ and hence $h(f) \leq h(g)$, by what we have just proved. □

This result shows that topological entropy is an invariant of topological conjugacy. The remainder of this section is devoted to the proof of a theorem of Bowen [42], which says that the topological entropy of a map is concentrated in its non-wandering set. The proof given here follows Xiong [132].

PROPOSITION 5 *If $f: X \to X$ is a continuous map and Y is a closed, invariant subset of X, then* $h(f|Y) \le h(f)$.

Proof Let α be an open cover of Y. For each $A \in \alpha$ there exists an open subset $\tilde{A}$ of X such that $A = \tilde{A} \cap Y$. The set of these $\tilde{A}$, together with the open set $X \setminus Y$, form an open cover $\tilde{\alpha}$ of X. If $g = f|Y$ and n is a positive integer, then

$$H(\alpha \vee g^{-1}\alpha \vee \ldots \vee g^{-n+1}\alpha) \le H(\tilde{\alpha} \vee f^{-1}\tilde{\alpha} \vee \ldots \vee f^{-n+1}\tilde{\alpha}).$$

It follows that $h(g,\alpha) \le h(f,\tilde{\alpha}) \le h(f)$ and hence, since α was arbitrary, $h(g) \le h(f)$. □

THEOREM 6 *If $f: X \to X$ is a continuous map, then* $h(f) = h(f|\Omega(f))$.

Proof It follows from Proposition 5 that $h(f|\Omega(f)) \le h(f)$. Thus we need only prove the reverse inequality.

We prove first that if an open cover α of X has the property that $\Omega(f) \subseteq A$ for some $A \in \alpha$, then $h(f,\alpha) = 0$.

For each $x \in X \setminus \Omega(f)$, let A_x be an open set in α containing x. Then there exists an open neighbourhood B_x of x such that $B_x \subseteq A_x$ and $f^k(B_x) \cap B_x = \emptyset$ for each positive integer k. The collection α' of all sets B_x together with A is an open cover of X which is a refinement of α. Let $\beta = \{A, B_{x_1}, \ldots, B_{x_t}\}$ be a finite subcover of α' with $t > 1$. Since $\alpha < \beta$, it is sufficient to prove that $h(f,\beta) = 0$.

For a given positive integer n, let β^n denote the collection of all sequences $(C_0, C_1, \ldots, C_{n-1})$, where each C_i is an open set of β. We define a map φ of β^n onto $\beta \vee f^{-1}\beta \vee \ldots \vee f^{-n+1}\beta$ by setting

$$\varphi(C_0, C_1, \ldots, C_{n-1}) = C_0 \cap f^{-1}(C_1) \cap \ldots \cap f^{-n+1}(C_{n-1}).$$

If for some $(C_0, C_1, \ldots, C_{n-1}) \in \beta^n$ the number of terms C_i different from A is greater than t, then there exist j and k with $j < k$ such that $C_j = C_k = B_{x_s}$ for some s. Then $\varphi(C_0, C_1, \ldots, C_{n-1}) = \emptyset$, because $x \in C_0 \cap f^{-1}(C_1) \cap \ldots \cap f^{-n+1}(C_{n-1})$ implies $f^j(x) \in B_{x_s}, f^k(x) \in B_{x_s}$, and hence $f^{k-j}(B_{x_s}) \cap B_{x_s} \ne \emptyset$, contrary to the definition of the sets B_x. Therefore the number of non-empty sets in the open cover $\beta \vee f^{-1}\beta \vee \ldots \vee f^{-n+1}\beta$ is not greater than the number of sequences $(C_0, C_1, \ldots, C_{n-1}) \in \beta^n$ with at most t terms different from A. Thus

$$\begin{aligned} N(\beta \vee f^{-1}\beta \vee \ldots \vee f^{-n+1}\beta) &\le 1 + \binom{n}{1}t + \ldots + \binom{n}{t}t^t \\ &\le n^t t^{t+1}, \end{aligned}$$

and

$$\begin{aligned} h(f,\beta) &= \lim_{n\to\infty} H(\beta \vee f^{-1}\beta \vee \ldots \vee f^{-n+1}\beta)/n \\ &\leq \lim_{n\to\infty} [\, t \log n + (t+1)\log t]/n \\ &= 0\,, \end{aligned}$$

as we wished to prove.

If α is any open cover of X and k any positive integer, then

$$k\alpha = \{A_1 \cup \ldots \cup A_k : A_i \in \alpha\}$$

is again an open cover of X. It is evident that

(i) $N(\alpha) \leq k\, N(k\alpha)$,
(ii) $f^{-1}(k\alpha) = k f^{-1}(\alpha)$,
(iii) if $\alpha_0, \ldots, \alpha_{n-1}$ are all open covers of X, then $k\alpha_0 \vee \ldots \vee k\alpha_{n-1}$ is a subcover of $k^n(\alpha_0 \vee \ldots \vee \alpha_{n-1})$.

It follows that

(iv) $h(f, k\alpha) \geq h(f,\alpha) - \log k$,

since

$$\begin{aligned} h(f,k\alpha) &= \lim_{n\to\infty} H(k\alpha \vee f^{-1}(k\alpha) \vee \ldots \vee f^{-n+1}(k\alpha))/n \\ &= \lim_{n\to\infty} H(k\alpha \vee kf^{-1}(\alpha) \vee \ldots \vee kf^{-n+1}(\alpha))/n \\ &\geq \lim_{n\to\infty} H(k^n(\alpha \vee f^{-1}\alpha \vee \ldots \vee f^{-n+1}\alpha))/n \\ &\geq \lim_{n\to\infty} [\log (N(\alpha \vee f^{-1}\alpha \vee \ldots \vee f^{-n+1}\alpha)/k^n)]/n \\ &= \lim_{n\to\infty} H(\alpha \vee f^{-1}\alpha \vee \ldots \vee f^{-n+1}\alpha)/n - \log k \\ &= h(f,\alpha) - \log k\,. \end{aligned}$$

If α is any open cover of X and if Y is an invariant closed subset of X, we denote by $\alpha\,|Y$ the open cover $\{A \cap Y : A \in \alpha\}$ of Y. It follows at once that

(v) $f^{-1}\alpha\,|Y = (f\,|Y)^{-1}(\alpha\,|Y)$,
(vi) if β is any other open cover of X, then

$$(\alpha \vee \beta)\,|Y = (\alpha\,|Y) \vee (\beta\,|Y)\,,$$

(vii) if Z is an invariant closed subset of Y, then $N(\alpha\,|Z) \leq N(\alpha\,|Y)$.

We show next that if α is any open cover of X, then

(viii) $h(f,\alpha) \leq H(\alpha\,|\Omega(f))$.

Let $k = N(\alpha\,|\Omega(f))$ and let $\{A_1 \cap \Omega(f), ..., A_k \cap \Omega(f)\}$, where $A_1, ..., A_k$ are in α, be a subcover of $\alpha\,|\Omega(f)$ containing k elements. Then

$$\alpha'' = \{A_1 \cup ... \cup A_{k-1} \cup A : A \in \alpha\}$$

is a subcover of $k\alpha$. Hence, by (iv),

$$h(f,\alpha'') \geq h(f, k\alpha) \geq h(f, \alpha) - \log k .$$

But $h(f,\alpha'') = 0$ by the first part of the proof, since $\Omega(f) \subseteq A_1 \cup ... \cup A_k$. Thus $h(f, \alpha) \leq \log k$, in agreement with (viii).

Finally, let α be any open cover of X and m a positive integer. Then, putting $\gamma = \alpha \vee f^{-1}\alpha \vee ... \vee f^{-m+1}\alpha$, we have

$$\begin{aligned} h(f, \alpha) &= \lim_{n\to\infty} H(\alpha \vee f^{-1}\alpha \vee ... \vee f^{-n+1}\alpha)/n \\ &= \lim_{n\to\infty} H(\alpha \vee f^{-1}\alpha \vee ... \vee f^{-mn+1}\alpha)/mn \\ &= \lim_{n\to\infty} H(\gamma \vee f^{-m}\gamma \vee ... \vee f^{-(n-1)m}\gamma)/mn \\ &= h(f^m,\gamma)/m . \end{aligned}$$

But, by (viii),

$$\begin{aligned} h(f^m,\gamma) &\leq H(\gamma\,|\Omega(f^m)) \\ &\leq H(\gamma\,|\Omega(f)) , \end{aligned}$$

and, by (v) and (vi),

$$H(\gamma\,|\Omega(f)) = H(\alpha\,|\Omega(f) \vee (f\,|\Omega(f))^{-1}(\alpha\,|\Omega(f)) \vee ... \vee (f\,|\Omega(f))^{-m+1}(\alpha\,|\Omega(f))) .$$

It follows that

$$h(f, \alpha) \leq h(f\,|\Omega(f), \alpha\,|\Omega(f)) .$$

Hence

$$h(f) \leq h(f\,|\Omega(f)) .$$

As stated at the outset, this is all we needed to prove. □

COROLLARY 7 *If $f: I \to I$ is a continuous map of a compact interval to itself, then*

$$h(f) = h(f\,|\overline{P}(f)) .$$

Proof This follows at once from Theorem 6 and Proposition IV.15. □

It may be noted also that if $f: X \to X$ is a continuous map of a compact metric space X into itself, then $h(f) = h(f\,|\overline{R}(f))$. This follows in the same way from Theorem 6 and Birkhoff's characterization of the centre of f.

2 RESULTS FOR A COMPACT INTERVAL

From now on we restrict attention to the case where $X = I$ *is a compact interval*. Moreover, although we naturally assume that I itself has non-empty interior, throughout the rest of this chapter we will use the word 'interval' to mean any connected subset of I. Thus an interval may now consist of a single point.

We consider first some ways of estimating the topological entropy $h(f)$ of a continuous map $f: I \to I$.

PROPOSITION 8 *Let $f: I \to I$ be a continuous map. If there exist disjoint closed intervals $J_1, \ldots, J_p$ such that*

$$J_1 \cup \ldots \cup J_p \subseteq f(J_i) \quad (i = 1, \ldots, p),$$

then $h(f) \geq \log p$.

Proof We can choose pairwise disjoint open intervals $G_1, \ldots, G_p$ with $J_i \subset G_i$ for $i = 1, \ldots, p$. By adjoining further open intervals $G_{p+1}, \ldots, G_q$, satisfying $G_i \cap J_k = \emptyset$ for $p+1 \leq i \leq q$ and $1 \leq k \leq p$, we obtain a finite open cover α.

For any positive integer n and any i_k with $1 \leq i_k \leq p$ the set

$$J_{i_1 \ldots i_n} = \{x : x \in J_{i_1}, f(x) \in J_{i_2}, \ldots, f^{n-1}(x) \in J_{i_n}\}$$

is non-empty. Each point in this set is contained in a unique element of $\alpha \vee f^{-1}\alpha \vee \ldots \vee f^{-n+1}\alpha$, namely

$$G_{i_1} \cap f^{-1}(G_{i_2}) \cap \ldots \cap f^{-n+1}(G_{i_n}).$$

It follows that

$$H(\alpha \vee f^{-1}\alpha \vee \ldots \vee f^{-n+1}\alpha) \geq n \log p.$$

Hence $h(f, \alpha) \geq \log p$ and, *a fortiori*, $h(f) \geq \log p$. □

Proposition 8 is a rather crude result which can be considerably improved. However, we will require some knowledge of the properties of non-negative matrices. Proofs of the following three lemmas may be found in Berman and Plemmons [15], for example.

LEMMA 9 *Let $A = (a_{ik})$ be a $p \times p$ matrix of non-negative real numbers. Then there exist $\lambda \geq 0$ and a non-zero vector $x = (x_k)$ with $x_k \geq 0$ $(k = 1, ..., p)$ such that $Ax = \lambda x$ and $|\mu| \leq \lambda$ for every other eigenvalue μ of A.*

We will refer to λ as the *maximal* eigenvalue of A. A non-negative matrix A is said to be *reducible* if there exists a permutation matrix P such that

$$P^t A P = \begin{bmatrix} B & 0 \\ C & D \end{bmatrix},$$

where B and D are square matrices of smaller size, and *irreducible* otherwise.

LEMMA 10 *For any non-negative matrix A there exists a permutation matrix P such that*

$$P^t A P = \begin{bmatrix} A_{11} & 0 & \dots & 0 \\ A_{21} & A_{22} & \dots & 0 \\ \dots & & \dots & \\ A_{r1} & A_{r2} & \dots & A_{rr} \end{bmatrix},$$

where each diagonal block A_{kk} $(k = 1, ..., r)$ is irreducible.

LEMMA 11 *Let A be a non-negative matrix with maximal eigenvalue λ. If A is irreducible, then there exists a positive integer h such that the eigenvalues of A with absolute value λ are*

$$\lambda, \lambda\omega, ..., \lambda\omega^{h-1},$$

where $\omega = \exp(2\pi i/h)$.

We define the *norm* of a real or complex matrix $A = (a_{ik})$ to be

$$|A| = \sum_{i,k} |a_{ik}|.$$

The maximal eigenvalue is related to the norm in the following way.

LEMMA 12 *Let A be a non-negative matrix with maximal eigenvalue λ. Then*

$$\lambda = \lim_{n\to\infty} |A^n|^{1/n}.$$

Proof Let $x = (x_k)$ be the non-negative eigenvector corresponding to the eigenvalue λ, so that

$$\lambda x_i = \Sigma_k a_{ik} x_k .$$

If we choose i so that $x_i = \max x_k$ we obtain

$$\lambda \leq \Sigma_k a_{ik} \leq |A|.$$

Applying this inequality to A^n, instead of A, we obtain $\lambda^n \leq |A^n|$ and hence

$$\lambda \leq \underline{\lim}_{n\to\infty} |A^n|^{1/n}.$$

Let T be a non-singular matrix such that $J = T^{-1}AT$ is in Jordan normal form. If $\rho > \lambda$ then $J^n/\rho^n \to 0$ as $n \to \infty$ and hence also $A^n/\rho^n \to 0$. Thus $|A^n| < \rho^n$ for all large n and

$$\overline{\lim}_{n\to\infty} |A^n|^{1/n} \leq \rho.$$

Since this holds for any $\rho > \lambda$, the result follows. □

Essentially the same argument shows that, for *any* matrix A,

$$\lim_{n\to\infty} |A^n|^{1/n} = \mu,$$

where μ is the maximum absolute value of any eigenvalue of A.

The maximal eigenvalue can also be characterized in terms of the trace.

LEMMA 13 *Let A be a non-negative matrix with maximal eigenvalue λ. Then*

$$\lambda = \overline{\lim}_{n\to\infty} (\operatorname{tr} A^n)^{1/n}.$$

Proof If A is a $p \times p$ matrix then $\operatorname{tr} A^n \leq p\lambda^n$ and hence

$$\overline{\lim}_{n\to\infty} (\operatorname{tr} A^n)^{1/n} \leq \lambda.$$

It remains to prove the reverse inequality.

Suppose first that A is irreducible. Then, by Lemma 11, the eigenvalues of A with absolute value λ are $\lambda, \lambda\omega, \ldots, \lambda\omega^{h-1}$, where $\omega = \exp(2\pi i/h)$. If we denote the remaining eigenvalues of A by $\lambda_{h+1}, \ldots, \lambda_p$ then

$$\mathrm{tr}\, A^n = \lambda^n + \lambda^n \omega^n + \ldots + \lambda^n \omega^{(h-1)n} + \lambda_{h+1}^n + \ldots + \lambda_p^n .$$

In particular if we take $n = kh$ to be a multiple of h then $\omega^n = 1$ and

$$\begin{aligned}(\mathrm{tr}\, A^n)^{1/n} &= \lambda[h + \textstyle\sum_{i=h+1}^{p} (\lambda_i / \lambda)^n]^{1/n} \\ &\to \lambda \ \text{ as } k \to \infty.\end{aligned}$$

In the general case we appeal to Lemma 10. For some k the irreducible diagonal block A_{kk} has maximal eigenvalue λ. Since

$$\mathrm{tr}\, A^n \geq \mathrm{tr}(A_{kk})^n ,$$

it follows from what we have just proved that

$$\overline{\lim}\, (\mathrm{tr}\, A^n)^{1/n} \geq \lambda. \quad \square$$

Suppose, in particular, that every element of the matrix A is 0 or 1. We define a *path* of length n to be a finite sequence $\{i_1, i_2, \ldots, i_{n+1}\}$ such that

$$a_{i_1 i_2}\, a_{i_2 i_3} \ldots a_{i_n i_{n+1}} \neq 0.$$

Then $|A^n|$ is just the number of paths of length n, since

$$|A^n| = \sum a_{i_1 i_2}\, a_{i_2 i_3} \ldots a_{i_n i_{n+1}} .$$

Similarly $(A^n)_{ik}$ is the number of paths of length n with $i_1 = i$, $i_{n+1} = k$.

We can now sharpen Proposition 8 in the following way.

PROPOSITION 14 *Let $f: I \to I$ be a continuous map. Let $J_1, \ldots, J_p$ be closed intervals with pairwise disjoint interiors and let $A = (a_{ik})$ be the $p \times p$ matrix defined by*

$$a_{ik} = 1 \ \text{if } J_k \subseteq f(J_i),\ a_{ik} = 0 \ \text{otherwise} .$$

Then $h(f) \geq \log \lambda$, *where λ is the maximal eigenvalue of A.*

Proof Evidently we may suppose that $\lambda > 1$. By Lemma 13 we have

$$\log \lambda = \overline{\lim}\ [\log \mathrm{tr}(A^n)]/n .$$

Since

$$\max_i (A^n)_{ii} \leq \mathrm{tr}(A^n) \leq p \max_i (A^n)_{ii} ,$$

it follows that for some i $(1 \le i \le p)$

$$\log \lambda = \overline{\lim}\ [\log (A^n)_{ii}]/n .$$

Thus for any μ with $1 < \mu < \lambda$ there exist arbitrarily large n such that there are more than μ^n paths of length n from J_i back to J_i. Evidently J_i must have non-empty interior. Hence there exist more than μ^n closed intervals K_j with pairwise disjoint interiors such that $K_j \subseteq J_i$, $f^n(K_j) = J_i$. By omitting the two intervals K_j which are furthest to the left and to the right and slightly shrinking the remaining intervals we obtain at least $[\mu^n] - 1$ disjoint closed intervals L_j such that

$$\bigcup_j L_j \subseteq \operatorname{int} f^n(L_m) \text{ for every } m.$$

Hence, by Propositions 2 and 8,

$$\begin{aligned} h(f) &= h(f^n)/n \ge [\log (\mu^n - 2)]/n \\ &= \log \mu + [\log(1 - 2/\mu^n)]/n . \end{aligned}$$

Since n can be arbitrarily large, it follows that $h(f) \ge \log \mu$. Since this holds for any $\mu < \lambda$, it follows that $h(f) \ge \log \lambda$. □

COROLLARY 15 *Let $f: I \to I$ be a continuous map. If f is turbulent, then* $h(f) \ge \log 2$. □

We are going to consider next some cases where the lower bound for $h(f)$ given by Proposition 14 is actually attained.

Let α be a collection of finitely many disjoint intervals. We will say that a finite sequence $\{A_1, \dots, A_n\}$ of intervals $A_k \in \alpha$ is a *chain* of *length n* if there exists a point x such that

$$x \in A_1, f(x) \in A_2, \dots, f^{n-1}(x) \in A_n,$$

i.e. if the set

$$A_1 \cap f^{-1}(A_2) \cap \dots \cap f^{-n+1}(A_n)$$

is non-empty. We will denote by α^n the set of all chains of length n and by $c_n(\alpha)$ its cardinality, i.e. the number of different chains of length n. Then

$$c_{m+n}(\alpha) \le c_m(\alpha)\, c_n(\alpha),$$

since if $\{A_1, \dots, A_{m+n}\}$ is a chain of length $m+n$ then $\{A_1, \dots, A_m\}$ is a chain of length m and $\{A_{m+1}, \dots, A_{m+n}\}$ is a chain of length n. Thus $\{\log c_n(\alpha)\}$ is a subadditive sequence. Therefore, by Lemma 1,

$$h^*(f, \alpha) = \lim_{n\to\infty} [\log c_n(\alpha)]/n$$

exists and $h^*(f, \alpha) \leq (1/n) \log c_n(\alpha)$ for every n.

Clearly we cannot expect the limit $h^*(f, \alpha)$ to have a connection with topological entropy unless α is a *cover*, i.e. unless the intervals of α have union I. However, in this case there is a connection.

PROPOSITION 16 *Let $f: I \to I$ be a continuous map. For any open cover β, there exists a cover α consisting of finitely many disjoint intervals such that*

$$h(f, \beta) \leq h^*(f, \alpha).$$

Conversely, for any cover α consisting of finitely many disjoint intervals there exists an open cover β such that

$$h^*(f, \alpha) \leq h(f, \beta) + \log 3.$$

Proof Let β be any open cover. We may suppose that β is the union of finitely many open intervals, since replacing β by a refinement does not decrease the entropy of f relative to the cover. Let α be a cover consisting of finitely many disjoint intervals such that each interval $A \in \alpha$ is contained in some open interval $B \in \beta$. For any chain $\{A_1, ... A_n\}$ in α^n pick some $B_k \in \beta$ such that $A_k \subseteq B_k$ $(1 \leq k \leq n)$. Since the collection of all open sets

$$B_1 \cap f^{-1}(B_2) \cap ... \cap f^{-n+1}(B_n)$$

obtained in this way covers I, it follows that

$$H(\beta \vee f^{-1}\beta \vee ... \vee f^{-n+1}\beta) \leq \log c_n(\alpha)$$

and hence $h(f, \beta) \leq h^*(f, \alpha)$.

Now let α be any cover consisting of finitely many disjoint intervals. Let β be a finite open cover of I, consisting of the interiors of the intervals in α together with small open intervals surrounding the endpoints of these intervals, chosen so that every open interval in β is contained in the union of at most three intervals in α (three intervals being needed if an interval in α contains only one point). Let σ_n denote a subcover of $\beta \vee f^{-1}\beta \vee ... \vee f^{-n+1}\beta$ of minimum cardinality. For any chain $\{A_1, ..., A_n\}$ in α^n there exists a point x with $f^{j-1}(x) \in A_j$ for $j = 1,...,n$. Then x belongs to some element

$$B_1 \cap f^{-1}(B_2) \cap ... \cap f^{-n+1}(B_n)$$

of σ_n. Evidently $A_k \cap B_k \neq \emptyset$ $(1 \leq k \leq n)$. Since the number of different chains $\{A_1, ..., A_n\}$ which correspond in this way to the same sequence $\{B_1, ..., B_n\}$ is at most 3^n, it follows that

$$\log c_n(\alpha) \leq H(\beta \vee f^{-1}\beta \vee ... \vee f^{-n+1}\beta) + n \log 3$$

and hence

$$h^*(f, \alpha) \leq h(f, \beta) + \log 3 . \quad \square$$

It follows as a corollary that

$$h(f) \leq \sup_\alpha h^*(f, \alpha) \leq h(f) + \log 3 .$$

There is one important case in which we can say much more. We say that f is *monotonic* on an interval J if f is either nondecreasing or nonincreasing on J. It should be noted that a map may be monotonic and yet not piecewise monotone, as defined in Chapter II.

PROPOSITION 17 *Let $f: I \to I$ be a continuous map, and let α be a cover consisting of finitely many disjoint intervals. If f is monotonic on each interval of α, then*

$$h^*(f, \alpha) = h(f) .$$

Proof We show first that $h(f) \geq h^*(f, \alpha)$. Let n be a fixed positive integer. For any chain $\{A_1, ..., A_n\}$ in α^n the set of all x such that $x \in A_1, f(x) \in A_2, ..., f^{n-1}(x) \in A_n$ is an interval, since f is monotonic on each A_j. Moreover the intervals corresponding to two distinct chains in α^n are disjoint, and every point of I belongs to such an interval. Thus we obtain a cover α_n of I consisting of finitely many disjoint intervals.

Put $\beta = \alpha_n$, $g = f^n$ and let k be any positive integer. For any chain $\{A_1, ..., A_{kn}\}$ in α^{kn} the sequences $\{A_1, ..., A_n\}, ..., \{A_{(k-1)n+1}, ..., A_{kn}\}$ are chains in α^n. Let $B_1, ..., B_k$ be the corresponding intervals in $\beta = \alpha_n$. Also let C be the interval in α_{kn} corresponding to the chain $\{A_1, ..., A_{kn}\}$ in α^{kn}. If $x \in C$ then $x \in B_1, f^n(x) \in B_2, ..., f^{(k-1)n}(x) \in B_k$ and hence $\{B_1, ..., B_k\}$ is in $\beta^k(g)$. Evidently to distinct chains in α^{kn} correspond distinct chains in $\beta^k(g)$. Conversely, for any chain $\{B_1, ..., B_k\}$ in $\beta^k(g)$ there exist intervals $A_1, ..., A_{kn}$ in α such that B_1 is the set of all x for which $x \in A_1, f(x) \in A_2, ..., f^{n-1}(x) \in A_n$; B_2 is the set of all x such that $x \in A_{n+1}, f(x) \in A_{n+2}, ..., f^{n-1}(x) \in A_{2n}$; and so on. But there exists a y such that

$$y \in B_1, f^n(y) \in B_2, ..., f^{(k-1)n}(y) \in B_k .$$

Hence $f^{j-1}(y) \in A_j$ $(1 \le j \le kn)$ and $\{A_1, \dots, A_{kn}\}$ is in α^{kn}.

Thus $c_{kn}(\alpha, f) = c_k(\beta, g)$ for every k and hence

$$h^*(f, \alpha) = h^*(g, \beta)/n.$$

Therefore, by Propositions 16 and 2,

$$\begin{aligned} h^*(f, \alpha) &\le [h(g) + \log 3]/n \\ &= h(f) + (\log 3)/n. \end{aligned}$$

Since this holds for arbitrary n, it follows that $h^*(f, \alpha) \le h(f)$.

To complete the proof we show next that $h(f) \le h^*(f, \alpha)$. It is sufficient to show that if $\tilde{\alpha}$ is any cover consisting of finitely many disjoint intervals which refines α then

$$h^*(f, \tilde{\alpha}) = h^*(f, \alpha).$$

For let β be any cover consisting of finitely many open intervals and let δ be a cover consisting of finitely many disjoint intervals which refines both α and β. Then it will follow that $h(f, \beta) \le h^*(f, \delta) = h^*(f, \alpha)$. Hence, since β is arbitrary, $h(f) \le h^*(f, \alpha)$.

Since $\tilde{\alpha}$ refines α, for any chain $\{A_1, \dots, A_n\}$ in α^n there is a chain $\{\tilde{A}_1, \dots, \tilde{A}_n\}$ in $\tilde{\alpha}^n$ with $\tilde{A}_k \subseteq A_k$ $(1 \le k \le n)$. Hence $c_n(\alpha) \le c_n(\tilde{\alpha})$ and $h^*(f, \alpha) \le h^*(f, \tilde{\alpha})$.

Consider a fixed chain $\{A_1, \dots, A_n\}$ in α^n and let Γ be the collection of all chains $\gamma = \{\tilde{A}_1, \dots, \tilde{A}_n\}$ in $\tilde{\alpha}^n$ with $\tilde{A}_k \subseteq A_k$ $(1 \le k \le n)$. The union of all intervals $\tilde{A}_k$ which appear in chains $\gamma \in \Gamma$ is again an interval B_k. Suppose B_k contains m_k intervals $\tilde{A}_k$. There is a chain $\{\tilde{E}_1, \dots, \tilde{E}_n\}$ in Γ which is *extreme* in the sense that $\tilde{E}_k$ is an end-interval of B_k for every k. For any chain $\gamma = \{\tilde{A}_1, \dots, \tilde{A}_n\}$ in Γ let μ_k $(0 \le \mu_k < m_k)$ denote the number of intervals in B_k on the same side of $\tilde{A}_k$ as E_k (excluding $\tilde{A}_k$ itself) and let $\mu(\gamma) = \mu_1 + \mu_2 + \dots + \mu_n$. If $\gamma' = \{\tilde{A}'_1, \dots, \tilde{A}'_n\}$ is any other chain in Γ then, since f is monotonic on each A_k, either $\mu'_k \le \mu_k$ for all k or $\mu_k \le \mu'_k$ for all k, with at least one inequality strict. Hence $\mu(\gamma') \ne \mu(\gamma)$. Thus the total number of chains in Γ is at most

$$1 + \textstyle\sum_{k=1}^{n} (m_k - 1) = \sum_{k=1}^{n} m_k + 1 - n.$$

If m is the maximum number of intervals in $\tilde{\alpha}$ which are contained in any one interval of α it follows that $c_n(\tilde{\alpha}) \le (mn + 1 - n)\, c_n(\alpha)$ for each n. Hence $h^*(f, \tilde{\alpha}) \le h^*(f, \alpha)$. □

From Proposition 17 we can derive a formula for the topological entropy of, in particular, any piecewise monotone map.

PROPOSITION 18 *Let $f: I \to I$ be a continuous map such that I is the union of finitely many intervals on each of which f is monotonic. Then*

$$h(f) = \lim (\log k_n)/n ,$$

where k_n is the least number of disjoint intervals with union I on each of which f^n is monotonic.

Proof There is a cover $\alpha = \alpha(n)$ of I consisting of exactly k_n disjoint intervals on each of which f^n is monotonic. For each positive integer j,

$$c_j(\alpha, f^n) \leq (k_n)^j$$

and hence

$$(\log c_j(\alpha, f^n))/j \leq \log k_n .$$

It follows from Proposition 17 that

$$h(f^n) = h^*(f^n, \alpha) \leq \log k_n .$$

Thus $h(f) = h(f^n)/n \leq (\log k_n)/n$ for each positive integer n, and hence

$$h(f) \leq \underline{\lim} (\log k_n)/n .$$

On the other hand, let γ denote the collection of maximal open intervals on which f is monotonic, and let δ denote the collection of endpoints of these intervals. Then $\beta = \gamma \cup \delta$ is a cover of I consisting of $2k_1 + 1$ disjoint intervals. Let n be a positive integer and let J be a maximal interval on which f^n is monotonic. For each $i = 1, \ldots ,n$ there is a unique element A_i of γ such that $f^{i-1}(J) \subseteq \overline{A}_i$. Then $\{A_1, \ldots , A_n\}$ is a chain in β^n. It follows from their maximality that distinct maximal intervals on which f^n is monotonic yield distinct elements of β^n in this way. Thus $k_n \leq c_n(\beta, f)$ for each positive integer n, and hence

$$\overline{\lim} (\log k_n)/n \leq h^*(f, \beta) = h(f). \quad \square$$

From Proposition 17 we obtain also

PROPOSITION 19 *Let $f: I \to I$ be a continuous map and suppose*

$$I = J_1 \cup \ldots \cup J_p ,$$

where $J_1, \dots, J_p$ are closed intervals with non-empty and pairwise disjoint interiors such that f is monotonic on each interval J_i and maps the set of endpoints of all intervals J_i into itself.

Let $A = (a_{ik})$ be the $p \times p$ matrix defined by

$$a_{ik} = 1 \ \text{ if } \ J_k \subseteq f(J_i)\,, \quad a_{ik} = 0 \ \text{ otherwise},$$

and let λ denote the maximal eigenvalue of A. Then $h(f) = \max(0, \log \lambda)$.

If in addition f is not constant on each interval J_i $(i = 1,\dots,p)$, then $\lambda \geq 1$ and $h(f) = \log \lambda$.

Proof Let α be the cover of I, consisting of disjoint intervals $K_1, \dots, K_{2p+1}$, which is formed by the endpoints of the intervals J_i and their interiors. Then $h(f) = h^*(f, \alpha)$, by Proposition 17.

Let $B = (b_{ij})$ be the $(2p+1) \times (2p+1)$ matrix defined by

$$b_{ij} = 1 \ \text{ if } \ K_j \subseteq f(K_i)\,, \quad b_{ij} = 0 \ \text{ otherwise}\,.$$

It follows from the hypotheses that $K_j \cap f(K_i) \neq \emptyset$ only if $K_j \subseteq f(K_i)$. Hence the number of chains in α^n is equal to the number of paths in B of length n, i.e. $c_n(\alpha) = |B^n|$. Consequently,

$$\log |B^n|^{1/n} \to h(f) \quad \text{as } n \to \infty.$$

If μ is the maximal eigenvalue of B, it now follows from Lemma 12 that $\mu \geq 1$ and $h(f) = \log \mu$.

We may suppose the notation chosen so that $K_1, \dots, K_{p+1}$ are the endpoints of the intervals J_i. Then B has the form

$$B = \begin{bmatrix} C & 0 \\ D & A \end{bmatrix},$$

and hence $\operatorname{tr} B^n = \operatorname{tr} C^n + \operatorname{tr} A^n$. Since $0 \leq \operatorname{tr} C^n \leq p+1$ for every n it follows from Lemma 13, and from Lemma 23 below, that $\log \mu = \max(0, \log \lambda)$. If f is not constant on any interval J_i then $\lambda \geq 1$, since the Markov graph of f contains a cycle. □

Although Propositions 14 and 19 are known, cf. Lemma 1.5 of Block *et al.* [37] and Theorem 1 of Misiurewicz and Szlenk [88], direct proofs have not been available in the literature. The following result is a simple consequence.

Proposition 20 *Let P be a finite subset of the real line, and let π:P $\to$ P be any map of this subset into itself. Let L denote the linearization of π, let I be any compact interval containing P, and let f:$I \to I$ be any continuous function such that $f\,|$P = π. Then*

$$h(f) \geq h(L),$$

with equality if I is the convex hull of P and f is monotonic between adjacent points of P.

Proof Let A be the matrix defined as in Proposition 19, using the intervals J_i whose endpoints are adjacent points of P. If I is the convex hull of P and f is monotonic between adjacent points of P, in particular if f = L, then by Proposition 19, h(f) = max(0, log λ), where λ is the maximal eigenvalue of A. In the general case h(f) $\geq$ max(0, log λ), by Proposition 14. □

A more substantial application of Propositions 14 and 19 is the next result, due to Block and Coppel [26]. It shows that, among all periodic orbits with a given period, the strongly simple orbits represent those with minimum entropy.

Proposition 21 *Let f: $I \to I$ be a continuous map with an orbit of period $n = qm$, where $q \geq 1$ is odd and $m = 2^s$. Then*

$$h(f) \geq (\log \lambda_q)/m, \qquad (*)$$

where λ_q is the unique positive root of the polynomial

$$L_q(\lambda) = \lambda^q - 2\lambda^{q-2} - 1$$

if $q > 1$ and $\lambda_q = 1$ if $q = 1$. Moreover, if equality holds in (*) then every orbit of period n is strongly simple.

Conversely, equality holds in (*) if the periodic orbit P = $\{x_1 < ... < x_n\}$ is strongly simple, $I = [x_1, x_n]$ and f is monotonic on each subinterval $J_i = [x_i, x_{i+1}]$ $(1 \leq i < n)$.

Proof We first note that the polynomial $L_q(\lambda)$ has at most one positive root, by Descartes' rule of signs, and that $1 < \lambda_q < 2$, since $L_q(1) < 0 < L_q(2)$.

We next prove the last assertion in the statement of the proposition. Thus we assume that f is monotonic on each subinterval J_i $(1 \leq i < n)$. Let A_P be the adjacency matrix corresponding to the permutation P. We will show that the characteristic polynomial of A_P is $\chi_m(\lambda)$ if $q = 1$ and $\psi_q(\lambda^m)\,\chi_m(\lambda)$ if $q > 1$, where

$$\chi_m(\lambda) = (\lambda^{m/2} - 1)(\lambda^{m/4} - 1) \ldots (\lambda^2 - 1)(\lambda - 1),$$
$$\psi_q(\lambda) = (\lambda^q - 2\lambda^{q-2} - 1)/(\lambda + 1).$$

We use the formulae (1), (2), (3) for the characteristic polynomials of adjacency matrices which were established in Chapter I. From (3) it follows at once by induction that if $n = m$ is a power of 2 then the characteristic polynomial of A_P is $\chi_m(\lambda)$. By (1) the characteristic polynomial of A_P is $\psi_q(\lambda)$ if $n = q$ is odd. It now follows from (2) that the characteristic polynomial is $\psi_q(\lambda^m)\,\chi_m(\lambda)$ in the general case $n = qm$. Hence the maximal eigenvalue of A_P is 1 if $q = 1$ and $\lambda_q^{1/m}$ if $q > 1$. Thus if $I = J_1 \cup \ldots \cup J_{n-1}$ then equality holds in (*), by Proposition 19.

If the periodic orbit $P = \{x_1 < \ldots < x_n\}$ is strongly simple but f is not assumed monotonic on each subinterval J_i, then the corresponding matrix A satisfies $A \geq A_P$ and hence its maximal eigenvalue $\lambda(A)$ satisfies $\lambda(A) \geq \lambda(A_P)$. Consequently the inequality (*) holds, by Proposition 14. Moreover if f has any orbit of period n then it has a strongly simple orbit of period n, by Theorem VII.18, and so again the inequality (*) holds.

Suppose finally that equality holds in (*). For any odd $r > 1$

$$\lambda_r^2 = 2 + 1/\lambda_r^{r-2} > 2 > \lambda_q.$$

It follows from what we have just proved that if $m > 1$ then f cannot have an orbit of period $rm/2$. Also, if $q > 1$ then for any odd $r < q$ we have $\lambda_r > \lambda_q$, since

$$L_q(\lambda_r) = \lambda_r^{q-r}(\lambda_r^r - 2\lambda_r^{r-2}) - 1 = \lambda_r^{q-r} - 1 > 0.$$

It follows in the same way that f cannot have an orbit of period rm. Hence f has no orbit of period k if k precedes n in the Šarkovskii ordering and $1 < k < n$. Therefore, by Theorem VII.11, every orbit of period n is strongly simple, except possibly when $q = 3$. But if $q = 3$ then f^m is not turbulent, since $h(f) < (\log 2)/m$, and hence every orbit of period n is strongly simple, by Corollary VII.13. □

It is natural to conjecture, as a generalization of Proposition 20, that a piecewise linear map has minimal topological entropy among all continuous maps which join the endpoints of its linear pieces. Rather surprisingly, this conjecture is false, as the following example shows.

EXAMPLE 22 Let $I = [0,1]$ and let $g: I \to I$ be the piecewise linear map defined by

$$g(0) = 0, \quad g(1/2) = \lambda/2, \quad g(1) = 0,$$

where $\lambda = (1+\sqrt{5})/2$. Then 1/2 is a periodic point of period 3 and $h(g) \geq \log \lambda$, by Proposition 21.

On the other hand, let $f:I \to I$ be the piecewise linear map defined by

$$f(0) = 0,\ f(1/2) = \lambda/2,\ f(\mu) = \mu,\ f(1) = 0,$$

where $\mu = (\lambda + 2)/4$. Since $\lambda/2 < \mu < 1$, we have $\omega(x) = \{\mu\}$, for every $x \in (0,1)$. Thus $\Omega(f) = \{0,\mu\}$, and hence $h(f) = 0$.

In this example we actually have $h(g) = \log \lambda$. Indeed, any piecewise linear map whose linear pieces all have slope $\pm\rho$, where $\rho \geq 1$, has topological entropy $\log \rho$. This follows from Theorem 3 of Misiurewicz and Szlenk [88], which relates the topological entropy of a map to its total variation.

Some interesting results on periodic orbits with *maximal* entropy have recently been obtained by Geller and Tolosa [64]. Let $n > 1$ be odd and let ℓ be the greatest integer $\leq (n-1)/4$. Thus $n = 4\ell + 1$ if $n \equiv 1 (\text{mod } 4)$ and $n = 4\ell + 3$ if $n \equiv 3 \ (\text{mod } 4)$. Let O_n be the cyclic permutation of $\{1,2,\ldots,n\}$ defined for odd j by

$$j \to n - 2\ell - j \quad \text{if } 1 \leq j < n - 2\ell,$$
$$j \to j - n + 2\ell + 1 \quad \text{if } n - 2\ell \leq j \leq n,$$

and for even j by

$$j \to n - 2\ell + j - 1 \quad \text{if } 1 \leq j \leq 2\ell,$$
$$j \to n + 2\ell - j + 2 \quad \text{if } 2\ell < j \leq n.$$

Geller and Tolosa show that, for each odd n, the linearizations of the cycle O_n and its dual $\overline{O}_n$, obtained by a reversal of orientation, have maximal entropy among the linearizations, not only of all cycles of period n, but even of all permutations of period n.

3 MISIUREWICZ'S THEOREM

We now embark on the proof of the theorem of Misiurewicz [83], mentioned at the beginning of this chapter. (Actually this theorem was already proved for piecewise monotone maps by Misiurewicz and Szlenk [88].) The treatment given here is based on the more extended account in Misiurewicz [82].

Let α be a collection of finitely many disjoint intervals. For any interval $A \in \alpha$, let $\alpha^n | A$ denote the set of all chains $\{A_1, \dots, A_n\}$ of length n with $A_1 = A$ and let $c_n(\alpha | A)$ denote the cardinality of $\alpha^n | A$. Then

$$c_n(\alpha) = \Sigma_{A \in \alpha} c_n(\alpha | A).$$

Since $c_n(\alpha | A) \leq c_n(\alpha)$ we evidently have

$$\overline{\lim}_{n\to\infty} [\log c_n(\alpha | A)]/n \leq h^*(f, \alpha) \text{ for every } A \in \alpha.$$

In fact equality must hold for some $A \in \alpha$, by virtue of the following elementary result.

LEMMA 23 *If $\{a_n\}$ and $\{b_n\}$ are sequences of positive numbers, then*

$$\overline{\lim}_{n\to\infty} [\log(a_n + b_n)]/n = \max\{\overline{\lim}_{n\to\infty} (\log a_n)/n, \overline{\lim}_{n\to\infty} (\log b_n)/n\}.$$

Proof Let λ and μ denote the left and right sides of the equality to be proved. Evidently $\lambda \geq \mu$. On the other hand, for any $\sigma > \mu$ there exists an integer p such that

$$a_n < e^{n\sigma}, \ b_n < e^{n\sigma} \quad \text{for all } n \geq p.$$

It follows that

$$[\log(a_n + b_n)]/n < \sigma + (\log 2)/n \quad \text{for } n \geq p,$$

and hence $\lambda \leq \sigma$. Since this holds for any $\sigma > \mu$, we must actually have $\lambda \leq \mu$. □

We will also require another result of the same type.

LEMMA 24 *If $\{a_n\}$, $\{b_n\}$ are two sequences of non-negative real numbers, then*

$$\overline{\lim}_{n\to\infty} [\log \Sigma_{k=0}^{n} \exp(a_k + b_{n-k})]/n \leq \max\{\overline{\lim}_{n\to\infty} a_n/n, \overline{\lim}_{n\to\infty} b_n/n\}.$$

Proof Denote the right side of the inequality by μ. Evidently we may suppose that $\mu < +\infty$. If we choose any $\lambda > \mu$ then there exists $p > 0$ such that

$$a_n/n \leq \lambda, \ b_n/n \leq \lambda \quad \text{for } n \geq p.$$

We can also choose $\sigma \geq \lambda$ so that $a_n/n \leq \sigma$, $b_n/n \leq \sigma$ for all $n \geq 0$.

We will show that if $n \geq 2p$ and $0 \leq k \leq n$ then

$$a_k + b_{n-k} \leq p\sigma + n\lambda.$$

In fact we must have either $k \geq p$ or $n-k \geq p$ or both:

if $k \geq p$ and $n-k \geq p$ then

$$a_k + b_{n-k} \leq k\lambda + (n-k)\lambda = n\lambda;$$

if $k < p$ and $n-k \geq p$ then, since $\sigma \geq \lambda > 0$,

$$a_k + b_{n-k} \leq k\sigma + (n-k)\lambda \leq p\sigma + n\lambda;$$

if $k \geq p$ and $n-k < p$ then

$$a_k + b_{n-k} \leq k\lambda + (n-k)\sigma \leq p\sigma + n\lambda.$$

Consequently if $n \geq 2p$ then

$$\log \Sigma_{k=0}^{n} \exp(a_k + b_{n-k}) \leq \log(n+1) + p\sigma + n\lambda,$$

and hence

$$\overline{\lim}_{n\to\infty} [\log \Sigma_{k=0}^{n} \exp(a_k + b_{n-k})]/n \leq \lambda.$$

Since $\lambda > \mu$ is arbitrary, the result follows. □

These two results will now be used to prove

PROPOSITION 25 *Let $f: I \to I$ be a continuous map. Let α be a collection of finitely many disjoint intervals and let γ be the* (necessarily non-empty) *collection of those intervals $A \in \alpha$ for which*

$$\overline{\lim}_{n\to\infty} [\log c_n(\alpha \,|A)]/n = h^*(f, \alpha).$$

Then, for any interval $A \in \gamma$,

$$\overline{\lim}_{n\to\infty} [\log c_n(\gamma \,|A)]/n = h^*(f, \alpha).$$

Proof Choose any $A \in \gamma$. Since $c_n(\gamma \,|A) \leq c_n(\alpha \,|A)$ we certainly have

$$\overline{\lim}_{n\to\infty} [\log c_n(\gamma \,|A)]/n \leq h^*(f, \alpha).$$

Let $\{A_1, \ldots, A_n\}$ be any chain in $\alpha^n \,|A$ and let k $(1 \leq k \leq n)$ be the greatest positive integer such that $A_j \in \gamma$ for $1 \leq j \leq k$. Then $\{A_1, \ldots, A_k\}$ is a chain in $\gamma^k \,|A$ and $\{A_{k+1}, \ldots, A_n\}$ is a chain in $\alpha^{n-k} \,|B$ for some $B \in \alpha \backslash \gamma$. Thus if g_k is the number of chains in $\alpha^n \,|A$ formed in this way for a given value of k, then

$$g_k \leq c_k(\gamma \,|A) \,.\, \Sigma_{B \in \alpha\backslash\gamma}\, c_{n-k}(\alpha \,|B).$$

Hence if we set $a_0 = b_0 = 0$ and

$$a_n = \log c_n(\gamma | A), \quad b_n = \log \Sigma_{B \in \alpha\vee\gamma} c_n(\alpha | B) \quad (n \geq 1),$$

then

$$c_n(\alpha | A) = g_1 + g_2 + \dots + g_n \leq \Sigma_{k=0}^{n} \exp(a_k + b_{n-k}).$$

It follows from Lemma 24 that

$$h^*(f, \alpha) \leq \max\{\overline{\lim}_{n\to\infty} [\log c_n(\gamma | A)]/n, \ \overline{\lim}_{n\to\infty} [\log \Sigma_{B \in \alpha\vee\gamma} c_n(\alpha | B)]/n\}.$$

But, by Lemma 23,

$$\begin{aligned} \overline{\lim}_{n\to\infty} [\log \Sigma_{B \in \alpha\vee\gamma} c_n(\alpha | B)]/n &= \max_{B \in \alpha\vee\gamma} \overline{\lim}_{n\to\infty} \log c_n(\alpha | B)]/n \\ &< h^*(f, \alpha). \end{aligned}$$

Hence

$$h^*(f, \alpha) \leq \overline{\lim}_{n\to\infty} [\log c_n(\gamma | A)]/n .$$

As already observed, the reverse inequality is trivial. □

To avoid interrupting the argument later we now prove the following result, which generalizes Lemma I.3.

LEMMA 26 *Let $f: I \to I$ be a continuous map. If J, K are intervals such that $f(J) \cap K \neq \emptyset$, then there exists an interval $L \subseteq J$ such that*

$$f(L) = f(J) \cap K .$$

Proof Evidently $M = f(J) \cap K$ is an interval. Since the result is obvious if M contains only a single point we may suppose that $\bar{M} = [a, b]$, where $a < b$. Choose $c, d \in \bar{J}$, and if possible $c, d \in J$, so that $f(c) = a$, $f(d) = b$. If $c < d$ let c' be the greatest value less than d such that $f(c') = a$ and let d' be the least value greater than c' such that $f(d') = b$. Then we can take L to be an appropriate (closed, open, or half-open) interval with endpoints c' and d'. If $c > d$ the proof is analogous. □

Again let α be a finite collection of disjoint intervals and let γ be defined as in the statement of Proposition 25. We now define inductively a sequence $\{\delta_n\}$, where each δ_n is a finite collection of disjoint intervals, in the following way:

(i) $\delta_1 = \gamma$,

(ii) if δ_n is defined, then for each interval $D \in \delta_n$ and each interval $A \in \gamma$ with $f^n(D) \cap A \neq \emptyset$ we choose, by Lemma 26, an interval $E(D, A)$ such that

$$E(D, A) \subseteq D\,, \quad f^n(E(D, A)) = f^n(D) \cap A\,.$$

The intervals $E(D, A)$ are necessarily disjoint and we take δ_{n+1} to be the collection of all such intervals.

By induction we see that if $\{A_1, \ldots, A_n\}$ is a chain in γ^n, then there exists a unique $D_n \in \delta_n$ such that

$$D_n \subseteq A_1\,,\ f(D_n) \subseteq A_2\,, \ldots, f^{n-2}(D_n) \subseteq A_{n-1},\ f^{n-1}(D_n) \subseteq A_n.$$

Then

$$f^{n-1}(D_n) = \{y_n \in A_n: \exists \text{ points } y_1,\ldots,y_{n-1} \text{ with } y_i \in A_i \text{ and } f(y_i) = y_{i+1} \text{ for } i = 1,\ldots,n-1\}\,.$$

Conversely, each interval $D_n \in \delta_n$ is contained in a unique interval $D_k \in \delta_k$ $(1 \leq k < n)$ and there is a unique interval $A_k \in \gamma$ such that $f^{k-1}(D_k) \subseteq A_k$ $(1 < k \leq n)$. Since $D_1 = A_1 \in \delta_1$, it follows that $f^{k-1}(D_n) \subseteq A_k$ $(1 \leq k \leq n)$. Thus $\{A_1, \ldots, A_n\}$ is a chain in γ^n.

Consequently the number of intervals in δ_n is exactly $c_n(\gamma)$, and the number of intervals in δ_n which are contained in an interval $A \in \gamma$ is $c_n(\gamma\,|A)$.

For any $A, B \in \gamma$ let $g(A, B, n)$ denote the number of intervals $D \in \delta_n$ such that

$$D \subseteq A,\ f^n(D) \supseteq B\,.$$

This notation, and the definitions of α and γ, are understood in the statements of the next two results.

PROPOSITION 27 *For any* $A, B, C \in \gamma$ *and any positive integers* m, n

$$g(A, B, m)\, g(B, C, n) \leq g(A, C, m+n)\,.$$

Proof Let D be an interval in δ_m such that $D \subseteq A$ and $f^m(D) \supseteq B$, and let E be an interval in δ_n such that $E \subseteq B$ and $f^n(E) \supseteq C$. Then there exists a chain $\{A_1, \ldots, A_m\}$ in $\gamma^m\,|A$ such that $f^{k-1}(D) \subseteq A_k$ $(1 \leq k \leq m)$, and a chain $\{B_1, \ldots, B_n\}$ in $\gamma^n\,|B$ such that $f^{k-1}(E) \subseteq B_k$ $(1 \leq k \leq n)$. Hence $\{A_1, \ldots, A_m, B_1, \ldots, B_n\}$ is a chain in $\gamma^{m+n}\,|A$ and there exists $D' \in \delta_{m+n}$ such that

$$f^{k-1}(D') \subseteq A_k \ (1 \leq k \leq m),\quad f^{m+k-1}(D') \subseteq B_k \ (1 \leq k \leq n)\,.$$

Moreover $D' \subseteq D$, since D' is contained in a unique interval of δ_m and D is the interval of δ_m with the itinerary $\{A_1, ..., A_m\}$. Since

$$f^m(D) \supseteq E, \ f^{k-1}(D) \subseteq A_k \ \ (1 \leq k \leq m),$$

and

$$f^n(E) \supseteq C, \ f^{k-1}(E) \subseteq B_k \ \ (1 \leq k \leq n),$$

it follows that for every $y_{m+n+1} \in C$ there are points y_i with $y_i \in A_i$ for $i = 1,...,m$ and $y_i \in B_{i-m}$ for $i = m+1,...,m+n$ and $f(y_i) = y_{i+1}$ for $i = 1,...,m+n$. Thus $f^{m+n}(D') \supseteq C$. Since D' is the only interval of δ_{m+n} with the itinerary $\{A_1, ..., A_m, B_1,..., B_n\}$, the result follows. □

PROPOSITION 28 *If* $h^*(f, \alpha) > \log 3$, *then there exists* $A \in \gamma$ *such that*

$$\overline{\lim}_{n\to\infty} [\log g(A, A, n)]/n \ = \ h^*(f, \alpha).$$

Proof Choose any $A \in \gamma$ and let μ be a real number such that $\log 3 < \mu < h^*(f, \alpha)$. We will show first that the following condition is satisfied:

(#) For every p there exists an integer $n \geq p$ such that

$$c_{n+1}(\gamma\,|A)/3 \ \geq \ c_n(\gamma\,|A) \ > e^{n\mu}.$$

Assume on the contrary that there exists a p such that, for $n \geq p$,

$$[\log c_n(\gamma\,|A)]/n \ > \ \mu \ \ \text{implies} \ \ c_{n+1}(\gamma\,|A) \ < 3c_n(\gamma\,|A).$$

If for some $q \geq p$ we have $\log c_n(\gamma\,|A) > \mu n$ for all $n \geq q$, then $c_{n+q}(\gamma\,|A) < 3^n c_q(\gamma\,|A)$ for all $n \geq 1$ and hence

$$\overline{\lim}_{n\to\infty} [\log c_n(\gamma\,|A)]/n \ \leq \ \log 3,$$

which is a contradiction. Therefore $(1/N) \log c_N(\gamma\,|A) \leq \mu$ for infinitely many positive integers N. Suppose that for such an N we have

$$\mu \ < \ [\log c_n(\gamma\,|A)]/n \quad \text{for} \ \ n = N+1, ..., N+r,$$

where $r \geq 1$. Since in general $c_{n+1}(\gamma\,|A) \leq s\, c_n(\gamma\,|A)$, where s is the number of intervals in γ, it follows that

$$c_{N+r}(\gamma\,|A) \ < \ s.3^{r-1} c_N(\gamma\,|A) \,.$$

Taking logarithms, we obtain

$$(\mathrm{N}+r)\mu < \log s + (r-1)\log 3 + \mathrm{N}\mu ,$$

i.e.,

$$r(\mu - \log 3) < \log s - \log 3 .$$

Thus $r \leq t$, for some positive integer t independent of N. It follows that

$$c_n(\gamma\,|A) < s.3^{t-1}e^{n\mu}$$

for all large n, and hence

$$\overline{\lim}_{n\to\infty} [\log c_n(\gamma\,|A)]/n \leq \mu,$$

which is again a contradiction. This establishes (#).

Now fix $D \in \delta_n$ with $D \subseteq A$. By the definition of δ_{n+1}, the number q of intervals in δ_{n+1} which are contained in D is equal to the number of intervals $C \in \gamma$ such that $f^n(D) \cap C \neq \emptyset$. Since $f^n(D)$ is an interval, at most two intervals of γ have non-empty intersection with $f^n(D)$ but are not contained in it. Hence the number of intervals $B \in \gamma$ such that $f^n(D) \supseteq B$ is at least $q-2$. Summing over all $D \in \delta_n$ with $D \subseteq A$ we obtain

$$\Sigma_{B\in\gamma}\, g(A,B,n) \geq c_{n+1}(\gamma\,|A) - 2c_n(\gamma\,|A).$$

Combining this with (#), we see that for infinitely many n we have

$$\Sigma_{B\in\gamma}\, g(A,B,n) \geq c_n(\gamma\,|A) > e^{n\mu}.$$

Hence, since $\mu < h^*(f,\alpha)$ is arbitrary,

$$\overline{\lim}_{n\to\infty} [\log \Sigma_{B\in\gamma}\, g(A,B,n)]/n \geq h^*(f,\alpha).$$

It follows from Lemma 23 that for each $A \in \gamma$ there exists $B = \varphi(A) \in \gamma$ such that

$$\overline{\lim}_{n\to\infty} [\log g(A,\varphi(A),n)]/n \geq h^*(f,\alpha).$$

Since γ is finite, the map $\varphi\colon \gamma \to \gamma$ has a periodic point A_0. Let m be its period. By repeated applications of Proposition 27 we obtain, for any positive integers n_i $(0 \leq i < m)$,

$$g(A_0, A_0, \Sigma_{i=0}^{m-1} n_i) \geq \Pi_{i=0}^{m-1}\, g(\varphi^i(A_0), \varphi^{i+1}(A_0), n_i) .$$

But for any $\lambda < h^*(f,\alpha)$ we can choose arbitrarily large n_i so that

$$\log g(\varphi^i(A_0), \varphi^{i+1}(A_0), n_i) \geq \lambda n_i .$$

Then, putting $n = \Sigma\, n_i$, we have

$$\log g(A_0, A_0, n) \geq \lambda n .$$

Thus

$$\overline{\lim}_{n\to\infty} [\log g(A_0, A_0, n)]/n \geq h^*(f, \alpha).$$

Since $g(A_0, A_0, n) \leq c_n(\gamma) \leq c_n(\alpha)$, the reverse inequality is obvious. □

After these preparations the main result can now be proved without difficulty.

THEOREM 29 *Let* $f: I \to I$ *be a continuous map. If* f *has topological entropy* $h(f) > 0$ *then, for any* λ *with* $0 < \lambda < h(f)$ *and any* $N > 0$, *there exist pairwise disjoint closed intervals* $J_1, \dots, J_p$ *and an integer* $n > N$ *such that* $(1/n) \log p > \lambda$ *and*

$$J_1 \cup \dots \cup J_p \subseteq \operatorname{int} f^n(J_i) \quad (i = 1, \dots, p) .$$

Proof Suppose first that $h(f) < +\infty$. Choose $\varepsilon > 0$ so small that $h(f) > \lambda + 2\varepsilon$ and let r be a positive integer such that $r\, h(f) > \varepsilon + \log 3$. There exists a finite open cover β of I such that

$$h(f^r, \beta) \geq h(f^r) - \varepsilon .$$

By Proposition 16, there exists a cover α consisting of finitely many disjoint intervals such that

$$h^*(f^r, \alpha) \geq h(f^r, \beta) .$$

Hence

$$h^*(f^r, \alpha) \geq r\, h(f) - \varepsilon > \log 3 .$$

By Proposition 28, applied to f^r and α, there exists an interval $A \in \alpha$ and arbitrarily large integers m such that

$$[\log g(A, A, m)]/m \geq h^*(f^r, \alpha) - \varepsilon \geq r\, h(f) - 2\varepsilon.$$

Thus if we put $n = mr$ there exist p_n disjoint intervals D_i such that $D_i \subseteq A$ and $f^n(D_i) \supseteq A$, where

$$(\log p_n)/n \geq h(f) - 2\varepsilon/r > \lambda .$$

Since $p_n \to \infty$ as $n \to \infty$, the interval A has non-empty interior and so have the intervals D_i . If we omit the two intervals which are furthest to the left and to the right and replace the remaining intervals D_i by slightly smaller closed intervals J_i then

$$J_1 \cup \dots \cup J_{p_n - 2} \subseteq \operatorname{int} f^n(J_i) \quad (i = 1, \dots, p_n - 2)$$

and, for all large n,

$$[\log(p_n-2)]/n \geq [\log(1-2/p_n)]/n + h(f) - 2\varepsilon/r > \lambda .$$

Suppose next that $h(f) = +\infty$. If we choose $\varepsilon = r = 1$ and a finite open cover β such that $h(f, \beta) > \lambda + 2$ then the preceding argument carries through. □

Theorem 29 has the following important corollaries.

PROPOSITION 30 *The topological entropy, regarded as a function* $h\colon C(I, I) \to \mathbb{R}^+ \cup \{+\infty\}$, *is lower semi-continuous.*

Proof We wish to show that if $\lambda < h(f)$ then $h(g) > \lambda$ for all g sufficiently close to f. Evidently we may suppose $h(f) > 0$. Then, in the statement of Theorem 29, for all g sufficiently close to f we will have

$$J_1 \cup \ldots \cup J_p \subseteq \operatorname{int} g^n(J_i) \quad (i = 1, \ldots, p).$$

By Proposition 8, this implies $h(g) \geq (1/n) \log p > \lambda$. □

PROPOSITION 31 *The set of all maps* $f \in C(I, I)$ *with* $h(f) = \infty$ *is dense in* $C(I, I)$.

Proof Since $C(I, I)$ is a complete metric space, the intersection of any countable collection of open, dense subsets is again dense, by the Baire Category Theorem. Thus, it suffices to show that the set $K_n = \{f \in C(I, I) : h(f) > \log n\}$ is open and dense in $C(I, I)$ for each integer $n > 1$.

It follows at once from Proposition 30 that K_n is open. We now show that K_n is dense. Let $g \in C(I, I)$ and let $\varepsilon > 0$. We will construct an $f \in C(I, I)$ with $d(f,g) < \varepsilon$ and $f \in K_n$.

Let x be a fixed point of g. First assume that x is not the right endpoint of I. Choose δ, with $0 < \delta < \varepsilon$, so that $[x, x+\delta] \subseteq I$ and $|g(y) - x| < \varepsilon/2$ for all $y \in [x, x+\delta]$. Next choose any points $a_0 < a_1 < \ldots < a_{2n}$ with $a_0 = x$, $a_{2n} = x + \delta/2$. Let $f(y) = g(y)$ for all y in the complement of $(x, x+\delta)$, let $f(a_i) = x$ for i even and $f(a_i) = x + \delta/2$ for i odd, and let f be linear on each interval $[a_i, a_{i+1}]$ $(i = 0,\ldots,2n-1)$ and on $[x + \delta/2, x + \delta]$. Then $|f(y) - x| < \varepsilon/2$ for all $y \in [x, x+\delta]$, and hence $d(f,g) < \varepsilon$. Moreover, it follows from Proposition 14 that $h(f) \geq \log 2n > \log n$. If x is the right endpoint of I, we make the same construction on an interval $[x - \delta, x]$. □

The proof of Proposition 31 can easily be modified to make it independent of Proposition 30. From Propositions 31 and 30 we immediately obtain

COROLLARY 32 *The topological entropy, regarded as a function* $h: C(I, I) \to \mathbb{R}^+ \cup \{+\infty\}$, *is continuous at* $f \in C(I, I)$ *if and only if* $h(f) = +\infty$. □

Our next corollary of Theorem 29 shows that the topological entropy of a continuous map of a compact interval is determined by its periodic orbits. For a periodic orbit P, let L_P denote the linearization of P. Note that $h(L_P)$ is easily determined from Proposition 19, and that $h(L_P) = h(L_Q)$ if P and Q have the same type.

PROPOSITION 33 *For any continuous map* $f: I \to I$,

$$h(f) = \sup h(L_P),$$

where the supremum is taken over all periodic orbits P *of* f.

Proof It follows from Proposition 20 that $h(f) \geq \sup h(L_P)$. We now show that $h(f) \leq \sup h(L_P)$.

We may assume that $h(f) > 0$, or else the conclusion is immediate. Let ε be any number such that $0 < \varepsilon < h(f)$ and choose a positive integer N so large that $\log 3 < \varepsilon N/2$. Then $h(f) - \varepsilon/2 > (\log 3)/N$. By Theorem 29 there exist positive integers n and k, with $n > N$ and $(\log k)/n > h(f) - \varepsilon/2$, and pairwise disjoint closed intervals $J_1, \ldots, J_k$ such that

$$J_1 \cup \ldots \cup J_k \subseteq f^n(J_i) \quad (i = 1,\ldots,k).$$

Evidently $k > 3$, and hence

$$\log(1 - 2/k) > -\log 3 > -\varepsilon n/2.$$

We may suppose the intervals J_i numbered according to their location on the real line, so that J_1 is furthest to the left and J_k furthest to the right. By forming from these intervals the f^n-cycle

$$J_2 \to J_1 \to J_2 \to J_k \to J_3 \to J_1 \to J_3 \to J_k \to$$
$$J_4 \to J_1 \to J_4 \to J_k \to \ldots \to J_{k-1} \to J_1 \to J_{k-1} \to J_k \to J_2$$

we see that f^n has a periodic orbit Q such that, for each positive integer i with $2 \leq i \leq k-1$, there are unique points x_i and y_i in $Q \cap J_i$ for which $f^n(x_i) \in J_1$ and $f^n(y_i) \in J_k$. It follows from Proposition 8, using the closed intervals $\langle x_i, y_i \rangle$ $(2 \leq i \leq k-1)$, that $h(L_Q) \geq \log(k-2)$.

There is a unique periodic orbit P of f with $Q \subset P$. Since Q is a periodic orbit of $(L_P)^n$, it follows from the first paragraph of this proof that $h((L_P)^n) \geq h(L_Q)$. We have

$$h(L_P) = h((L_P)^n)/n \geq h(L_Q)/n \geq [\log(k-2)]/n$$
$$> [\log k]/n - \varepsilon/2 > h(f) - \varepsilon.$$

Since ε was arbitrary, $\sup h(L_P) \geq h(f)$. □

Proposition 33 was stated by Takahashi [119], although his proof appears to be valid only for piecewise monotone maps. A complete proof was given by Block and Coven [30]. The proof given here is based on Misiurewicz and Nitecki [86].

Finally we use Theorem 29 to derive another characterization of chaotic maps.

PROPOSITION 34 *A continuous map* $f: I \to I$ *is chaotic if and only if it has topological entropy* $h(f) > 0$.

Proof If f is chaotic then f^n is strictly turbulent for some $n > 0$. Hence, by Corollary 15, $h(f^n) \geq \log 2$ and $h(f) \geq (\log 2)/n$.

Conversely, suppose $h(f) > 0$. Then in the statement of Theorem 29 we must have $p > 1$. Hence f^n is turbulent and f is chaotic. □

Proposition 34 is surely the most fundamental justification for our use of the word 'chaotic'. A proof based on Corollary 7, rather than on Theorem 29, has been given by Xiong [133]. The 'easy half' of Proposition 34, that f chaotic implies $h(f) > 0$, was already proved by Bowen and Franks [44].

IX
Maps of the Circle

1 LIFTS

In this chapter we discuss some of the basic ideas involved in studying the dynamics of continuous maps of the circle to itself, and we state, without proof, a selection of the many interesting results. The purpose of this chapter is to present a guide for the reader familiar with the material on maps of the interval in the previous chapters to the related literature on maps of the circle, as well as to present the necessary background.

We let S^1 denote the unit circle in the plane, i.e. $S^1 = \{(x, y) \in \mathbb{R} \times \mathbb{R} : x^2 + y^2 = 1\}$. Throughout the chapter f denotes a continuous map of S^1 to itself. The map $\prod : \mathbb{R} \to S^1$ defined by $\prod(t) = (\cos 2\pi t, \sin 2\pi t)$ is clearly continuous and onto, and $\prod(t_1) = \prod(t_2)$ if and only if $t_1 - t_2$ is an integer. Using the map $\prod$, we can associate to f a map F of the real line to itself.

PROPOSITION 1 *There is a continuous map $F : \mathbb{R} \to \mathbb{R}$ and a unique integer $k = k(f)$ such that*

(i) *for each $x \in \mathbb{R}$, $F(x + 1) = F(x) + k$,*

(ii) $\prod \circ F = f \circ \prod$, *i.e. the diagram*

$$\begin{array}{ccc} \mathbb{R} & \xrightarrow{\;F\;} & \mathbb{R} \\ {\scriptstyle \prod}\downarrow & & \downarrow{\scriptstyle \prod} \\ S^1 & \xrightarrow[\;f\;]{} & S^1 \end{array}$$

commutes.

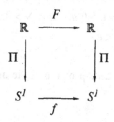

The integer $k = k(f)$ in Proposition 1 is called the *degree* of f. Any map F satisfying the conditions of Proposition 1 is called a *lift* of f.

PROPOSITION 2 *Lifts have the following properties:*

(i) *if F_1 is a lift of f and j is an integer, then $F_2 : \mathbb{R} \to \mathbb{R}$ is also a lift if*

$$F_2(x) = F_1(x) + j \text{ for all } x \in \mathbb{R},$$

(ii) *if F_1 and F_2 are both lifts of f, then there is an integer j such that*

$$F_2(x) = F_1(x) + j \text{ for all } x \in \mathbb{R}.$$

It follows from Proposition 2 that there are countably many lifts of a given map f, but if the value of the lift at a single real number is specified, then the lift is completely determined.

An elementary discussion of lifts which includes proofs of Propositions 1 and 2 may be found in Chapter 6 of Wall [124]. In that text the lift of f is defined to be a map $\hat{F} : I \to \mathbb{R}$, where $I = [0,1]$. As we will see later, in Proposition 7 and the discussion that follows, one may easily extend $\hat{F}$ to a map $F : \mathbb{R} \to \mathbb{R}$ with the appropriate properties.

Many properties of orbits under f may be described in terms of a lift F of f. For example, the following proposition follows easily from the definitions.

PROPOSITION 3 *Let F be a lift of f. Let $x, y \in S^1$, and let $\tilde{x}, \tilde{y} \in \mathbb{R}$ satisfy $\Pi(\tilde{x}) = x$, $\Pi(\tilde{y}) = y$. Then*

(i) *for any positive integer n, $f^n(x) = x$ if and only if $F^n(\tilde{x}) = \tilde{x} + k$ for some integer k,*

(ii) *$y \in \omega(x, f)$ if and only if, for any $\varepsilon > 0$ and any $N > 0$, there is an integer $n > N$ and an integer k such that $|F^n(\tilde{x}) - (\tilde{y} + k)| < \varepsilon$.*

In Proposition 1 we started with a map of the circle and obtained a lift. One can also proceed in the other direction.

PROPOSITION 4 *Suppose $F : \mathbb{R} \to \mathbb{R}$ is continuous, k is an integer, and*

$$F(x+1) = F(x) + k \quad \text{for each } x \in \mathbb{R}.$$

Then there is a unique continuous map $f : S^1 \to S^1$ such that F is a lift of f.

The map f in Proposition 4 is constructed as follows. For $x \in S^1$ let $\tilde{x} \in \mathbb{R}$ satisfy $\Pi(\tilde{x}) = x$. Then we set

$$f(x) = \Pi(F(\tilde{x})).$$

It is easy to verify that f is well-defined and continuous. Furthermore, f must be defined in this way for F to be a lift of f. By definition, f has degree k. We will refer to f as the degree k map obtained from F.

Let $\alpha \in \mathbb{R}$, and define $F : \mathbb{R} \to \mathbb{R}$ by $F(x) = x + \alpha$. Then $F(x + 1) = F(x) + 1$ for each $x \in \mathbb{R}$. Let f be the degree one map obtained from F. Then f is called a *rotation*; in fact f rotates each point of the circle by $2\pi\alpha$ radians. The map f is called a *rational* or *irrational* rotation according as α is rational or irrational.

PROPOSITION 5 *Suppose f is a rational rotation with lift $F(x) = x + \alpha$, and $\alpha = p/q$, where p and q are relatively prime integers with $q > 0$. Then each point of the circle is a periodic point of f with period q.*

Proposition 5 follows easily from statement (i) of Proposition 3.

PROPOSITION 6 *If f is an irrational rotation, then the orbit of each point is dense in the circle.*

Proposition 6 may be proved using statement (ii) of Proposition 3. It follows from Proposition 6 that for an irrational rotation the entire circle is a minimal set, and hence each point is strongly recurrent.

PROPOSITION 7 *If $\tilde{F} : [0,1] \to \mathbb{R}$ is a continuous map such that $\tilde{F}(1) - \tilde{F}(0)$ is an integer k, then $\tilde{F}$ may be extended in a unique way to a map $F : \mathbb{R} \to \mathbb{R}$ such that $F(x + 1) = F(x) + k$ for every $x \in \mathbb{R}$.*

The map F in Proposition 7 is obtained from $\tilde{F}$ as follows. Define F on $[1,2]$ by $F(x) = \tilde{F}(x - 1) + k$, and define F inductively in a similar way on each interval $[n, n + 1]$, where n is a positive integer. Also, define F on $[-1, 0]$ by $F(x) = \tilde{F}(x + 1) - k$, and define F inductively in a similar way on each interval $[n-1, n]$, where n is a negative integer.

Suppose a continuous map $g : [0,1] \to \mathbb{R}$ is given, and $g(1) - g(0)$ is an integer k. Let F be the extension of g given in Proposition 7. If f is the degree k map obtained from F, we will also refer to f as the degree k map obtained from g. This provides another method of constructing examples.

By a closed interval on S^1 we mean the image under Π of a closed interval in $\mathbb{R}$. If K is a proper closed interval on S^1 and K is the image under Π of an interval $[c,d]$ we write $K = [a,b]$, where $a = \Pi(c)$ and $b = \Pi(d)$. Note that a and b are uniquely determined, even though c and d are not.

PROPOSITION 8 *Let g_0 be a continuous map of a compact interval I to itself. There is a continuous map f of the circle to itself and a proper, closed, invariant interval K on S^1 such that $f|K$ is topologically conjugate to g_0.*

One method of constructing the map f in Proposition 8 is as follows. First, we construct a map g_1 of the interval $[1/3,2/3]$ to itself which is topologically conjugate to g_0. Then we extend g_1 to a map $g : [0,1] \to [1/3,2/3]$ with $g(0) = g(1)$. Finally, we let f be the degree zero map of the circle to itself obtained from g. If K is the image under Π of the interval $[1/3,2/3]$, then $f|K$ is topologically conjugate to g_1, and hence also to g_0. The restriction of Π to $[1/3,2/3]$ provides the conjugacy.

Proposition 8 implies that any phenomena which occur for maps of the interval also occur for maps of the circle. Of course the converse is not true. For example, we have seen that for irrational rotations the entire circle is a minimal set. For maps of the interval the entire interval cannot be a minimal set, since there must exist a fixed point.

2 PERIODIC POINTS

The theorem of Šarkovskii (Theorem I.1) specifies, for continuous maps of an interval, which sets of positive integers may occur as the set of periods. Results along these lines have also been obtained for maps of the circle.

Let J and K be proper closed intervals on S^1. We say J *f-covers* K if there is a closed interval $L \subseteq J$ such that $f(L) = K$. Note that for a continuous map $g : I \to \mathbb{R}$, if J and K are compact intervals in I and $g(J) \supseteq K$, then J g-covers K, by Lemma I.3. However, for maps of the circle the corresponding statement is not true. This is why the concept of one interval covering another must be formulated in this way for maps of the circle. We have the following three elementary results. Proposition 9 follows from Lemma I.3 and the definitions, whereas the proof of Lemma 10 is similar to that of Lemma I.3. Lemma 11 is proved in Block [23].

PROPOSITION 9 *Let F be a lift of f, and let J and K be proper closed intervals on S^1. Then J f-covers K if and only if there are compact intervals $\tilde{J}$ and $\tilde{K}$ in $\mathbb{R}$ such that $\Pi(\tilde{J}) = J$, $\Pi(\tilde{K}) = K$, and $F(\tilde{J}) \supseteq \tilde{K}$.*

LEMMA 10 *Let J and K be proper closed intervals on S^1. Suppose $f(J) \supseteq K$ and $f(J)$ is a proper subset of S^1. Then J f-covers K.*

LEMMA 11 *Let $J = [a,b]$ be a proper closed interval on S^1. Suppose $f(a) = c$, $f(b) = d$, and $c \neq d$. Then J f-covers either $[c,d]$ or $[d,c]$.*

The following result is analogous to Lemma I.4 and may be proved in a similar way.

LEMMA 12 *Suppose $J_1, J_2, \ldots, J_n$ are proper closed intervals on S^1 such that J_i f-covers J_{i+1} for $i = 1, \ldots, n-1$ and J_n f-covers J_1. Then there is a fixed point x of f^n such that $x \in J_1$, $f(x) \in J_2, \ldots, f^{n-1}(x) \in J_n$.*

Given a finite *f*-invariant set *L* on the circle, one may define the *Markov graph* associated to *L* as in Chapter I. However, the Markov graph is not determined by the action of *f* on *L* but by the action of *F* on $\tilde{L}$, where *F* is a lift of *f* and $\tilde{L}$ is a set of points contained in a half open interval of length 1 such that $\Pi(\tilde{L}) = L$. Once the Markov graph is determined, the existence of periodic orbits corresponding to cycles in the graph may be deduced from Lemma 12.

Using these basic ideas, one may prove the following result of Block [25]. Related results, depending on the degree of *f*, were obtained in Block *et al.* [37].

THEOREM 13 *Suppose f has a fixed point and a point of period $n > 1$. Then at least one of the following holds:*

(i) *for every integer k with $n < k$, f has a periodic point of period k,*
(ii) *for every integer k with $n \prec k$, f has a periodic point of period k.*

In Theorem 13, $<$ denotes the usual ordering of the positive integers, while $\prec$ denotes the Šarkovskii ordering given in Theorem I.1. By constructing appropriate examples one may obtain the following companion of Theorem 13, also in Block [25].

THEOREM 14 *Let S be a subset of the positive integers with $1 \in S$. Suppose that for each $n \in S$ with $n \neq 1$ at least one of the following holds:*

(i) *for every integer k with $n < k$, $k \in S$.*
(ii) *for every integer k with $n \prec k$, $k \in S$.*

Then there exists a continuous map f of the circle to itself, such that the set of periods of the periodic points of f is exactly S.

Theorems 13 and 14 characterize the set of periods which can occur for a continuous map of the circle to itself which has a fixed point. Of course, not every map f of the circle has a fixed point. However, it may be shown that f has a fixed point if the degree of f is not one. Partly for this reason, degree one maps of the circle require special attention in the study of periodic orbits and other properties. We will consider this class of maps in the last section of this chapter.

The proof that a map f of the circle of degree k has a fixed point if $k \neq 1$ proceeds in the following way. Let F be a lift of f. Then $F(n) = F(0) + kn$ for every integer n. If $k > 1$, then $F(n) - n \to +\infty$ as $n \to +\infty$ and $F(n) - n \to -\infty$ as $n \to -\infty$. Hence, by continuity, $F(x) - x = 0$ for some $x \in \mathbb{R}$, and $\Pi(x)$ is a fixed point of f. A similar argument applies if $k < 1$.

Finally we mention that *minimal* periodic orbits can be defined and described also for maps of the circle having a fixed point, using the two orderings in Theorem 13. This is carried out in Alseda and Llibre [3].

3 TOPOLOGICAL DYNAMICS

We begin this section with the following elementary result of Mulvey [90], which gives the same alternative characterization of nonwandering points for maps of the circle as Proposition IV.16 for maps of an interval.

PROPOSITION 15 *If $x \in \Omega(f)$, then there exists a sequence of points (x_n) and a sequence of positive integers (k_n) such that $x_n \to x$, $k_n \to \infty$, and $f^{k_n}(x_n) = x$.*

The next result of Block *et al.* [32] shows that the statement of Proposition IV.30 also holds for maps of the circle.

PROPOSITION 16 *For any odd positive integer n,* $\Omega(f) = \Omega(f^n)$.

For maps of the interval we have seen that $\overline{P}(f) = \overline{R}(f) = \Omega(f\,|\,\Omega(f))$. The first equality need not hold for maps of the circle in general; any irrational rotation yields a counterexample. However, we have the following results.

PROPOSITION 17 *If f has a periodic point, then* $\overline{P}(f) = \overline{R}(f)$.

PROPOSITION 18 *For any continuous map* $f: S^1 \to S^1$, $\overline{R}(f) = \Omega(f\,|\,\Omega(f))$.

Propositions 17 and 18 are due to Coven and Mulvey [56], although in the case where *f* has no periodic point Proposition 18 is implicitly contained in Auslander and Katznelson [8]. The equality $\overline{P}(f) = \overline{R}(f)$, if *f* has a periodic point, was also obtained in Mulvey [90] and Bae and Yang [9].

We also have the following analogues of Theorems VI.46 and VI.48. Theorem 19 is due to Coven and Mulvey [56] and Theorem 20 to Hidalgo [66].

THEOREM 19 *The following statements are equivalent:*

(i) *there is a positive integer m such that* f^m *is transitive and has both a fixed point and a point of odd period greater than one,*

(ii) *there is a positive integer m such that* f^{2m} *is transitive and* f^m *has a fixed point,*

(iii) f^n *is transitive for every positive integer n, and f has a periodic point,*

(iv) *f is topologically mixing.*

THEOREM 20 *Suppose* $\Omega(f) = S^1$. *Then there is a positive integer N such that* S^1 *admits a decomposition*

$$S^1 = K \cup J_1 \cup J_2 \cup \dots ,$$

where the terms of the decomposition have pairwise disjoint interiors, K is a closed set of fixed points of f^N, *each* J_i *is a nondegenerate closed* f^N*-invariant interval, and* $f^N\,|\,J_i$ *is topologically mixing for each i. The collection* $\{J_1, J_2, \dots\}$ *is countable, finite, or empty.*

Moreover, if f has a fixed point, then the above statement holds with $N = 1$ *or* $N = 2$.

For maps of the interval, if $P(f)$ is closed then $\Omega(f) = P(f)$ and in fact $CR(f) = P(f)$. These statements are clearly false for maps of the circle if $P(f)$ is empty. However, we have the following two results.

PROPOSITION 21 *If* $P(f)$ *is closed and non-empty, then* $\Omega(f) = P(f)$.

PROPOSITION 22 *If* $P(f)$ *is empty, then every point of the circle is chain recurrent.*

Proposition 21 was obtained by Block *et al.* [32] and Proposition 22 by Block and Franke [35]. Maps of the circle with $P(f)$ empty have been analysed by Auslander and Katznelson [8]. For such maps $\Omega(f)$ need not be the entire circle.

It is not true that if $P(f)$ is closed and non-empty then $CR(f) = P(f)$. For example, let $g : [0,1] \to [0,1]$ be defined by $g(x) = x^2$, and let f be the degree one map of the circle obtained from g. Then f is a homeomorphism, and $P(f)$ contains exactly one point, $\Pi(0) = \Pi(1)$. However, it follows from Corollary V.41 that each point of the circle is chain recurrent.

For homeomorphisms of the circle the chain recurrent set must be either $P(f)$ or the entire circle. This result was obtained by Block and Franke [35]. In general, for maps of the circle with $P(f)$ closed and non-empty the chain recurrent set may be neither $P(f)$ nor the entire circle. For example, let g be a continuous map of $[0,1]$ to itself with the following properties (see Figure 1):

(1) The points 0, 1/3, 1/2, 2/3, and 1 are fixed points of g,

(2) $2/3 < g(1/6) < 1$,

(3) each of the intervals $[1/3,1/2]$, $[1/2,2/3]$, and $[2/3,1]$ is invariant under g,

(4) g is strictly increasing on the intervals $[0,1/6]$ and $[1/3,1]$, and strictly decreasing on the interval $[1/6,1/3]$,

(5) $g(x) > x$ if $x \in (0,1/3) \cup (1/3,1/2) \cup (2/3,1)$,

(6) $g(x) < x$ if $x \in (1/2,2/3)$.

It follows that $P(g) = \{0, 1/3, 1/2, 2/3, 1\}$. Now, let f be the degree one map of the circle obtained from g. Then $P(f)$ consists of the four fixed points $\Pi(1/3)$, $\Pi(1/2)$, $\Pi(2/3)$, and $\Pi(0) = \Pi(1)$. Also, the only chain recurrent point in $\Pi(1/3,2/3)$ is $\Pi(1/2)$. This follows easily from the definition given in Chapter V. Finally, it follows from Corollary V.41 that $\Pi(1/6)$ is chain recurrent. Thus, the chain recurrent set is neither $P(f)$ nor the entire circle.

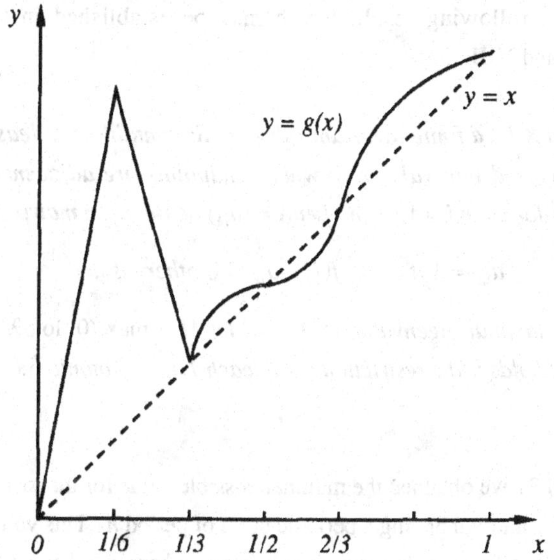

Fig. 1

To conclude this section, we recall that for a map f of the interval the set $\Lambda(f)$ of all ω-limit points is closed and moreover, by Proposition V.10,

$$\Lambda(f) = \bigcap_{n\geq 0} f^n(\Omega(f)).$$

We do not know if either of these results holds on the circle. However, we do have the following result, due to Bae and Yang [10].

PROPOSITION 23 $\overline{P}(f) \subseteq \Lambda(f)$.

4 TOPOLOGICAL ENTROPY AND CHAOTIC MAPS

Several important results on topological entropy given in Chapter VIII have analogues on the circle. Theorem VIII.29 of Misiurewicz [83] holds on the circle in the following form.

THEOREM 24 *If* $h(f) > 0$ *then, for any* λ *with* $0 < \lambda < h(f)$ *and any* $N > 0$, *there exist pairwise disjoint closed intervals* $J_1, \ldots, J_p$ *on the circle and an integer* $n > N$ *such that* $\log p > \lambda n$ *and* J_i f^n*-covers* J_k *for each* $i, k \in \{1, \ldots, p\}$.

We also have the following result, which may be established in the same way as Propositions VIII.14 and VIII.19.

PROPOSITION 25 *Let K be a finite invariant subset of S^1 containing at least two points. Let $J_1, \ldots, J_n$ denote the closed intervals on S^1 whose endpoints are adjacent points of K, and suppose that $f(J_i) \neq S^1$ for each $i = 1, \ldots, n$. Let $A = (a_{ik})$ be the $n \times n$ matrix defined by*

$$a_{ik} = 1 \text{ if } J_k \subseteq f(J_i), \quad a_{ik} = 0 \text{ otherwise.}$$

Let λ denote the maximal eigenvalue of A, and let $M = \max(0, \log \lambda)$. Then $h(f) \geq M$. Furthermore, equality holds if the restriction of f to each J_i is monotonic (i.e. *nondecreasing or nonincreasing*).

In Proposition VIII.21 we obtained the minimal possible value for the topological entropy of a continuous map of the interval having a periodic point of period n. This value may be denoted by $\log \sigma_n$, where σ_n is defined as follows. Let $n = qm$, where $q \geq 1$ is odd and $m = 2^s$. For $q \geq 3$, let λ_q denote the largest root of $L_q(x) = x^q - 2x^{q-2} - 1$. Then $\sigma_n = 1$ if $q = 1$ and $\sigma_n = \lambda_q^{1/m}$ if $q \geq 3$.

There is an analogous result for maps of the circle.

THEOREM 26 *Let μ_n denote the largest root of the polynomial*

$$M_n(x) = x^{n+1} - x^n - x - 1.$$

If a continuous map f of the circle to itself has a fixed point and a periodic point of period $n \geq 2$, then $h(f) \geq \min(\log \mu_n, \log \sigma_n)$.

PROPOSITION 27 *Let $n = qm$, where $q \geq 1$ is odd, $m = 2^s$ and $n > 1$. Then the following inequalities hold:*

(i) *if $s = 0$, then $\mu_n < \sigma_n$ for $n \neq 3$ but $\mu_3 = \sigma_3$,*
(ii) *if $1 \leq s \leq 6$, then $\sigma_n < \mu_n$ when $q \leq 2s + 3$ and $\mu_n < \sigma_n$ when $q \geq 2s + 5$,*
(iii) *if $s \geq 7$, then $\sigma_n < \mu_n$ when $q \leq 2s + 5$ and $\mu_n < \sigma_n$ when $q \geq 2s + 7$.*

These two results, obtained in Block *et al.* [33], together give the minimal possible value of the topological entropy for a map of the circle having a fixed point and a point of period n. Of course, the number $\log \sigma_n$ is just the topological entropy associated to a strongly simple orbit (of a map of the interval) of period n. Thus, using the construction of Proposition 8, one easily obtains examples of maps of the circle having a fixed point and a point of period n with topological entropy equal to $\log \sigma_n$.

The number $\log \mu_n$ is obtained as follows. Let $x_1, x_2, \ldots, x_n$ be points in (0,1) with $x_1 < x_2 < \ldots < x_n$, and let $g : [0,1] \to \mathbb{R}$ be the piecewise linear map defined by

$$g(0) = 0,\ g(x_i) = x_{i+1} \text{ for } 1 \le i \le n-1,\ g(x_n) = x_1 + 1,\ \text{and } g(1) = 1.$$

Let f be the degree one map of the circle obtained from g. Then f has a fixed point $\Pi(0)$ and a periodic orbit $\{\Pi(x_1), \Pi(x_2), \ldots, \Pi(x_n)\}$ of period n, and it may be shown using Proposition 25 that $h(f) = \log \mu_n$.

We say that f is *strictly turbulent* if there are disjoint closed intervals J and K on S^1 such that each of the intervals J and K f-covers both J and K. This is analogous to the definition for maps of the interval, so that some proofs given for maps of the interval may be adapted to maps of the circle. An example is the proof of Proposition II.15.

We may define the unstable manifold of a periodic point, and homoclinic points, exactly as in Chapter III. We have the following equivalent conditions, analogous to the corresponding conditions for maps of the interval.

THEOREM 28 *The following statements are equivalent:*

(i) *f has positive topological entropy,*
(ii) *f has periodic points with periods n and k, where $n < k$ and k/n is not a power of 2,*
(iii) *f^n is strictly turbulent for some positive integer n,*
(iv) *f has a non-periodic, nonwandering point with a finite orbit,*
(v) *f has a nonwandering homoclinic point,*
(vi) *there is a positive integer n, a closed f^n-invariant subset X of S^1, and a continuous surjection $h : X \to \Sigma$ such that the diagram*

$$\begin{array}{ccc} X & \xrightarrow{f^n} & X \\ h\downarrow & & \downarrow h \\ \Sigma & \xrightarrow[\sigma]{} & \Sigma \end{array}$$

commutes, where Σ *is the space of* (0,1)*-sequences and* $\sigma : \Sigma \to \Sigma$ *is the shift defined in Chapter II, Section 3.*

The equivalence of (i), (ii), (iii), (iv), and (v) is proved in Block *et al.* [32], using Theorem 24 to obtain (i) $\Rightarrow$ (iii) and Theorem 26 to obtain (ii) $\Rightarrow$ (i); (vi) $\Rightarrow$ (i) follows from Propositions VIII.2 and VIII.4 and the fact that $h(\sigma) = \log 2$, and (iii) $\Rightarrow$ (vi) may be established as in the proof of Proposition II.15.

As for maps of the interval, the equivalent conditions of Theorem 28 imply the following:

(vii) *for some* $x \in S^1$, $\omega(x,f)$ *properly contains a periodic orbit.*

This can be established as in the proof of Proposition II.17. However, in contrast to the interval case, it is not known if, conversely, (vii) implies the conditions of Theorem 28.

Finally, we recall that if a map g of the interval is transitive, then g^2 is turbulent, and hence g has a periodic point of period 6 and topological entropy at least $(\log 2)/2$. Results along these lines have been obtained for maps of the circle by Hidalgo [66] in various cases depending on the degree of the map.

5 DEGREE ONE MAPS

In this section we deal with the important special case of degree one maps of the circle. As noted earlier, a map of the circle of degree different from one has a fixed point, so that results such as Theorems 13 and 26 are relevant. We will describe some results along the same lines as these, and also some different results, which hold for degree one maps.

We begin with a result of Bernhardt [16].

THEOREM 29 *Suppose* f *does not have a fixed point. Let* t *and* s *denote the two smallest positive integers which are periods of periodic points of* f*, and suppose* t *and* s *are relatively prime. Then for any positive integers* m *and* n, f *has a periodic point of period* $mt + ns$.

An important concept in the study of degree one maps is the rotation number. Let f be a degree one map, and let F be a lift of f. If $y \in \mathbb{R}$, then the *rotation number* of y under F is defined by

$$\rho_F(y) = \overline{\lim}_{n\to\infty} [F^n(y) - y]/n.$$

Basic properties of the rotation number for degree one maps were obtained and used in Newhouse *et al.* [92]. The classical case of an orientation preserving homeomorphism was already considered by Poincaré; see Poincaré [98] or Nitecki [93].

PROPOSITION 30 *Let F and F' be lifts of f, and let $y, z \in \mathbb{R}$. Then the following properties hold:*

(i) *if $\Pi(y) = \Pi(z)$, then $\rho_F(y) = \rho_F(z)$,*
(ii) *if $F' = F + m$, where m is an integer, then $\rho_{F'}(y) = \rho_F(y) + m$,*
(iii) *if $F^n(y) = y + k$, where n is a positive integer and k is an integer, then $\rho_F(y) = k/n$,*
(iv) *if f is a homeomorphism, then $\lim_{n\to\infty} [F^n(y) - y]/n$ exists and is independent of y.*

We will mainly be concerned with rotation numbers for periodic orbits. In this case one may simply use (iii) of Proposition 30, with n the period of the orbit, to define the rotation number.

For $x \in S^1$ we may define $\rho_F(x)$ to be $\rho_F(y)$, where $y \in \mathbb{R}$ and $\Pi(y) = x$. This is well-defined, by (i) of Proposition 30.

It follows from Ito [70] that the set $\{\rho_F(x) : x \in S^1\}$ is a closed interval or a single point. This set is called the *rotation interval* of F.

We now present a result of Misiurewicz [84], which describes the possible set of periods of the periodic points of a continuous degree one map of the circle. This result complements Theorems 13 and 14.

For any real numbers a and b, let $M(a,b)$ denote the set of positive integers n such that there is an integer k with $a < k/n < b$. If $a \in \mathbb{R}$ and ℓ is either a positive integer or the symbol 2^∞, we define a subset $S(a,\ell)$ of the positive integers $\mathbb{Z}^+$ as follows. If a is irrational, then $S(a,\ell) = \emptyset$. If a is rational we can write $a = k/n$, where n is a positive integer, k is an integer, and k and n are relatively prime. If ℓ is a positive integer, then $S(a,\ell)$ denotes the set of positive integers of the form ns, where $\ell \prec s$ (in the Šarkovskii ordering); if ℓ is the symbol 2^∞, then $S(a,\ell)$ denotes the set of positive integers of the form ns, where s is a power of 2.

THEOREM 31 *Let f be a continuous map of the circle to itself of degree one. Then there exist $a,b \in \mathbb{R}$, with $a \le b$, and $\ell, r \in \mathbb{Z}^+ \cup \{2^\infty\}$ such that the set of periods of all periodic points of f is*

$$M(a,b) \cup S(a,\ell) \cup S(b,r).$$

Conversely, for every subset A of $\mathbb{Z}^+$ of the form

$$A = M(a,b) \cup S(a,\ell) \cup S(b,r)$$

there is a continuous map of the circle to itself of degree one whose set of periods is exactly A.

As suggested by the definition of $M(a,b)$, the rotation number of periodic points plays a role in the proof of Theorem 31, even though it does not appear in the statement.

We wish to describe a result analogous to Theorem VII.19, which characterized primary periodic orbits for maps of the interval. However, instead of looking at periodic orbits on the circle, we must lift these orbits to the real line. We will refer to the lift of a periodic orbit as a periodic lift. Formally, a (degree one) *periodic lift* of *period n* is a map $\varphi : P \to P$, where $P = \{... < x_{-1} < x_0 < x_1 < ...\}$ is a subset of $\mathbb{R}$, such that

(a) $x_{i+n} = x_i + 1$ for each i,
(b) for all $x \in P$, $\varphi(x + 1) = \varphi(x) + 1$,
(c) for all $x,y \in P$, there are integers $m > 0$ and s such that $\varphi^m(x) = y + s$.

If $\varphi : P \to P$ is a periodic lift of period n, it may be shown that there is an integer r such that, for all $x \in P$, $\varphi^n(x) = x + r$. The number $\rho(P) = r/n$ is called the *rotation number* of the periodic lift.

We say that periodic lifts $\varphi : P \to P$ and $\psi : Q \to Q$ of the same period are equivalent, in symbols $\varphi \approx \psi$, if we can write $P = \{... < x_{-1} < x_0 < x_1 < ...\}$ and $Q = \{... < y_{-1} < y_0 < y_1 < ...\}$ in such a way that $\varphi(x_i) = x_j$ if and only if $\psi(y_i) = y_j$. By a *degree one lift*, we mean a continuous map $F : \mathbb{R} \to \mathbb{R}$ such that $F(x + 1) = F(x) + 1$ for all $x \in \mathbb{R}$ (i.e. F is the lift of a continuous degree one map of the circle to itself). If P and Q are not equivalent we say that P *forces* Q, in symbols $P \prec Q$, if every degree one lift which has a periodic lift equivalent to P also has a periodic lift equivalent to Q.

We say a periodic lift $\varphi : P \to P$ is a *twist lift* if φ is strictly increasing. The word 'twist' comes from the fact that, for a rational rotation, if one lifts a periodic orbit to the real line one obtains a twist lift. The next two results follow from Alseda *et al.* [5], Misiurewicz [85], or Chenciner *et al.* [47].

THEOREM 32 *Every periodic lift is either a twist lift or forces a twist lift with the same rotation number.*

THEOREM 33 *The following statements about a periodic lift* P *are equivalent:*

(i) P *is a twist lift,*
(ii) *no periodic lift with rotation number* $\rho(P)$ *is forced by* P,
(iii) *no periodic lift is forced by* P,
(iv) *there is a degree one lift whose only periodic lift is* P.

In view of Theorem 33, one may think of a twist lift for a lift of a degree one map of the circle as being analogous to a fixed point for a map of the interval.

We say a periodic lift P is *primary* if no periodic lift of the same period as P is forced by P. To present a characterization of primary periodic lifts we introduce the notion of an extension.

Let $\varphi : P \to P$ and $\psi : Q \to Q$ be periodic lifts. Moreover, suppose that P has period n, and that Q has period m and rotation number r/m. Let X be a periodic orbit, as defined at the beginning of Chapter VII. We say that P is an *X-extension* of Q if each of the following holds:

(1) $n = sm$ for some positive integer s,
(2) we can write $Q = \{\ldots < y_{-1} < y_0 < y_1 < \ldots\}$ and $P = \{\ldots < z_{-1} < z_0 < z_1 < \ldots\}$ so that if we set $P_i = \{z_{is}, \ldots, z_{(i+1)s-1}\}$ for each integer i, then $\varphi(P_i) = P_j$ if and only if $\psi(y_i) = y_j$,
(3) there is an integer k with $0 \le k \le m-1$ such that φ is monotone on P_i, except possibly when $i \equiv k \pmod{m}$,
(4) P_k is a periodic orbit of $\varphi^m - r$ of the same *type* as X, in the sense of Chapter VII.

If X is an orbit of period two, we call an X-extension a *double cover*; note that in this case condition (3) is automatically satisfied. If X is a Stefan orbit, we call an X-extension a *Stefan extension.*

Note that we could make an analogous definition of a Stefan extension for periodic orbits of maps of the interval. It would then follow from the definitions that a periodic orbit is strongly simple if and only if it can be obtained from a fixed point by taking a finite number (possibly zero) of double covers and then at most one Stefan extension. Hence the following theorem is analogous to Theorem VII.19.

THEOREM 34 *The following statements about a periodic lift* P *are equivalent*:

(i) P *is primary,*
(ii) *no periodic lift with the same period and rotation number as* P *is forced by* P,
(iii) P *can be obtained from a twist lift by taking a finite number (possibly zero) of successive double covers and then at most one Stefan extension,*
(iv) *there is a degree one lift whose only periodic lift with the same period as* P *is* P *itself.*

We remark that the proof of Theorem 34 given by Block *et al.* [31] uses Proposition VII.20.

In conclusion, we consider the topological entropy of degree one maps of the circle. The following result of Ito [71] complements Theorem 26.

THEOREM 35 *Let f be a continuous, degree one map of the circle to itself. Suppose f has periodic orbits of periods q and s, where $q,s \geq 2$ and q and s are relatively prime. Let $\gamma_{q,s}$ denote the largest root of the equation*

$$z^{q+s} - z^q - z^s - 1 = 0.$$

Then $h(f) \geq \log \gamma_{q,s}$.

Moreover, for every pair of integers $q,s \geq 2$ with q and s relatively prime, there exists a continuous, degree one map of the circle to itself having periodic orbits of periods q and s and topological entropy exactly $\log \gamma_{q,s}$.

One can also give a lower bound for the topological entropy in terms of the rotation interval. Our final theorem is a sharp result of this form, due to Alseda *et al.* [4].

For $x > 1$ and real numbers c,d with $c < d$, set

$$R_{c,d}(x) = \sum_{n=1}^{\infty} K(n)\, x^{-n},$$

where $K(n)$ denotes the number of integers p for which $c < p/n < d$. The power series defining $R_{c,d}(x)$ is convergent for $x > 1$. Furthermore, it can be verified that the equation $R_{c,d}(x) = 1/2$ has a unique solution with $x > 1$. Denote this unique solution by $\beta_{c,d}$.

THEOREM 36 *Let f be a continuous, degree one map of the circle to itself, and suppose that $[c,d]$ is the rotation interval of a lift of f, where $c < d$. Then* $h(f) \geq \log \beta_{c,d}$.

Moreover, for every pair of real numbers c,d with $c < d$, there exists a continuous, degree one map f of the circle to itself such that $[c,d]$ is the rotation interval of a lift of f and $h(f) = \log \beta_{c,d}$.

References

[1] R.L. Adler, A.G. Konheim and M.H. McAndrew, Topological entropy, *Trans. Amer. Math. Soc.*, **114** (1965), 309-319. MR 30#5291

[2] S.J. Agronsky, A.M. Bruckner, J.G. Ceder and T.L. Pearson, The structure of ω-limit sets for continuous functions, *Real Anal. Exchange*, **15** (1989/90), 483-510.

[3] L. Alsedà and J. Llibre, Orbites périodiques minimales des applications continues du cercle dans lui-même ayant un point fixe (English summary), *C.R. Acad. Sci. Paris Sér. I Math.*, **301** (1985), no.11, 601-604. MR 87c:58099

[4] L. Alsedà, J. Llibre, F. Mañosas and M. Misiurewicz, Lower bounds of the topological entropy for continuous maps of the circle of degree one, *Nonlinearity*, **1** (1988), 463-479. MR 89m:58119

[5] L. Alsedà, J. Llibre, M. Misiurewicz and C. Simó, Twist periodic orbits and topological entropy for continuous maps of the circle of degree one which have a fixed point, *Ergodic Theory Dynamical Systems*, **5** (1985), 501-517. MR 87h:58185

[6] L. Alsedà, J. Llibre and R. Serra, Minimal periodic orbits for continuous maps of the interval, *Trans. Amer. Math. Soc.*, **286** (1984), 595-627. MR 86c:58124

[7] D.V. Anosov, *Geodesic flows on closed Riemannian manifolds with negative curvature*, English translation, Amer. Math. Soc., Providence, R.I., 1969. MR 39#3527 [Russian original, 1967]

[8] J. Auslander and Y. Katznelson, Continuous maps of the circle without periodic points, *Israel J. Math.*, **32** (1979), 375-381. MR 81e:58048

[9] J.S. Bae and S.K. Yang, $\bar{P} = \bar{R}$ for maps of the circle, *Bull. Korean Math. Soc.*, **24** (1987), 151-157. MR 88m:58158

[10] J.S.Bae and S.K. Yang, ω-limit sets for maps of the circle, *Bull. Korean Math. Soc.*, **25** (1988), 233-242. MR 90m:58164

[11] S. Baldwin, Generalizations of a theorem of Sarkovskii on orbits of continuous real-valued functions, *Discrete Math.*, **67** (1987), 111-127. MR 89c:58057

[12] M. Barge and J. Martin, Chaos, periodicity, and snakelike continua, *Trans. Amer. Math. Soc.*, **289** (1985), 355-365. MR 86h:58079

[13] M. Barge and J. Martin, Dense periodicity on the interval, *Proc. Amer. Math. Soc.*, **94** (1985), 731-735. MR 87b:58068

[14] M. Barge and J. Martin, Dense orbits on the interval, *Michigan Math. J.*, **34** (1987), 3-11. MR 88c:58031

[15] A. Berman and R.J. Plemmons, *Nonnegative matrices in the mathematical sciences*, Academic Press, New York, 1979. MR 82b:15013

[16] C. Bernhardt, Periodic orbits of continuous mappings of the circle without fixed points, *Ergodic Theory Dynamical Systems*, **1** (1981), 413-417. MR 84h:58118

[17] C. Bernhardt, The ordering on permutations induced by continuous maps of the real line, *Ergodic Theory Dynamical Systems*, **7** (1987), 155-160. MR 88h:58099

[18] N.P. Bhatia, A.C. Lazer and G.P. Szego, On global weak attractors in dynamical systems, *J. Math. Anal. Appl.*, **16** (1966), 544-552. MR 34#5076

[19] G.D. Birkhoff, *Dynamical Systems*, Amer. Math. Soc. Colloquium Publications, Vol. IX, Providence, R.I., revised edition, 1966. MR 35#1 [Original edition, 1927]

[20] L.S. Block, Continuous maps of the interval with finite nonwandering set, *Trans. Amer. Math. Soc.*, **240** (1978), 221-230. MR 57#13887

[21] L.S. Block, Homoclinic points of mappings of the interval, *Proc. Amer. Math. Soc.*, **72** (1978), 576-580. MR 81m:58063

[22] L.S. Block, Simple periodic orbits of mappings of the interval, *Trans. Amer. Math. Soc.*, **254** (1979), 391-398. MR 80m:58031

[23] L.S. Block, Periodic orbits of continuous mappings of the circle, *Trans. Amer. Math. Soc.*, **260** (1980), 553-562. MR 83c:54057

[24] L.S. Block, Stability of periodic orbits in the theorem of Šarkovskii, *Proc. Amer. Math. Soc.*, **81** (1981), 333-336. MR 82b:58071

[25] L.S. Block, Periods of periodic points of maps of the circle which have a fixed point, *Proc. Amer. Math. Soc.*, **82** (1981), 481-486. MR 82h:58042

[26] L.S. Block and W.A. Coppel, Stratification of continuous maps of an interval, *Trans. Amer. Math. Soc.*, **297** (1986), 587-604. MR 88a:58164

[27] L.S. Block and E.M. Coven, ω-limit sets for maps of the interval, *Ergodic Theory Dynamical Systems*, **6** (1986), 335-344. MR 88a:58165

[28] L.S. Block and E.M. Coven, Maps of the interval with every point chain recurrent, *Proc. Amer. Math. Soc.*, **98** (1986), 513-515. MR 87j:54059

[29] L.S. Block and E.M. Coven, Topological conjugacy and transitivity for a class of piecewise monotone maps of the interval, *Trans. Amer. Math. Soc.*, **300** (1987), 297-306. MR 88c:58032

[30] L.S. Block and E.M. Coven, Approximating entropy of maps of the interval, *Dynamical Systems and Ergodic Theory*, pp. 237-242, Banach Center Publications, Vol. 23, Polish Scientific Publishers, Warsaw, 1989.

[31] L.S. Block, E.M. Coven, L. Jonker and M. Misiurewicz, Primary cycles on the circle, *Trans. Amer. Math. Soc.*, **311** (1989), 323-335. MR 90c:58082

[32] L. Block, E. Coven, I. Mulvey and Z. Nitecki, Homoclinic and nonwandering points for maps of the circle, *Ergodic Theory Dynamical Systems*, **3** (1983), 521-532. MR 86b:58101

[33] L. Block, E. Coven and Z. Nitecki, Minimizing topological entropy for maps of the circle, *Ergodic Theory Dynamical Systems*, **1** (1981), 145-149. MR 83h:58058

[34] L.S. Block and J.E. Franke, The chain recurrent set for maps of the interval, *Proc. Amer. Math. Soc.*, **87** (1983), 723-727. MR 84j:58103

[35] L.S. Block and J.E. Franke, The chain recurrent set, attractors, and explosions, *Ergodic Theory Dynamical Systems*, **5** (1985), 321-327. MR 87i:58107

[36] L.S. Block and J.E. Franke, Isolated chain recurrent points for one-dimensional maps, *Proc. Amer. Math. Soc.*, **94** (1985), 728-730. MR 86g:58078

[37] L. Block, J. Guckenheimer, M. Misiurewicz, and L.S. Young, Periodic points and topological entropy of one dimensional maps, *Global Theory of Dynamical Systems*, pp. 18-34, Lecture Notes in Mathematics, 819, Springer, Berlin, 1980. MR 82j:58097

[38] L.S. Block and D. Hart, The bifurcation of periodic orbits of one-dimensional maps, *Ergodic Theory Dynamical Systems*, **2** (1982), 125-129. MR 84h:58105

[39] L.S. Block and D. Hart, Orbit types for maps of the interval, *Ergodic Theory Dynamical Systems*, **7** (1987), 161-164. MR 89g:58172

[40] A.M. Blokh, On the limit behaviour of one-dimensional dynamical systems, *Russian Math. Surveys*, **37** (1982), 157-158. MR 83i:58082

[41] A.M. Blokh, An interpretation of a theorem of A.N. Šarkovskii (Russian), *Oscillation and stability of solutions of functional-differential equations*, pp. 3-8, Akad. Nauk Ukrain. SSR, Inst. Mat., Kiev, 1982. MR 85g:58074

[42] R. Bowen, Topological entropy and axiom A, *Global Analysis* (Proc. Sympos. Pure Math. XIV, Berkeley, Calif., 1968), pp. 23-41, Amer. Math. Soc., Providence, R.I., 1970. MR 41#7066

[43] R. Bowen, *On axiom A diffeomorphisms*, Regional Conference Series in Mathematics No. 35, Amer. Math. Soc., Providence, R.I., 1978. MR 58#2888

[44] R. Bowen and J.M. Franks, The periodic points of maps of the disk and the interval, *Topology*, **15** (1976), 337-342. MR 55#4283

[45] I.U. Bronštein and V.P. Burdaev, Chain recurrence and extensions of dynamical systems (Russian), *Mat. Issled.*, **55** (1980), 3-11. MR 82d:58058

[46] I.U. Bronštein and A.Ya. Kopanskii, Chain recurrence in dispersive dynamical systems (Russian), *Mat. Issled.*, **80** (1985), 32-47. MR 87a:58101

[47] A. Chenciner, J.-M. Gambaudo and C. Tresser, Une remarque sur la structure des endomorphismes de degré 1 du cercle (English summary), *C.R. Acad. Sci. Paris Sér.I Math.*, **299** (1984), no.5, 145-148. MR 86b:58102

[48] H. Chu and J.C. Xiong, A counter example in dynamical systems of the interval, *Proc. Amer. Math. Soc.*, **97** (1986), 361-366. MR 87i:58140

[49] P. Collet and J.-P. Eckmann, *Iterated maps on the interval as dynamical systems*, Progress in Physics, 1, Birkhäuser, Basel, 1980. MR 82j:58078

[50] C. Conley, The gradient structure of a flow: I, *Ergodic Theory Dynamical Systems*, **8*** (1988), 11-26. MR 90e:58136 [IBM Research Report, 1972]

[51] C. Conley, *Isolated invariant sets and the Morse index*, CBMS Regional Conference Series in Mathematics, 38, Amer. Math. Soc., Providence, R.I., 1978. MR 80c:58009

[52] W.A. Coppel, The solution of equations by iteration, *Proc. Cambridge Philos. Soc.*, **51** (1955), 41-43. MR 16-577

[53] W.A. Coppel, Šarkovskii-minimal orbits, *Math. Proc. Cambridge Philos. Soc.*, **93** (1983), 397-408. MR 84h:58120

[54] M.Y. Cosnard, On the behavior of successive approximations, *SIAM J. Numer. Anal.*, **16** (1979), 300-310. MR 81a:65027

[55] E.M. Coven and G.A. Hedlund, $\bar{P} = \bar{R}$ for maps of the interval, *Proc. Amer. Math. Soc.*, **79** (1980), 316-318. MR 81b:54042

[56] E.M. Coven and I. Mulvey, Transitivity and the centre for maps of the circle, *Ergodic Theory Dynamical Systems*, **6** (1986), 1-8. MR 87j:58074

[57] E.M. Coven and I. Mulvey, The Barge-Martin decomposition theorem for pointwise nonwandering maps of the interval, *Dynamical Systems* (College Park, Md., 1986-87), pp. 100-104, Lecture Notes in Mathematics, 1342, Springer-Verlag, Berlin, 1988. MR 90c:58147

[58] E.M. Coven and Z. Nitecki, Non-wandering sets of the powers of maps of the interval, *Ergodic Theory Dynamical Systems*, **1** (1981), 9-31. MR 82m:58043

[59] D.M. Cvetković, M. Doob and H. Sachs, *Spectra of graphs*, Academic Press, New York, 1980. MR 81i:05054

[60] J.-P. Delahaye, Fonctions admettant des cycles d'ordre n'importe quelle puissance de 2 et aucun autre cycle, *C.R. Acad. Sci. Paris Sér. A-B*, **291** (1980), A323-325. MR 83e:58073a,b

[61] P. Erdös and A.H. Stone, Some remarks on almost periodic transformations, *Bull. Amer. Math. Soc.*, **51** (1945), 126-130. MR 6-165

[62] M.J. Evans, P.D. Humke, C.M. Lee and R.J. O'Malley, Characterizations of turbulent one-dimensional mappings via ω-limit sets, *Trans. Amer. Math. Soc.*, to appear.

[63] V.V. Fedorenko and A.N. Šarkovskii, Continuous mappings of an interval with closed sets of periodic points (Russian), *An investigation of differential and differential-difference equations*, pp. 137-145, Akad. Nauk Ukrain. SSR, Inst. Mat., Kiev, 1980. MR 83i:58083

[64] W. Geller and J. Tolosa, Maximal entropy odd orbit types, *Trans. Amer. Math. Soc.*, to appear.

[65] W.H. Gottschalk and G.A. Hedlund, *Topological Dynamics*, Amer. Math. Soc. Colloquium Publications, Vol. 36, Providence, R.I., 1955. MR 17-650

[66] M. Hidalgo, *Periodic points and topological entropy of transitive maps of the circle*, PhD Thesis, Wesleyan University, Middletown, Connecticut, 1989.

[67] C.W. Ho, On the structure of the minimum orbits of periodic points for maps of the real line, Preprint No. 67, Southern Illinois University, 1982.

[68] C.W. Ho and C. Morris, A graph theoretic proof of Sharkovsky's theorem on the periodic points of continuous functions, *Pacific J. Math.*, **96** (1981), 361-370. MR 83d:58056

[69] J.G. Hocking and G.S. Young, *Topology*, Addison-Wesley, Reading, Mass., 1961. MR 23#A2857

[70] R. Ito, Rotation sets are closed, *Math. Proc. Cambridge Philos. Soc.*, **89** (1981), 107-111. MR 82i:58061

[71] R. Ito, Minimal entropy for endomorphisms of the circle, *Proc. Amer. Math. Soc.*, **86** (1982), 321-327. MR 84b:58093

[72] M.S. Izman, The asymptotic stability of sets in dispersive dynamical systems (Russian), *Differencial'nye Uravnenija*, **7** (1971), 615-621. MR 45#4387

[73] K. Janková and J. Smítal, A characterization of chaos, *Bull. Austral. Math. Soc.*, **34** (1986), 283-292. MR 87k:58178

[74] I. Jungreis, Some results on the Šarkovskii partial ordering of permutations, *Trans.*

Amer. Math. Soc., to appear.

[75] A. Lasota and J.A. Yorke, On the existence of invariant measures for transformations with strictly turbulent trajectories, *Bull. Acad. Polon. Sci. Ser. Sci. Math. Astronom. Phys.*, **25** (1977), 233-238. MR 56#12228

[76] T.-Y. Li, M. Misiurewicz, G. Pianigiani and J.A. Yorke, No division implies chaos, *Trans. Amer. Math. Soc.*, **273** (1982), 191-199. MR 83i:28024

[77] T.-Y. Li and J.A. Yorke, Period three implies chaos, *Amer. Math. Monthly*, **82** (1975), 985-992. MR 52#5898

[78] Liao Gongfu, A condition for a self-map of an interval to have a prime periodic point (Chinese), *Acta Sci. Nat. Univ. Jilin*, 1985, no.3, 55-58. MR 87c:58096

[79] Liao Gongfu, Chain recurrent orbits of mapping of the interval, *Dongbei Shuxue*, **2** (1986), 240-244. MR 88b:58113

[80] Liao Gongfu, A note on a chaotic map with topological entropy 0, *Dongbei Shuxue*, **2** (1986), 379-382. MR 88i:58112

[81] J. Milnor and W. Thurston, On iterated maps of the interval, *Dynamical Systems* (College Park, Md., 1986-87), pp. 465-563, Lecture Notes in Mathematics, 1342, Springer-Verlag, Berlin, 1988. [Preprint, Princeton, 1977] MR 90a:58083

[82] M. Misiurewicz, Horseshoes for continuous mappings of an interval, *Dynamical Systems* (Bressanone, 1978), pp. 127-135, Liguori, Naples, 1980. MR 83h:58076

[83] M. Misiurewicz, Horseshoes for mappings of the interval, *Bull. Acad. Polon. Sci. Sér. Sci. Math.*, **27** (1979), 167-169. MR 81b:58033

[84] M. Misiurewicz, Periodic points of maps of degree one of a circle, *Ergodic Theory Dynamical Systems*, **2** (1982), 221-227. MR 84j:58101

[85] M. Misiurewicz, Twist sets for maps of the circle, *Ergodic Theory Dynamical Systems*, **4** (1984), 391-404. MR 86m:58135

[86] M. Misiurewicz and Z. Nitecki, Extensions and representatives of cycles in maps of the interval, Preprint, 1987.

[87] M. Misiurewicz and J. Smítal, Smooth chaotic maps with zero topological entropy, *Ergodic Theory Dynamical Systems*, **8** (1988), 421-424. MR 90a:58118

[88] M. Misiurewicz and W. Szlenk, Entropy of piecewise monotone mappings, *Studia Math.*, **67** (1980), 45-63. MR 82a:58030

[89] M. Morse, Recurrent geodesics on a surface of negative curvature, *Selected Papers*, pp. 21-37, Springer-Verlag, New York, 1981. MR 83e:01091 [Original, 1921]

[90] I. Mulvey, *Periodic, recurrent, and non-wandering points for maps of the circle*, PhD Thesis, Wesleyan University, Middletown, Connecticut, 1982.

[91] V.V. Nemytskii and V.V. Stepanov, *Qualitative theory of differential equations*, Princeton University Press, Princeton, N.J.,1960. MR 22#12258 [Russian original, 1947-52]

[92] S. Newhouse, J. Palis and F. Takens, Bifurcations and stability of families of diffeomorphisms, *Inst. Hautes Etudes Sci. Publ. Math.*, **57** (1983), 5-71. MR 84g:58080

[93] Z. Nitecki, *Differentiable Dynamics*, MIT Press, Cambridge, Mass., 1971. MR 58#31210

[94] Z. Nitecki, Periodic and limit orbits and the depth of the center for piecewise monotone interval maps, *Proc. Amer. Math. Soc.*, **80** (1980), 511-514. MR 81j:58068

[95] Z. Nitecki, Maps of the interval with closed periodic set, *Proc. Amer. Math. Soc.*, **85** (1982), 451-456. MR 83k:58067

[96] Z. Nitecki, Topological dynamics on the interval, *Ergodic theory and dynamical systems, II* (College Park, Md., 1979-80), pp. 1-73, Progr. Math. 21, Birkhäuser, Boston, Mass., 1982. MR 84g:54051

[97] M. Osikawa and Y. Oono, Chaos in C^0-endomorphism of interval, *Publ. Res. Inst. Math. Sci.*, **17** (1981), 165-177. MR 82e:58068

[98] H. Poincaré, Sur les courbes définies par les équations différentielles, Chapitre XV. Étude particulière du tore, *Oeuvres, Tome* 1, pp. 137-158, Gauthier-Villars, Paris, 1951. [Original, 1885]

[99] H. Poincaré, *Les méthodes nouvelles de la mécanique céleste, t.III*, Dover, New York, 1957. MR 19-414 [Original edition, 1899]

[100] W. Rudin, *Real and Complex Analysis*, 2nd. ed., McGraw-Hill, New York, 1974. MR 49#8783

[101] W. Rudin, *Principles of Mathematical Analysis*, 3rd. ed., McGraw-Hill, New York, 1976. MR 52#5893

[102] A.N. Šarkovskii, Coexistence of cycles of a continuous mapping of the line into itself (Russian), *Ukrain. Mat. Ž.*, **16** (1964), no.1, 61-71. MR 28#3121

[103] A.N. Šarkovskii, Non-wandering points and the centre of a continuous mapping of the line into itself (Ukrainian), *Dopovidi Akad. Nauk Ukrain. RSR*, 1964, 865-868. MR 29#2467

[104] A.N. Šarkovskii, On cycles and the structure of a continuous mapping (Russian), *Ukrain. Mat. Ž.*, **17** (1965), no.3, 104-111. MR 32#4213

[105] A.N. Šarkovskii, Attracting and attracted sets (Russian), *Dokl. Akad. Nauk SSSR*, **160** (1965), 1036-1038. MR 32#6419

[106] A.N. Šarkovskii, Continuous mapping on a set of ω-limit points (Ukrainian), *Dopovidi Akad. Nauk Ukrain. RSR*, 1965, 1407-1410. MR 33#1847

[107] A.N. Šarkovskii, The behaviour of a mapping in the neighbourhood of an attracting set (Russian), *Ukrain. Mat. Ž.*, **18** (1966), no.2, 60-83. MR 35#3649

[108] A.N. Šarkovskii, Continuous mapping on the set of limit points of an iteration sequence (Russian), *Ukrain. Mat. Ž.*, **18** (1966), no.5, 127-130. MR 34#6732

[109] A.N. Šarkovskii, On a theorem of G.D. Birkhoff (Ukrainian), *Dopovidi Akad. Nauk Ukrain. RSR Ser.A*, 1967, 429-432. MR 35#3646

[110] A.N. Šarkovskii, Structural theory of differentiable dynamical systems, and weakly non-wandering points (Russian), *VII. Internationale Konferenz über Nichtlineare Schwingungen* (Berlin, 1975), Band I, Teil 2, pp. 193-200, Akademie-Verlag, Berlin, 1977. MR 58#24370

[111] A.N. Šarkovskii, On some properties of discrete dynamical systems, *Iteration theory and its applications* (Toulouse, 1982), pp. 153-158, Colloq. Internat. CNRS 33, Paris, 1982. MR 86k:58058

[112] A.N. Šarkovskii, S.F. Kolyada, A.G. Sivak and V.V. Fedorenko, *Dynamics of one-dimensional mappings* (Russian), Naukova Dumka, Kiev, 1989.

[113] A.N. Šarkovskii, Yu.L. Maistrenko and E.Yu. Romanenko, *Difference equations and their applications* (Russian), Naukova Dumka, Kiev, 1986. MR 88k:39007

[114] G.R. Sell, *Topological dynamics and ordinary differential equations*, Van Nostrand, London, 1971. MR 56#1283

[115] K.S. Sibirskii, *Introduction to topological dynamics*, Noordhoff, Leiden, 1975. MR 50#10452 [Russian original, 1970]

[116] J. Smítal, Chaotic functions with zero topological entropy, *Trans. Amer. Math. Soc.*, **297** (1986), 269-282. MR 87m:58107

[117] P. Štefan, A theorem of Šarkovskii on the existence of periodic orbits of continuous endomorphisms of the real line, *Comm. Math. Phys.*, **54** (1977), 237-248. MR 56#3894

[118] P.D. Straffin, Periodic points of continuous functions, *Math. Mag.*, **51** (1978), no.2, 99-105. MR 80h:58043

[119] Y. Takahashi, A formula for topological entropy of one-dimensional dynamics, *Sci. Papers College Gen. Ed. Univ. Tokyo*, **30** (1980), no.1, 11-22. MR 82i:58057

[120] Tang Junjie, ω-limit sets for a class of self-mappings of an interval (Chinese), *J. Shanghai Jiaotong Univ.*, 1987, no.4, 91-98. MR 89g:58111

[121] A. Thue, Ueber unendliche Zeichenreihen, *Selected Mathematical Papers*, pp. 139-158, Universitetsforlaget, Oslo, 1977. MR 57#46 [Original, 1906]

[122] M.B. Vereikina and A.N. Šarkovskii, Recurrence in one-dimensional dynamical systems (Russian), *Approximate and qualitative methods of the theory of functional-differential equations*, pp. 35-46, Akad. Nauk Ukrain. SSR, Inst. Mat., Kiev, 1983. MR 85m:58149

[123] M.B. Vereikina and A.N. Šarkovskii, The set of almost recurrent points of a dynamical system (Russian), *Dokl. Akad. Nauk Ukrain. SSR Ser.A*, 1984, no.1, 6-9. MR 85i:58099

[124] C.T.C. Wall, *A geometric introduction to topology*, Addison-Wesley, Reading, Mass., 1972. MR 57#17617

[125] Xiong Jincheng, Continuous self-maps of the closed interval whose periodic points form a closed set, *J. China Univ. Sci. Tech.*, **11** (1981), no.4, 14-23. MR 84h:58124a

[126] Xiong Jincheng, Nonwandering sets of continuous interval self-maps, *Kexue Tongbao* (English Ed.), **29** (1984), 1431-1433. MR 86j:58131

[127] Xiong Jincheng, Sets of recurrent points of continuous maps of the interval, *Proc. Amer. Math. Soc.*, **95** (1985), 491-494. MR 87d:58114

[128] Xiong Jincheng, A note on chain recurrent points of continuous self-maps of the interval (Chinese), *J. China Univ. Sci. Tech.*, **15** (1985), 385-389. MR 87h:58179

[129] Xiong Jincheng, Set of almost periodic points of a continuous self-map of an interval, *Acta Math. Sinica (N.S.)*, **2** (1986), no.1, 73-77. MR 88d:58093

[130] Xiong Jincheng, A chaotic map with topological entropy, *Acta Math. Sci.* (English Ed.), **6** (1986), 439-443. MR 89b:58143

[131] Xiong Jincheng, The topological structure of nonwandering sets of self-maps of an interval (Chinese), *Acta Math. Sinica*, **29** (1986), 691-696. MR 88a:58170

[132] Xiong Jincheng, A note on topological entropy, *Chinese Sci. Bull.*, **34** (1989), 1673-1676.

[133] Xiong Jincheng, A simple proof of a theorem of Misiurewicz (Chinese), *J. China Univ. Sci. Tech.*, **19** (1989), 21-24. MR 90j:58121

[134] Xiong Jincheng, Chaoticity of interval self-maps with positive entropy, Preprint IC/88/385, International Centre for Theoretical Physics, Trieste, 1988.

[135] Xiong Jincheng, The closure of periodic points of a piecewise monotone map of the interval. [Manuscript]

[136] L.S. Young, A closing lemma on the interval, *Invent. Math.*, **54** (1979), 179-187. MR 80k:58084

[137] Zhou Zuoling, Self-mappings of the interval without homoclinic points (Chinese), *Acta Math. Sinica*, **25** (1982), 633-640. MR 85c:58090

Index

Notations